# Introduction to Botany

## James Schooley

**Delmar Publishers**

*an International Thomson Publishing company* I(T)P®

Albany • Bonn • Boston • Cincinnati • Detroit • London • Madrid
Melbourne • Mexico City • New York • Pacific Grove • Paris • San Francisco
Singapore • Tokyo • Toronto • Washington

## NOTICE TO THE READER

Publisher does not warrant or guarantee any of the products described herein or perform any independent analysis in connection with any of the product information contained herein. Publisher does not assume, and expressly disclaims, any obligation to obtain and include information other than that provided to it by the manufacturer.

The reader is expressly warned to consider and adopt all safety precautions that might be indicated by the activities herein and to avoid all potential hazards. By following the instructions contained herein, the reader willingly assumes all risks in connection with such instructions.

The publisher makes no representation or warranties of any kind including but not limited to the warranties of fitness for particular purpose or merchantability, nor are any such representations implied with respect to the material set forth herein, and the publisher takes no responsibility with respect to such material. The publisher shall not be liable for any special, consequential, or exemplary damages resulting, in whole or part, from the readers' use of, or reliance upon, this material.

Cover illustration by Laurette Richin

**Delmar Staff**
Publisher: Tim O'Leary
Acquisitions Editor: Cathy L. Esperti
Senior Project Editor: Andrea Edwards Myers

Production Manager: Wendy A. Troeger
Production Editor: Carolyn Miller
Marketing Manager: Maura Theriault

COPYRIGHT © 1997
By Delmar Publishers
a division of International Thomson Publishing Inc.

The ITP logo is a trademark under license.

Printed in the United States of America

For more information, contact:

Delmar Publishers
3 Columbia Circle, Box 15015
Albany, New York 12212-5015

International Thomson Publishing Europe
Berkshire House 168-173
High Holborn
London, WC1V 7AA
England

Thomas Nelson Australia
102 Dodds Street
South Melbourne, 3205
Victoria, Australia

Nelson Canada
1120 Birchmount Road
Scarborough, Ontario
Canada M1K 5G4

International Thomson Editores
Campos Eliseos 385, Piso 7
Col Polanco
11560 Mexico D F Mexico

International Thomson Publishing GmbH
Königswinterer Strasse 418
53227 Bonn
Germany

International Thomson Publishing Asia
221 Henderson Road
#05-10 Henderson Building
Singapore 0315

International Thomson Publishing — Japan
Hirakawacho Kyowa Building, 3F
2-2-1 Hirakawacho
Chiyoda-ku, Tokyo 102
Japan

All rights reserved. No part of this work covered by the copyright hereon may be reproduced or used in any form or by any means—graphic, electronic, or mechanical, including photocopying, recording, taping, or information storage and retrieval systems—without the written permission of the publisher.

1 2 3 4 5 6 7 8 9 10 XXX 03 02 01 00 99 98 97

**Library of Congress Cataloging-in-Publication Data**
Schooley, James.
    Introduction to botany / James Schooley.
      p.    cm.
    Includes bibliographical references (p.    ) and index.
    ISBN 0-8273-7378-3
    1. Botany.  2. Horticulture.  I. Title.
QK47.S37  1996
581—dc20
                                96-4945
                                   CIP

# Contents

Introduction .................................................... ix

**Chapter 1**     **The Origin of Life** .................................. 1
The Theory of Spontaneous Generation • Life Viewed Through the Microscope • A Modern-day Theory

**Chapter 2**     **The Cell** ............................................. 5
The Cell Doctrine • Mitochondria • Golgi Bodies • Endoplasmic Reticulum • Nuclear Membrane • Cell Membrane • Cell Walls • Chloroplasts • Cilia • Plastids • Vacuoles • A Single-celled Imposter

**Chapter 3**     **Mitosis and Meiosis** ............................... 21
The Basics of Mitosis • Plant Mitosis • The Basics of Meiosis • Aberrations in Mitosis and Cellular Composition

**Chapter 4**     **Chemistry** .......................................... 37
Empirical and Structural Formulas • Alcohols • Organic Acids • Amino Acids • Polymers • Proteins • Carbohydrates • Lipids

**Chapter 5**     **Mendelian Genetics** ............................... 49
Pre-Mendelian Theorists and Theories • Gregor Mendel • Mendel's Laws • Mendel's Experiments • Applying Genetics

**Chapter 6**     **DNA** ................................................ 59
The Search for the Substance of Heredity • The Structure of DNA • The Functions of DNA • Amino Acids • Transfer RNA • Enzymes • Mutations in DNA • Gene Repression

**Chapter 7**     **Viruses** ............................................. 75
T-2 Bacteriophage • Plant Viruses

**Chapter 8**     **Physical Properties of Protoplasm** ............. 81
Composition of Protoplasm • Colloids • Diffusion

| Chapter | 9 | **Photosynthesis** | 89 |
|---|---|---|---|

Early Research • Modern-day Research • Chlorophyll • Light • Electron Transfer • The Calvin Cycle • The $C_4$ Plants • The CAM Carbon Pathway

| Chapter | 10 | **Respiration and Fermentation** | 101 |
|---|---|---|---|

The ATP Molecule • Respiration and Photosynthesis • The Anaerobic and Aerobic Pathways • Hydrogenation • The Carbon Cycle

| Chapter | 11 | **Bacteria** | 109 |
|---|---|---|---|

Is Bacteria a Plant? • The Original Bacteria • Modern, Aerobic Bacteria • Characteristics of Bacteria • Benefits of Bacteria • Hazards of Bacteria • Identifying Bacteria • Bacterial Growth

| Chapter | 12 | **The Blue-green Algae** | 119 |
|---|---|---|---|

Primordial Ooze • Characteristics of the Blue-green Algae • Types of Blue-green Algae

| Chapter | 13 | **The Green Algae** | 125 |
|---|---|---|---|

The Volvacine Line • The Tetrasporine Line • The Siphonous Line • Green Alga or Bryophyta?

| Chapter | 14 | **Phaeophyta: The Brown Algae** | 143 |
|---|---|---|---|

Products from Brown Algae • Reproduction in Brown Algae

| Chapter | 15 | **Rhodophyceae: The Red Algae** | 149 |
|---|---|---|---|

Bangiophycidae • Floridiophycidae

| Chapter | 16 | **Other Algae** | 155 |
|---|---|---|---|

Xanthophyta: The Yellow-green Algae • Euglenophyta • Chrysophyta • Pyrrophyta • *Acetabularia:* A Green Alga

| Chapter | 17 | **Fungi** | 169 |
|---|---|---|---|

Fungi Classification • Myxomycetes

| Chapter | 18 | **Phycomycetes** | 173 |
|---|---|---|---|

Chytridiomycetes • Zygomycetes • Oomycetes

| Chapter | 19 | **The Ascomycetes** | 183 |
|---|---|---|---|

Reproduction in Ascomycetes • Fruiting Bodies • Yeasts • Pathogenic Ascomycetes • *Penicillium* and *Aspergillus* • Morels and Truffles • Ergot

| Chapter | 20 | **The Basidiomycetes** | 195 |
|---|---|---|---|

Rusts • Smuts • Puffballs • Mycorrhiza

| Chapter | 21 | **Fungi Imperfecti** ................................................. 207 |
|---|---|---|
| | | Problems in Classification • Moniliales |

| Chapter | 22 | **Lichens** ......................................................... 211 |
|---|---|---|
| | | Reproduction in Lichens • The Members of a Lichen • Growth of Lichens • Products from Lichens • Lichens and the Doctrine of Signatures |

| Chapter | 23 | **Plant Classification** ............................................. 215 |
|---|---|---|
| | | Early Efforts at Plant Classification • Carolus Linnaeus • The Theory of Evolution • Problems in Classification • Monophyletic or Polyphyletic? • System of Plant Classification • What Is a Species? |

| Chapter | 24 | **Bryophytes: The Liverworts, Hornworts, and Mosses** .. 221 |
|---|---|---|
| | | Liverworts • Hornworts • Mosses |

| Chapter | 25 | **Pteridophytes: The Ferns, Club Mosses, and Horsetails** .. 229 |
|---|---|---|
| | | Ferns • Club Mosses • Horsetails |

| Chapter | 26 | **Tissues** .......................................................... 235 |
|---|---|---|
| | | Meristematic Tissue • Simple Tissues and Complex Tissues |

| Chapter | 27 | **Gymnosperms** ................................................... 245 |
|---|---|---|
| | | The First Seed Plants • Classifying the Gymnosperms • Coniferales • *Ginkgo biloba* |

| Chapter | 28 | **Angiosperms** .................................................... 255 |
|---|---|---|
| | | Life Cycle • Lilies • Comparing Angiosperms to Gymnosperms |

| Chapter | 29 | **Hormones** ....................................................... 263 |
|---|---|---|
| | | In Search of Auxin • Other Plant Hormones |

| Chapter | 30 | **Biological Clocks** ............................................... 275 |
|---|---|---|
| | | Circadian Rhythms • *Gonyaulax polyedra* |

| Chapter | 31 | **Plant Nutrition** ................................................. 279 |
|---|---|---|
| | | Required Minerals • Determining Mineral Needs • Symptoms of Improper Nutrition |

| Chapter | 32 | **Stems** ........................................................... 283 |
|---|---|---|
| | | The Woody Dicot Stem • The Herbaceous Dicot Stem • The Monocot Stem • Modified Stems |

| Chapter 33 | **Roots** . . . . . . . . . . . . . . . . . . . . . . . . . . . . . . . . . . . . . . . . . . . . . 299 |
|---|---|
| | Contributors to Root Growth • Root Hairs • Structure of a Root • Casparian Strip • Root Growth |

| Chapter 34 | **Leaves** . . . . . . . . . . . . . . . . . . . . . . . . . . . . . . . . . . . . . . . . . . . . . 307 |
|---|---|
| | Simple versus Compound Leaves • Transpiration • Leaves and Transplanting • Guttation • Structure of a Leaf • Leaves and Plant Classification and Identification |

| Chapter 35 | **Flowers** . . . . . . . . . . . . . . . . . . . . . . . . . . . . . . . . . . . . . . . . . . . 317 |
|---|---|
| | How Flowers Are Formed • Variations in Flowers • Evolutionary Modifications in Flowers |

| Chapter 36 | **Fruits and Seeds** . . . . . . . . . . . . . . . . . . . . . . . . . . . . . . . . . . 327 |
|---|---|
| | Forms of Fruit • Seed Structure and Characteristics • Functions of Seeds • Variations in Seed Composition • Seed Longevity • Seed Germination • Reproduction |

| Chapter 37 | **Other Methods of Propagation** . . . . . . . . . . . . . . . . . . . . . 337 |
|---|---|
| | Division • Layering • Cuttings • Grafting |

| Chapter 38 | **Evolution** . . . . . . . . . . . . . . . . . . . . . . . . . . . . . . . . . . . . . . . . . 343 |
|---|---|
| | Early Changes in Thought • Charles Darwin • The Tenets of Darwinian Theory • Other Theories of Evolution • The First Organisms • Prokaryotic Life • Eukaryotic Life • The Emergence of Seed Plants • Grasses • Human Life • Life over Time |

| Chapter 39 | **Ecology** . . . . . . . . . . . . . . . . . . . . . . . . . . . . . . . . . . . . . . . . . . . 353 |
|---|---|
| | Plant Ecology • Adaptation • Environment • Climate • The Global-Warming Controversy • Ecological Interrelationships • Natural Recycling • Plant Succession |

| Chapter 40 | **Plants and Human Welfare** . . . . . . . . . . . . . . . . . . . . . . . . . . 359 |
|---|---|
| | Feeding an Increasing Population • Other Human Uses for Plants • Cultivated Plants • Viruses, Bacteria, and Fungi |

| **Glossary** | . . . . . . . . . . . . . . . . . . . . . . . . . . . . . . . . . . . . . . . . . . . . . . . . . . . . . . . . . . 365 |
|---|---|
| **Index** | . . . . . . . . . . . . . . . . . . . . . . . . . . . . . . . . . . . . . . . . . . . . . . . . . . . . . . . . . . 399 |

# Introduction

The study of biology historically has been divided into two realms: botany, for plants, and zoology, for animals. This suggests that all living organisms are either plants or animals, a theory that presented little problem when applied to giraffes and elm trees. But when bacteria were discovered, there resulted some puzzlement regarding to which realm these organisms should be relegated. Further research and discoveries only increased the uncertainty until, in 1959, Professor R.H. Whittaker proposed a five-kingdom system as follows: Monera, Protista, Fungi, Plants, and Animals. Members of the Monera kingdom are prokaryotic (having no definite nuclei) cells such as bacteria and blue-green algae. Members of the Protista kingdom are eukaryotic (having true nuclei) cells. Members of the Fungi kingdom are plantlike but lack chlorophyll. Such organisms, therefore, do not manufacture carbohydrate as do green plants, and must therefore live as either parasites or saprophytes (organisms that live on dead matter). Because people are so accustomed to classifying organisms as either plant or animal, this system has been slow to take hold. And while this five-kingdom system does not solve all problems relating to classification, it does constitute a step forward. It is thus the system of classification employed in this text.

## Acknowledgments

The author wishes to thank Dr. Knut Norstog, formerly editor of the *American Journal of Botany*, first, for his friendship, and second, for helping to put in clear language some comments regarding the origin of seed plants. The author also wishes to express appreciation to the following, all at Delmar Publishers: Cathy Esperti, acquisitions editor, for fine-tuning the manuscript; Wendy Troeger, who worked on art and book manufacturing; and Maura Theriault and Suzanne Fronk, for their work in marketing. Appreciation is also expressed to those other professionals at Delmar Publishers who aided this work without even making themselves known, and to Thomas J. Gagliano, Gagliano Graphics, Albuquerque, New Mexico for the illustrations. Finally, the author wishes to thank the following reviewers, who provided constructive comments and input:

Cheryl Carney
Iowa Lakes Community College

Alan Smith
University of Minnesota

Connie Fox
Tarleton State University

# Dedication

This book is dedicated to the memory of Barbara McClintock (1902–1992). In 1983 she was awarded the Nobel Prize in Genetics for the discovery of "jumping genes" (genes that regularly change their positions on chromosomes). In 1944 she served as President of the Genetics Society of America, and in 1945 was President of the National Academy of Sciences. She once confided to me that one reason for her devotion to the study of corn was to combat loneliness.

# 1

# The Origin of Life

<small>❦ Notes ❦</small>

The study of botany very properly begins with a few comments about the origin of life. How did life come about? We know there was a time when the planet had no life. Life may have begun nearly four billion years ago.

As we look around today, we are led to the conclusion that all life comes from previously existing life. Further, knowing that **organisms** are composed of cells, we concur with what Rudolf Virchow (1821–1902) said in 1858: that "all cells come from previously existing cells." Our common sense tells us so.

## The Theory of Spontaneous Generation

Yet, this is modern-day common sense. Common sense in other times told people quite a different thing. They saw earthworms arising from the mud, especially after a rain. They saw maggots coming out of the garbage. They saw evidence all around them of life arising from nonliving precursors—of the **spontaneous generation** of life. In fact, Jan van Helmont (1577–1644) passed on a recipe for making mice: put some old rags in a dark corner, sprinkle some grains of wheat on the rags, and in twenty-one days you have mice. The mice presumably generated spontaneously.

Francesco Redi (1626–1697) was the first to investigate the theory of spontaneous generation of life. He took two dishes of meat, covered one with gauze, left the other dish uncovered, and let both dishes stand for a time. While the meat in both dishes decayed, only the uncovered dish developed maggots. Redi's experiment did not disprove the spontaneous generation of life, however; it disproved only the spontaneous generation of maggots.

## Life Viewed Through the Microscope

In 1590 Zacharias Janssen invented the microscope. Johannes Kepler and Christoph Scheiner soon made improvements on this invention. Then, in

## Notes

1676, Anton van Leeuwenhoek, a Dutch dry-goods merchant who manufactured his own microscopes, presented a paper before the Royal Society of London in which he claimed to have discovered "wee beasties" under his microscope's lenses.

Here, then, was microscopically sized life; and while the spontaneous generation of maggots had been disproven, and the "creation" of mice by rags in a dark corner seemed unlikely, surely these extremely small creatures must have arisen from a nonliving precursor. In 1749 John Needham devised an experiment that seemed to confirm this theory. He prepared a broth in which grew great numbers of tiny organisms. He then brought the broth to a boil and determined that all the organisms were destroyed through boiling. After letting the broth stand for a day or two, he observed that the creatures reappeared. This was interpreted as a proof of the spontaneous generation of life.

Lazzaro Spallanzani (1729–1799) was a skilled experimenter who studied blood circulation, respiration, digestion, the senses of bats, and the breeding of eels. His name belongs here, however, because of his experiments with bacteria. He repeated John Needham's experiments. But after boiling the broth to destroy the microorganisms, he drew out the neck of the flask and sealed it against any further invasion by organisms. When he broke open the flask after several days and examined the contents for microorganisms, none could be found. This may appear to be a turning point in the study of the origin of life. Another hundred years would pass before Louis Pasteur's series of experiments in 1859 finally put to rest the concept of spontaneous generation of life.

## A Modern-day Theory

We are brought back, then, to modern-day thinking, which tells us that all living things are products of living things. At the same time, however, we realize that there was a time when life did not exist on Earth and that there had to be a beginning. This leads us to the following understanding: life does not arise spontaneously in the world as we know it today, but in the primordial world, when conditions were different, life arose from a nonliving precursor. It is important to emphasize that world conditions were different from those we know today. Specifically, the atmosphere at that time was either nearly or completely devoid of oxygen.

Before continuing, it is also important to make clear that theories regarding the origin of life reside in the realm of educated speculation. It is contended that organic molecules were formed in the primordial sea or in the atmosphere; that these molecules accumulated, persisted, and got together in clusters; and that molecules formed that were able to govern both their own replication and the formation of other molecules. If this sounds like the contention of an exalted imagination, keep in mind that this process occurred in a world having conditions different from those of the world today. There was no decay, because decay is the function of organisms; there was little or no oxygen; and molecules did not tend to break down in being oxidized. Here, an argument may be made that without oxygen, there was no ozone layer; and that without ozone in the upper atmosphere, ultraviolet light would have been able to penetrate to the Earth's surface; and that ultraviolet radiation is not compatible with life. This objection is countered by the fact that ultraviolet light is not able to penetrate water, and life is believed to have begun in the water. So long as there was no oxygen in the atmosphere, life was confined to the sea.

Stanley Miller's experiment is significant to this hypothesis. In 1953 he constructed an apparatus intended to simulate the ancient atmosphere in the neighborhood of a volcano. This apparatus included a mixture of gases, hydrogen, ammonia, methane, and water. He subjected this mixture to heat and electrical discharges, and, in time, determined that a number of **amino acids** had formed. This was significant because until this time, it was believed that amino acids were made only by organisms. Friedrich Wohlor's (1800–1882) successful synthesis of urea in 1828 provided another example of the synthesis of an organic molecule not made by an organism.

Laboratory observations such as those of Miller and Wohler are far from the creation of life. Nor do they fully explain how life came into being. Yet one must ask what we have to take the place of such observations. The theological explanation is simple enough: God made life. But that is not an explanation. It just makes an explanation unnecessary. Thus, it is not the approach taken in this text.

Laboratory observations, then, demonstrate neither the formation of **deoxyribonucleic acid (DNA)** nor the formation of **chlorophyll**, the green stuff that can trap light energy and use it in the manufacture of starch. All one can say is that these things did happen somewhere along the line. When the capacity of **photosynthesis** came into that primordial ooze that we call blue-green algae, oxygen was liberated to the atmosphere. The stage was thus set for the emergence of **terrestrial** life. It is difficult to conceive of the events that were to follow; events that would lead eventually to tears and laughter—events that unfolded over hundreds of millions of years.

## Questions for Review

1. What condition of the primordial world that does not exist in today's world could perhaps have allowed for the spontaneous generation of life?
2. Describe an experiment conducted by Francesco Redi, specifically, what the experiment demonstrated.
3. Who invented the microscope? When?
4. John Needham conducted an experiment that he claimed proved spontaneous generation of life. Describe this experiment.
5. It is asserted that in the primordial world there was no oxygen in the atmosphere, and, because of this, no _____, which prevents _____ _____ from reaching the Earth.
6. Recount the experiment carried out by Stanley Miller, specifically, what the experiment demonstrated.

## ❈ Notes ❈

## Suggestions for Further Reading

Alder, I. 1957. *How Life Began.* New York: New American Library, Signet Books.

Farley, J. 1977. *The Spontaneous Generation Controversy from Descartes to Oparin.* Baltimore: Johns Hopkins University Press.

Goldsmith, D., and T. Owen. 1980. *The Search for Life in the Universe.* Menlo Park, CA: Benjamin/Cummings.

Margulis, Lynn. *Early Life.* Boston: Science Books International, Inc.

Ponnamperuma, C. 1972. *The Origin of Life.* New York: E. P. Dutton and Co.

Smith, C.U.M. 1976. *The Problem of Life: An Essay in the Origin of Biological Thought.* New York: Halsted Press (Wiley).

# 2

# The Cell

**W**hy should a chapter regarding the cell begin with the name of Robert Hooke (1635–1703)? Did he discover cells? No. Robert Hooke liked to play with microscopes, and he wrote descriptions of the fly's compound eye, lice, fungi, and gnats. He was interested in and studied many things, including gravity, the motions of heavenly bodies, the nature of light, clocks, springs, and balances. But he is recalled in the study of botany for introducing the name **cell** to describe the minute units that are the building blocks of life. In 1665 he was peering through his microscope at a thin slice of cork and observed that the tissue was organized into little compartments, little boxes, which he named *cells*. Given that Zacharias Janssen invented the microscope in 1590, seventy-five years had passed since this invention before the name cell was introduced.

❋ **Notes** ❋

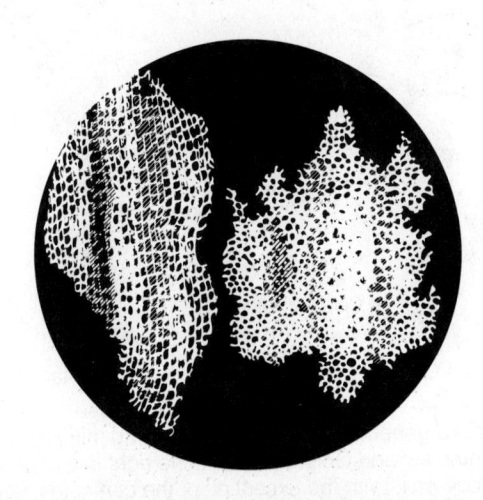

**Figure 2-1** Cork cells as seen by Robert Hooke in 1665. Among his comments was that it "seems to be like a kind of Mushrome."

We now define a cell as "a mass of **protoplasm** surrounded by a membrane and containing a **nucleus** or a number of nuclei". Thus, there were no cells present in the material examined by Robert Hooke; rather, Hooke observed empty spaces where cells previously resided.

## The Cell Doctrine

The doctrine that all living things are made up of cells was published in 1839. This doctrine is credited to two men: Matthias Schleiden (1804–1881), a botanist, and Theodor Schwann (1810–1882), an anatomist. Though working independently, Schleiden and Schwann came to this conclusion at nearly the same moment.

As it turns out the doctrine that all living things are composed of cells is not tenable. Some things do not have a cellular organization. Yet, most organisms, both plant and animal, are constituted of cells; and cells, no matter what their sources, share certain characteristics. The cells of cabbages and giraffes, for example, have much in common. This suggests that the great diversity of life comes from a common beginning; and as focus is directed to the minute **organelles** that reside in cells, the differences fade even more to where such structures as **mitochondria** and **Golgi bodies** appear to be quite the same whatever their sources.

As microscopes were improved, the structures contained within cells were revealed, and how these structures are involved in cell activity also became known. Our knowledge of cells has advanced along two fronts. On the one front, increasingly powerful microscopes have allowed the identification of the smaller aspects of the cell; on the other front, biochemical methods enabled

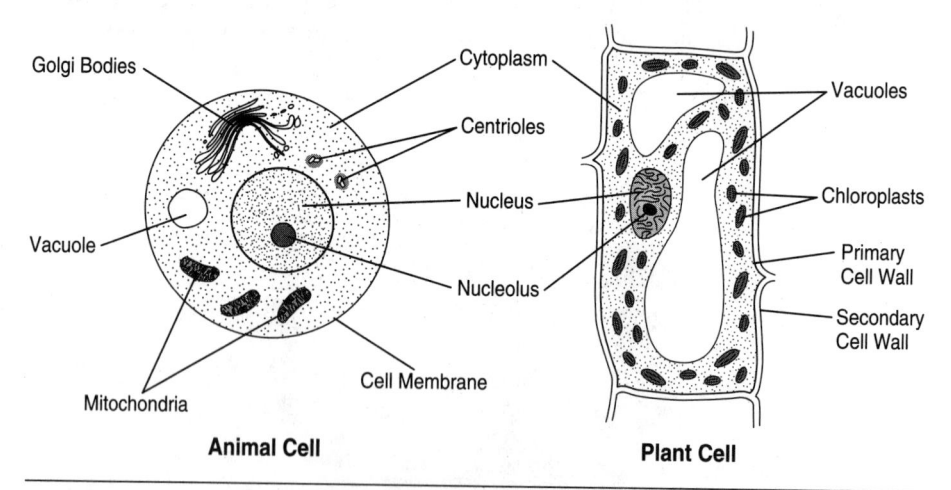

**Figure 2-2** At the left is a generalized animal cell showing mitochondrion, vacuole, nucleus, nucleolus, centrioles, and Golgi bodies. At the right is a plant cell, which conforms to the shape of a rigid cell wall. With the exception of the centrioles, which are generally not seen in plant cells, a plant cell possesses the same organelles as are shown for the animal cell. Shown for the plant cell are chloroplasts, a large vacuole, and both primary and secondary cell walls, which lie outside of the cell membrane.

**Figure 2-3** Matthias Schleiden (1804–1881) contributed to the theory that all life is composed of cells. (Illustration by Donna Mariano)

the discovery of how these delicate microstructures are involved in metabolism. Figure 2-4 shows what can generally be seen with a light microscope.

Now, consider for a moment the smallest imaginable cell having all the components necessary for the maintenance of life. A **mycoplasm** is an example. It is estimated that such a cell needs more than 1,000 different kinds

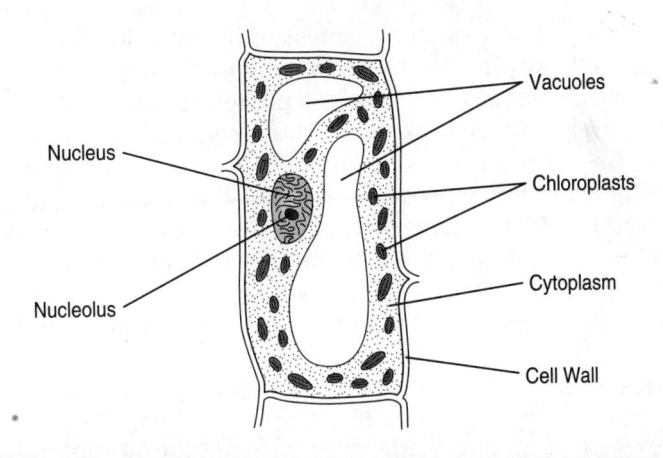

**Figure 2-4** Plant cell as seen through a light microscope: nucleolus, nucleus, cell wall, cytoplasm, chloroplast, and vacuole.

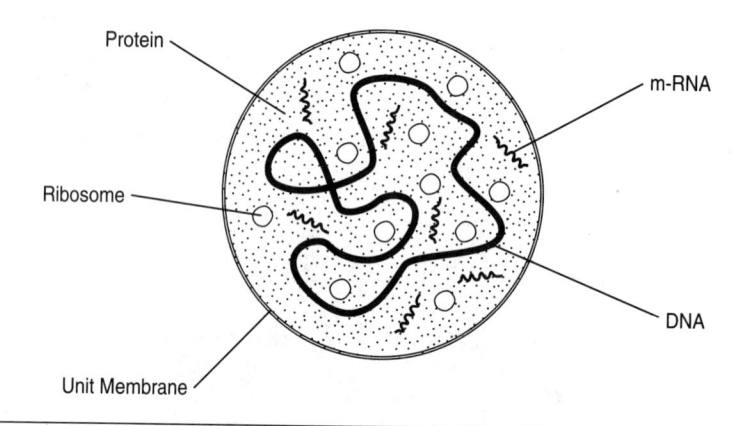

**Figure 2-5** A mycoplasm, the smallest possible cell containing all components necessary for life: protein, unit membrane, m-RNA, DNA, and ribosome.

of molecules in order to sustain and perpetuate its life. Such cells are smaller and less complex than the simplest bacteria. There is a delicate plasma membrane made of **proteins** and **lipids**. The cell is, of course, **prokaryotic**, that is, having no discrete nucleus. Recently, the electron microscope has allowed for magnification of such cells by much more than 100,000 diameters, giving us additional knowledge regarding cell structures.

Following is a discussion of the mitochondria, Golgi body, endoplasmic reticulum, nuclear membrane, cell membrane, cell walls, chloroplasts, cilia, plastids, and vacuoles.

## Mitochondria

Mitochondria are minute organelles measuring approximately 1 by 3 microns. Mitochondria are said to be the "power houses" of the cell because many of the chemical changes associated with respiration take place in these structures. As glucose is broken down in respiration, energy is released. This energy goes into the manufacture of **ATP** (adenosine triphosphate). This breakdown is achieved stepwise, and while some of the steps occur in the **cytoplasm**, most of the changes occur in the mitochondria. A much larger amount of ATP is manufactured in the mitochondria, and the ATP that is made there is stored there. The mitochondria, then, contain the energy reserves that are called upon to do the work of the cell; thus the nickname "power houses."

It is not the aim at this point to consider the chemistry of respiration but, rather, to consider the structure of the mitochondria. Whereas these structures are present in all **eukaryotic** cells, they are not present in prokaryotes. Structurally, they resemble chloroplasts in that they each possess a double-

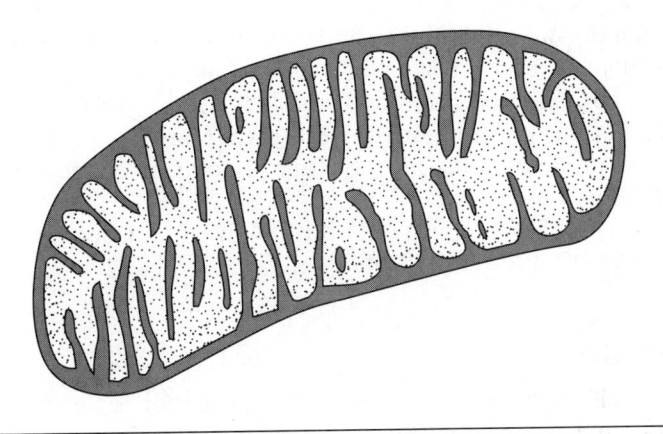

**Figure 2-6** A mitochondrion. The inner membrane of this double-membraned structure has folds that extend into the interior of the organelle. This infolding increases the inner surface area.

membrane system: an outer smooth membrane and an inner, much convoluted membrane. The outer membrane contains passageways. The inner membrane is impermeable. The inward projecting parts of the inner membrane are called **cristae**; in plants, they commonly appear as tubules. Many enzymes involved in the chemistry of respiration are aligned along these cristae. Every chemical change requires its own particular kind of enzyme, and the various enzymes appear to be arranged here in a proper sequencing. There are perhaps as many as seventy different kinds of enzymes in the mitochondria.

Given that mitochondria are associated with steps in respiration and energy harvesting, one might suppose that they would be found where the most energy is required. The cluster of mitochondria found at the bases of **flagella** appears to confirm this assumption. While not forgetting that our concern is botany, it is interesting to note that 500 times more mitochondria are found in heart muscle cells than are found in cells of other, less active muscles.

As already indicated, mitochondria from different kinds of cells, and even from different kinds of organisms, have quite similar structures; and, of course, they all perform the same functions.

As is true of chloroplasts, mitochondria have their own DNA, **RNA (ribonucleic acid)**, and **ribosomes**. Further, both mitochondria and chloroplasts are self-replicating structures; that is, all mitochondria come from previously existing mitochondria, and all chloroplasts come from previously existing chloroplasts.

Mitochondria are, then, semi-autonomous. They are only partially dependent on nuclear genes. Professor Lynn Margulis postulates that their presence in cells is a consequence of invasion; that is, mitochondria came to be in cells

❋ **Notes** ❋

by virtue of a prokaryotic cell crawling into another prokaryote and taking up residence there. The event is called **endosymbiosis** and is thought to have taken place approximately one and one-half billion years ago. The same is proposed regarding other organelles; Professor Margulis further suggests that the origin of mitochondria may be traced to **purple bacteria** entering a prokaryote in this manner.

Mitochondrial DNA replication and the division of mitochondria are not synchronized with nuclear division.

## Golgi Bodies

Eukaryotic cells regularly contain numerous flattened, saclike structures, which under an electron microscope appear as a stack of pancakes. These are the Golgi bodies. They derive their name from Camillo Golgi (1843–1926), an Italian physician who discovered these structures in 1883, while examining nerve cells of a barn owl.

Although they are present in all kinds of cells, Golgi bodies appear to be more prominent in cells that produce secretions; and they are always found in association with basal bodies of flagella and cilia, and with **centrioles**. Figure 2-7 shows that the flattened sacs seem to pinch off little vesicles, which are able to migrate through the cytoplasm and deliver their contents to specific sites. Golgi bodies contain enzymes, and in addition to their role of delivery, they may be involved in manufacturing. They are also believed to play a part in cell plate formation, making microtubules, and synthesizing enzymes. While they are more common in animal cells than in plant cells, they are

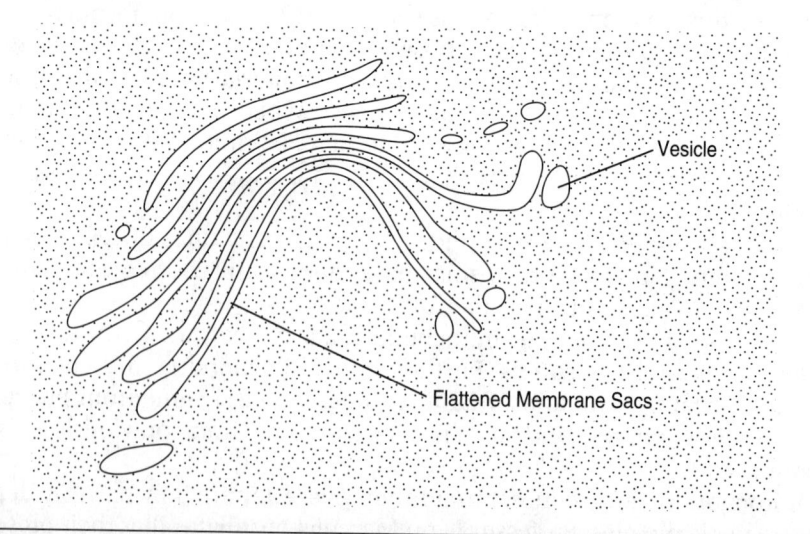

**Figure 2-7** Golgi bodies: flattened membranous sacs and vesicle.

found in both. They are versatile organelles, performing different functions in different kinds of cells. In an egg cell, for example, they are involved in the production of yolk; in an adrenal gland cell, they play a role in making a hormone; and in a salivary gland cell, they participate in making a digestive enzyme. Golgi bodies do what they do in accordance with instructions from nuclear DNA. We know that the nuclei of all kinds of cells are alike, and that, in fact, all nuclei come from previously existing nuclei. So it is a striking feature of nuclear DNA that it is able to give certain instructions and leave other instructions switched off. Thus, in differing cells, the instructions may be different by calling upon different DNA segments.

In plant cells, the Golgi bodies are commonly called **dictyosomes**. They may make **cellulose**, and the vesicles associated with them may deliver the cellulose to be deposited in the cell walls. In depositing cellulose in the secondary cell wall, the vesicles that carry the cellulose must pass through the cell membrane. This suggests that they also play a role in cell membrane repair. Proteins manufactured in the endoplasmic reticulum may be passed to the dictyosomes, where they are modified by the addition of sugars or fat groups. These dictyosomes do not reproduce themselves in the way that mitochondria and chloroplasts do. Rather, at the time of **mitosis**, the dictyosomes fragment into fine granules, which are then distributed to the daughter cells.

If the contents of the vesicles are digestive enzymes, the vesicles are called **lysosomes**. The enzymes are believed to be manufactured in the endoplasmic reticulum and passed on first to the Golgi bodies and then to the lysosome vesicles. The lysosome vesicles may then either rupture within the cell, where the release of enzymes would result in the dissolution of the cell, or migrate outside of the cell, where they will rupture and release the enzymes. Because such events appear to be more often associated with animal cells, they will not be elaborated on here.

## Endoplasmic Reticulum

The **endoplasmic reticulum** (**ER**, *endo* meaning "inner," *reticular* meaning "net") was an unknown constituent of cells until early in the 1950s when the electron microscope brought it into view. It is now known that the endoplasmic reticulum is present in all eukaryotic cells. It appears as a system of paired, parallel membranes running through the cytoplasm and taking the form of flattened tubes or bags. The bags are called cisternae. It has been suggested that the endoplasmic reticulum divides the cytoplasm into compartments and that it may be likened to a mass of soap bubbles continually changing form and position. There are two known kinds of endoplasmic reticulum: rough and smooth. **Rough endoplasmic reticulum** is so-called because it has ribosomes on its outer surface. (Ribosomes are involved with protein synthesis and secretions.) **Smooth endoplasmic reticulum** has no ribosomes and may be involved in the production of carbohydrate. The endoplasmic

❋ **Notes** ❋

※ **Notes** ※

reticulum, particularly the smooth form, may be associated with **plasmodesmata**, strands of cytoplasm that run through cell walls, creating the appearance of communicating linkages between cells. Evidence also indicates that the endoplasmic reticulum is contiguous with the outer nuclear membrane.

## Nuclear Membrane

The **nuclear membrane** is a double membrane much like the double membrane of the endoplasmic reticulum. An electron microscope reveals a light line sandwiched between two dark lines. This double-membrane system is called the *nuclear envelope*, and each unit membrane is believed to be composed of a central lipid layer sandwiched between layers of protein. Numerous pores can be seen in this envelope (see figure 2-8); perhaps one-third of the surface is taken up by these perforations. The holes of the outer membrane appear contiguous with the channels of the endoplasmic reticulum. These holes are believed to allow the passage of materials from the

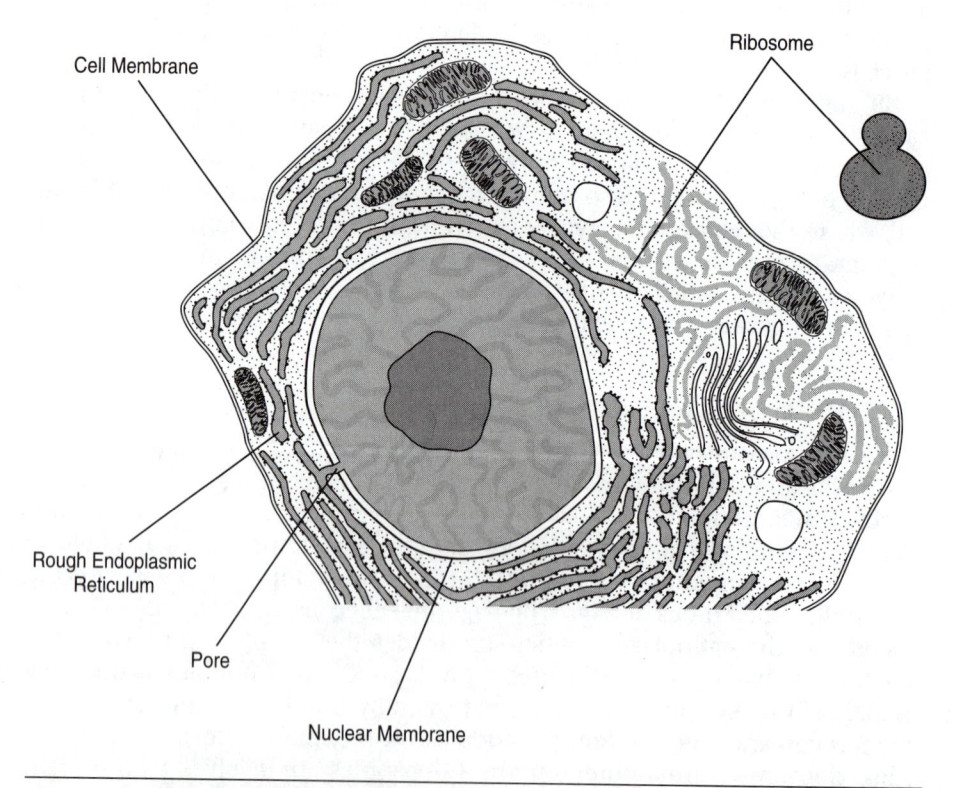

**Figure 2-8** Greatly magnified cell showing the endoplasmic reticulum (ER): nuclear membrane, ribosome, rough endoplasmic reticulum, and cell membrane.

nucleus to the cytoplasm and vice versa, although there is not universal agreement on this point. Some researchers claim that the pores of the nuclear envelope play no role in passage.

The inner and outer membranes of the envelope are connected to each other at the margins of the pores, and the pores appear to be lined. This lining is called the **annulus**, and while it fills much of the pore, a central channel remains. The channel may sometimes appear to be clogged with material. This material may be ribosomal, although this is not certain.

## Cell Membrane

All cells are bounded by **cell membranes**, which are similar in all cells. In prokaryotic cells, the membranes appear to be much-folded, the convolutions extending to the interior of the cell and having the effect of increasing surface area. Plant and animal cells are alike in this respect; however, a significant difference between the two is found in the cell wall. In plants, cell walls are secreted by plant cells and lie outside of the cell membrane. Animal cells for the most part do not exhibit this characteristic. (Cell walls are further described in an upcoming section of this chapter.)

The cell membrane, or **plasmalemma**, has three layers, as seen through an electron microscope. There is a light line in the center bounded by dark lines on each side. The center portion is made of phospholipids, and the dark lines are made of protein. Such a membrane is called a *unit membrane*. It is perforated by many holes.

While the cell wall is freely permeable to both water and dissolved materials, the cell membrane exercises selectivity; that is, it allows some materials to pass through and restricts others. This **selective permeability** may be thought to relate to pore size, with molecules and ions smaller than the openings being able to pass through, and molecules larger than the openings being restricted. This reasonable postulate, which depends on the constant motion of molecules and their tendency to diffuse, accords with some observations; but other factors come into play. Dependency on diffusion alone would be too slow a process. A cell membrane is a living structure and can exercise selectivity entirely separate from the presence of apertures. The movement of materials through a membrane involves the expenditure of energy and is called **active transport**. Enzymes are involved. The permeability of the membrane constantly changes. A substance that is allowed to pass through at one time may be disallowed at another time. Materials become dissolved in the membrane, migrate across it, and emerge on the other side. Certain molecules are moved across the cell membrane by carriers. They become attached to carrier molecules, are transported through the membrane, and are released on the other side. These migrations do not involve movement through holes. The capacity of a cell membrane to allow or disallow the passage of solutes depends on several factors. Ions having a charge of plus one (+1) tend to increase permeability. Ions having a charge of plus two (+2) tend to decrease permeability. Nitrates and phosphates tend to increase cellular metabolism and, hence, accelerate the movement of dissolved materials through the membrane. When calcium ions are deficient, the membrane tends to be damaged and develops leaks.

## Cell Walls

The presence of a wall secreted by the cell is a characteristic of plant cells. Animal cells do not produce walls. Many, but not all, **protista** have **cell walls**. Fungi and bacteria produce cell walls. The distinction between the cell membrane and cell wall is an important one. The cell membrane is a part of the cell and is a living structure. The cell wall is not part of the cell; rather, it is secreted by the cell, lies outside of it, and is not living. The cell wall may appear homogeneous when viewed through an ordinary light microscope, yet there are actually two forms of cell wall: the **primary wall**, which is produced when the cell is young and continuing to grow, and the **secondary wall**, which is produced after the cell has completed its growth. If a polarized light source and a polarizing microscope are used, a primary wall and a secondary wall can be distinguished. Both primary and secondary walls are composed largely of cellulose deposited in the form of microfibrils. **Pentosans** are also present in the wall. Whereas cellulose is composed of long chains of 6-carbon sugars (such as glucose) linked together, pentosans are composed of linkages of 5-carbon sugars. It is thought that the synthesis of these molecules is accomplished in the Golgi bodies. **Pectin** is another substance found in plant cell walls. Pectic substances also form a thin layer called the **middle lamella** and found between adjacent cells. Many cells, such as those of wood, contain **lignin**. The walls of cork cells and certain leaf cells possess a waxy material called **suberin**. Cell walls do not deter the passage of water and dissolved materials unless the walls are impregnated with suberin.

Soon after the completion of cell division, the primary cell wall is deposited by the daughter cells on each side of the middle lamella. As a result, the cell membrane rests against the primary cell wall rather than against the middle lamella. Because the secondary cell wall is secreted after the primary cell wall is formed, the secondary cell wall lies internal to the primary cell wall. Many perforations occur in the walls. Strands of cytoplasm commonly run through these perforations, producing linkages with the cytoplasm of adjacent cells. These strands are called the *plasmodesmata*.

The arrangement of cellulose fibers in the primary and secondary cell walls differs, being randomly oriented in the primary wall and spirally arranged in the secondary wall. Whereas the primary wall is flexible and can be stretched while the cell is growing, the secondary wall is more rigid. In many plants, the protoplast dies after secondary wall formation is complete. The constituents of the protoplast are then removed, leaving only the cell wall. This occurs in most wood cells.

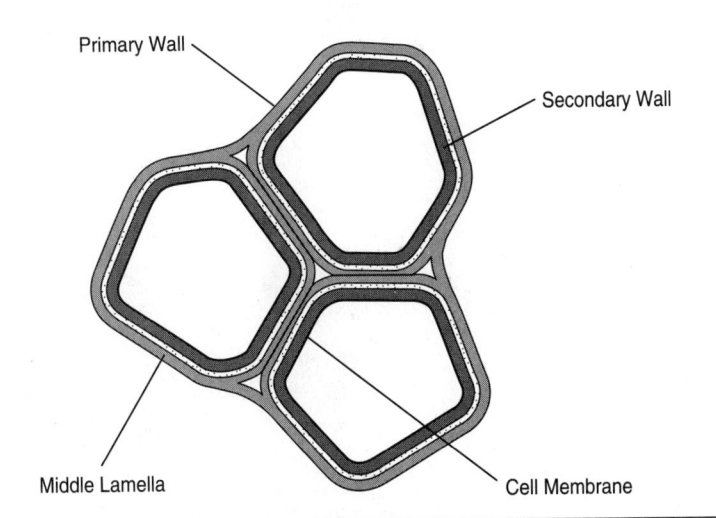

**Figure 2-9** The primary cell wall, which is the first formed, lies against the middle lamella. The middle lamella is the point of contact between the two cells. The secondary cell wall lies inward, adjacent to the cell membrane, and is thicker than the primary cell wall.

## Chloroplasts

Nearly all life on Earth runs on sunlight and, thus, depends on the processes that occur in **chloroplasts**. It is therefore fitting that these structures be examined in depth.

The most significant distinction between animal cells and plant cells is the presence of chloroplasts in plant cells. Under a light microscope, chloroplasts appear as uniformly green, often lens shaped, and commonly about 6 microns in diameter. A single leaf cell may contain 20 to 100 chloroplasts; each cell of a spinach leaf may have 500 chloroplasts; and a square millimeter of leaf surface may have one-half million chloroplasts. In these organelles, chlorophyll catalyzes the reactions of photosynthesis, thereby converting carbon dioxide and water to carbohydrate and oxygen. The oxygen is then liberated to the atmosphere.

Examination under an electron microscope reveals that chlorophyll is not uniformly dispersed in the chloroplast; rather, chlorophyll is concentrated in **grana** suspended in a clear **stroma**. The stroma contains protein. The chloroplast possesses a double-membrane structure similar to that of mitochondria, except that the inner membrane is not folded as it is in mitochondria. When grana are further magnified, it becomes apparent that chlorophyll is contained in compressed stacks of paired **lamellae**. These disc-like lamellae are called **thylakoids**. (*Thylakos* is a Greek word for "sac.") Grana are interconnected by

## ❧ Notes ❧

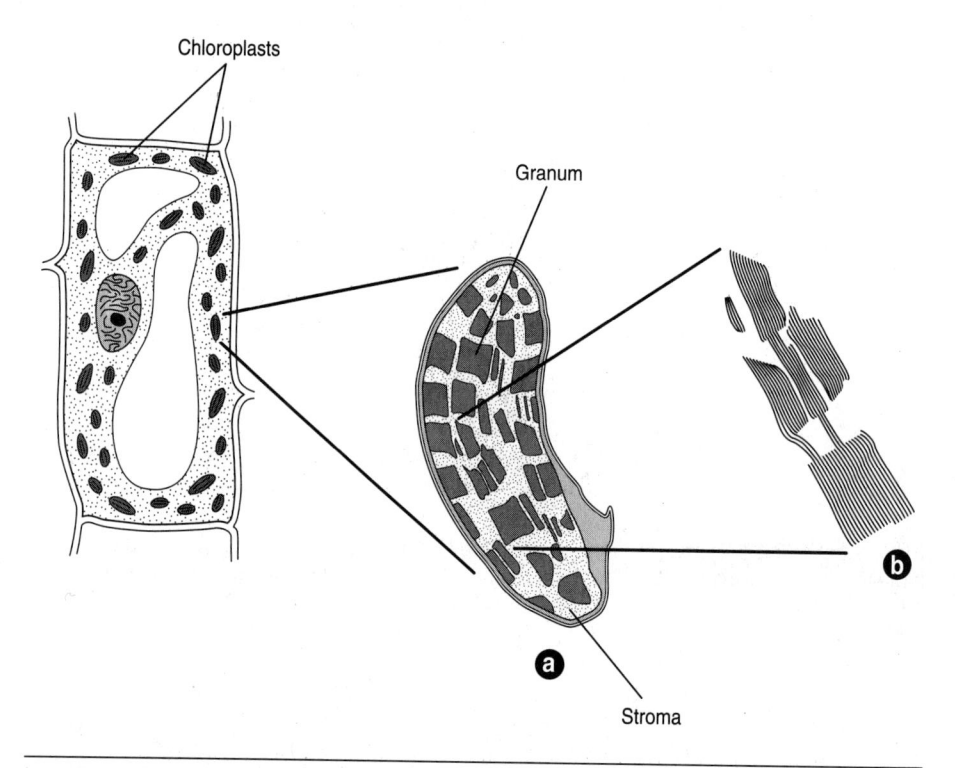

**Figure 2-10** Under low magnification a chloroplast appears uniformly green. High magnification (a) however, reveals that the chlorophyll is in discrete bodies, called grana, which are surrounded by a clear stroma. Higher magnification (b) reveals that the chlorophyll is arranged in a lamellar fashion.

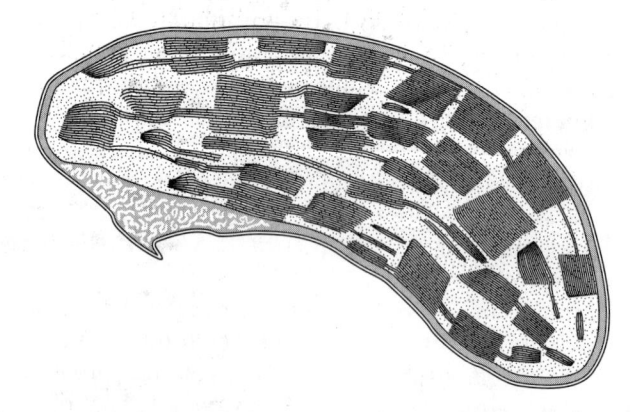

**Figure 2-11** A chloroplast from a cell of *Zea mays* (corn). The grana are distinct and appear to be constructed of a stack of lamellae, or thylakoids. The grana are interconnected by extensions of the lamellae.

**fret membranes**. The arrangement of chlorophyll in the structure of the lamellae is important to the chlorophyll's capacity to carry on photosynthesis.

Chlorophyll is linked to proteins, and the resemblance of chlorophyll's molecular structure to that of hemoglobin is remarkable. One significant difference between the two types of molecules is found at the centers of the molecules. A chlorophyll molecule possesses an atom of magnesium, while a hemoglobin molecule contains an atom of iron. There are several variations on the molecular structure of chlorophyll, the different forms being found in different groups of plants. All photosynthetic organisms (except several forms of bacteria) have chlorophyll a. Flowering plants have two forms of chlorophyll: chlorophyll a and chlorophyll b. Certain algae have chlorophylls c and d. Photosynthetic bacteria have their own type of chlorophyll.

Chloroplasts contain their own DNA and thus are able to make a number of their own components. Chloroplasts divide independently of the cells in which they reside, although the first formed chloroplasts arise from **proplastids**. Chloroplasts are not, however, totally autonomous; some of their components are supplied by the cell.

There are some interesting speculations regarding the origin of chloroplasts. As was mentioned earlier, Professor Lynn Margulis of Boston University proposes that eukaryotic cells arose from prokaryotic cells by invasion, or endosymbiosis. According to this theory, nuclei had their origins through a prokaryote entering another prokaryote and taking up residence there. The same reasoning has been applied to the origin of chloroplasts; that is, that they entered cells by being ingested. A number of researchers have voiced objection to this theory, however. While nuclei, chloroplasts, and mitochondria are all double membraned structures and all contain their own DNA, they do greatly different things. For this reason some researchers believe that these structures came about through evolutionary trends rather than through ingestion.

## Cilia

Many microscopically sized plants and certain fungi contain hairlike structures that project out from the cell surface. These structures are used to propel the cells through the water and are called **cilia** or flagella. In many plants, cilia or flagella are found only in sperm cells. There is little difference between cilia and flagella except for length (flagella tend to be longer), and method of movement. An electron microscope reveals the same structure for both. A cross-sectional view of a cilium shows a circle of nine pairs of microtubules, with two single microtubules in the center. Each microtubule possesses thirteen longitudinal filaments. This structure is universal for all cilia and flagella except those occurring in bacteria. Flagella and cilia grow out from an organelle called the **basal body**.

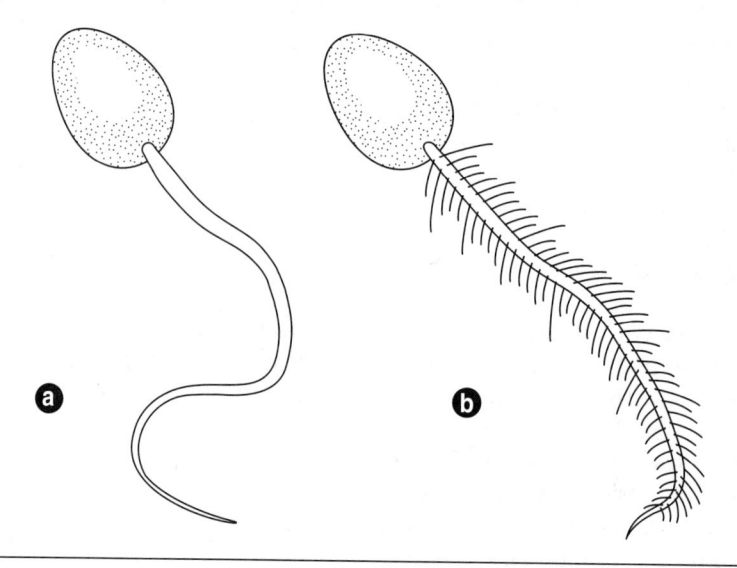

**Figure 2-12** Cilia: (a) whiplash and (b) tinsel type.

# Plastids

**Plastids** are cellular organelles that become specialized to serve different purposes. Surely the most significant plastids are the chloroplasts, which were discussed separately. Two other types of plastids are **leucoplasts** and **chromoplasts**. Leucoplasts are colorless and function in the storage of starch and oil. Chromoplasts are colored plastids, having various pigments primarily **xanthophylls** and **carotenes**. **Anthocyanins** are water soluble and found in vacuoles. Xanthophylls and carotenes are fat soluble and are found in plastids. Chromoplasts tend to be yellow, orange, or red. Tomatoes, oranges, and carrots owe their colors to chromoplasts.

Plastids are commonly disk shaped, sphere shaped or shaped like a double convex lens. Shape as well as color varies with the tissue and species. Plastids are interchangeable in form. Leucoplasts, such as those in a potato, will become chloroplasts if exposed to light. Chloroplasts can become chromoplasts. Granules of chromonucleoprotein have been found in plastids, suggesting that plastids have their own genetic mechanism.

Plastids frequently have their beginnings as proplastids in **meristematic** cells. These proplastids appear as minute bodies scarcely visible through a light microscope. They may divide by simple constriction, resulting in large numbers of them. They grow into mature plastids as the cells mature.

# Vacuoles

In addition to a nucleus, mitochondria, Golgi bodies, endoplasmic reticulum, and ribosomes, a plant cell contains small droplets of watery solution called the *cell sap*. The droplets are called **vacuoles**. Vacuoles are almost universally present in plant cells and are surrounded by a membrane that separates their contents from the rest of the cytoplasm. Most young, meristematic cells, which are continuously dividing, have numerous small, round or drawn out vacuoles (although the cells of **cambium**, also actively dividing cells, seem instead to have a large vacuole). These vacuoles coalesce as the cells mature so that older cells tend to have a single, large vacuole. The vacuole may occupy most of the volume of the cell.

The concentration of substance dissolved in the vacuolar sap is often far greater than that in the surrounding cytoplasm and may be great enough to come out of solution and to form minute crystals. For example, analysis of the vacuolar sap of Nitella, a green alga, discloses a tenfold greater concentration of calcium ion in the vacuole than in the surrounding cytoplasm, and a one hundredfold greater concentration of potassium ion.

What is suspended in the water of a vacuole? Sugars, salts, pigments, and some enzymes. The composition of cell sap varies in different plants and also varies under changing conditions. Water may compose up to 98 percent of cell sap. Sodium, potassium, magnesium, and calcium ions are also generally present. Carbohydrates, nitrogenous compounds, proteins, amino acids, and probably waste products from the cell metabolism are commonly present. There may also be dissolved gases, specifically oxygen and carbon dioxide. Enzymes such as **diastase** and **invertase** are also common, and both acetic acid and formic acid are thought to be present in all living cells. Finally, many vacuoles contain anthocyanins, or colored pigments. The yellow, pink, and blue of flowers are often imparted by anthocyanins residing in vacuoles. Red cabbage and beets also owe their color to pigments in vacuoles.

The membrane that bounds the vacuole is similar to the cell membrane. Both are selectively permeable; both allow the free passage of water; and both have a certain control over what solutes are allowed to pass.

**Chrysophyta** algae have unicellular, flagellated cells, each possessing a reservoir at the attached end of the flagella. Just beneath the reservoir are one or more contractile vacuoles, which can discharge their contents out of the cell. These pulsating vacuoles alternately contract and expand. In this way, they can both get rid of waste products and regulate water content.

Are vacuoles alive? While they are certainly not passive, the contents of vacuoles are not chemically active; vacuoles are therefore not considered living. Yet, another interesting question arises: Is any constituent of a cell alive when considered by itself, isolated from the rest of the cell?

Vacuoles play an important part in the sexual process observed in Spirogyra, a green alga. The conjugation of Spirogyra is described in Chapter 13. Two filaments of Spirogyra come to lie side by side when conjugation bridges form between adjacent cells. One of the filaments takes on the role of female (being a receiver), and the other adopts the role of male (giving its substance). The protoplasts of the male cells migrate across the bridges and unite with the protoplasts of the female cells. As a result the receiver cells have twice as much cytoplasm as before conjugation; but they are not twice as large. The vacuoles play a significant role here, functioning

### Notes

as pumping stations; they pump water out of the female cell. So much water is removed that when the two protoplasts unite, the resulting zygote does not even fill the cell.

## A Single-celled Imposter

Having discussed the characteristics of cells at some length, to mention *Caulerpa floridana* ("turtle grass") may seem to undo it all. Here is an organism that seems to have leaves, stem, and roots (or rhizoids); an organism that can reach a length of two feet or more. Yet it is a *single cell*! Although it contains great numbers of nuclei, mitochondria, and chloroplasts, it has no cross walls, and, hence, is not divided into a number of cells. This puts *Caulerpa* in the kingdom Protista; and to be faithful to the concept of five kingdoms, it is not a plant.

## Questions for Review

1. What distinguishes a living organism from a nonliving substance?
2. How does one distinguish between cytoplasm and a vacuole?
3. What is the difference between cytoplasm and protoplasm?
4. Characterize the properties of plant cell walls.
5. Describe the characteristics of each of the following: mitochondria, Golgi bodies, endoplasmic reticulum, and nuclear membrane.
6. Who gave the name to what we now call cells? Was he really looking at what we regard as cells? Explain your answer.
7. Who is credited with establishing the cell doctrine? What is the cell doctrine? Is it correct?
8. Describe the ultrastructure of chloroplasts.
9. Name some types of plastids other than chloroplasts.

## Suggestions for Further Reading

Avers, C.J. 1978. *Basic Cell Biology.* New York: D. Van Nostrand Company.

Jensen, W.A. 1978. *The Plant Cell.* Belmont, CA: Wadsworth Publishing Co.

Jensen, W.A., and R.B. Park. 1967. *Cell Ultrastructure.* Belmont, CA: Wadsworth Publishing Co.

Loewenstein, W.R. 1970. Intercellular communication. *Scientific American,* May, 222:10, 78–84.

# 3

# Mitosis and Meiosis

**W**hen mitosis occurs, cell division usually but not always takes place. Mitosis, therefore, is not the same as cell division. The term *mitosis* applies only to nuclear division. Two terms are commonly used to make the distinction between cellular and nuclear division. **Karyokinesis** refers to the division of the nucleus, and **cytokinesis** refers to the division of the cell. Karyokinesis can take place without cytokinesis.

More than one-half century before the details of mitosis were completely understood, cell division had already been observed many times. Wilhelm Hofmeister (1824–1877) described much of mitosis but apparently did not recognize its significance. The stages of mitosis were perhaps best described during this time period by Klein in 1880. Many hundreds of hours of study were undoubtedly involved in his study.

## The Basics of Mitosis

In mitosis, the daughter nuclei each retain the same number of chromosomes as the parent nuclei; in meiosis, by **reduction division**, the daughter nuclei each have half the chromosome number of the parent. We thus speak of the **diploid** number and the **haploid** number of chromosomes. Common practice is to represent the haploid number with the letter N. The diploid number, therefore, is 2N. In a general sense, all nonreproductive cells in a body have the same number of chromosomes—that is, the diploid number. This is clear because a body grows in size by mitosis and cell division. All members of the same species, with few exceptions, have the same number of chromosomes. Many know that the human chromosome number is 46 (2N = 46), the haploid number being 23. The chromosome number of cabbages is 18 (2N = 18); corn has 20 chromosomes; sunflowers have 34; and plums have 48. In the animal kingdom, cats have 38 chromosomes, dogs have 78, and crayfish have approximately 200. In the plant kingdom, the fern *Ophioglossum vulgatum* has 500 chromosomes.

❃ Notes ❃

## ❧ Notes ❧

The process of mitosis, once initiated, is a continuous flow. It is convenient, however, to designate the various stages with names. Figure 3-1 shows the events in animal mitosis, provided for purposes of comparison. The student who is already familiar with these details may wish to proceed to the section on plant mitosis (later in this chapter).

The top left illustration in figure 3-1 shows a cell in which there is no evidence of nuclear division. This is the **interphase** of mitosis. While this is sometimes called the **resting stage**, resting it is not. In fact, much synthetic activity is occurring. The Golgi bodies, endoplasmic reticulum, and ribosomes all increase to provide enough for two cells, should cell division occur. The DNA is replicated, and protein is synthesized. Mitochondria and chloroplasts also reproduce themselves at this time. Considering the entire time lapse from the beginning of interphase to the beginning of the next interphase, approximately 90 percent of the time of mitosis is spent in interphase.

Interphase   Early Prophase   Prophase   Metaphase

Anaphase   Telophase   Cell Division   Daughter Cells

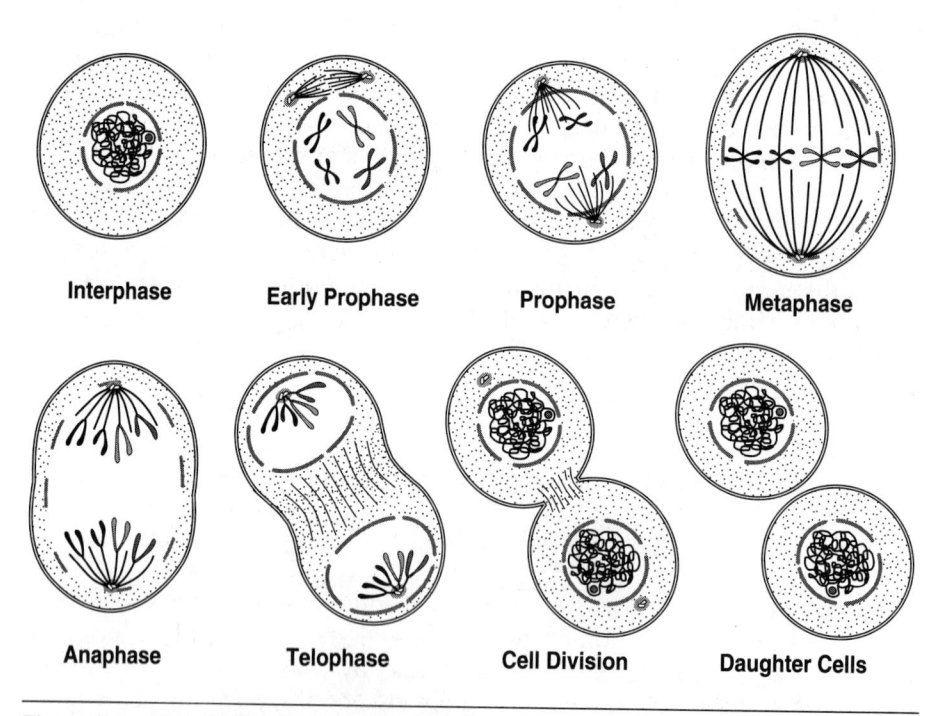

**Figure 3-1**  The events of mitosis in animal cells: interphase, showing an intact nuclear membrane and two centrioles lying side by side; early prophase, when the centrioles begin to move apart and a spindle begins to form; in prophase, the chromosomes replicate but remain attached at the centromeres; metaphase, only the centromeres of the chromosomes lie on the equatorial plate; anaphase; telophase, when the chromosomes have reached the centrioles at the poles; cell division, the process of mitosis having been completed; and daughter cells.

The next stage in mitosis is **prophase**. During prophase, the centrioles begin to move apart, the nuclear membrane disappears, and the nuclear material forms what first appear to be fine threads and then clearly takes the form of chromosomes. The chromosomes are then free in the cytoplasm. In prophase, the chromosomes replicate, and their number doubles. The example in figure 3-1 shows a cell with a diploid number of 4 ($2N = 4$). When chromosomes replicate, giving the appearance of splitting down the middle, they do so faithfully, particle by particle. The products of this process are two members of the same composition. These members remain temporarily attached to each other at a point called the **centromere**, or **kinetochore**.

The chromosomes migrate to a position along an equatorial plate, and the centrioles continue their movement until they lie at opposite **poles** of the cell, one on each side of the centrally placed chromosomes. This is called **metaphase**. Careful examination reveals **spindle** fibers running from one centriole to the other and from the centriole to the centromere of the replicated chromosomes.

The chromosomes which had been attached next become disjoined and begin to move away from the **equatorial plate** and toward the centrioles. At this time, it appears that the spindle fibers attached to the centromeres contract, thus pulling the chromosomes toward the centrioles. This is called **anaphase** (see figure 3-2).

The migration of chromosomes continues until they reach the centrioles, at which time the cell is in **telophase**. During this stage, the chromosomes seem to lose their organization, becoming indiscernible as chromosomes; the centrioles divide; and new nuclear membranes are formed. The nucleolus also reforms at this time, apparently at a specific locus of a chromosome called the **nucleolar organizer**. Two examples are shown in figure 3-3.

This marks the completion of mitosis, and the daughter nuclei return to interphase. If cell division is to occur, it will become evident at this time.

✺ **Notes** ✺

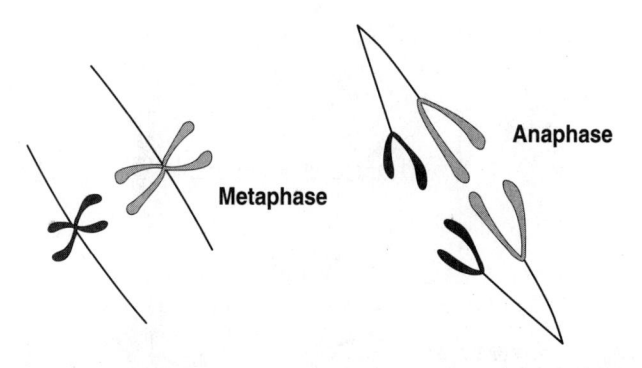

**Figure 3-2** Chromosomes replicate during prophase. At left, they are shown at metaphase. They then part beginning at the point of the centromeres and go into anaphase.

❋ **Notes** ❋

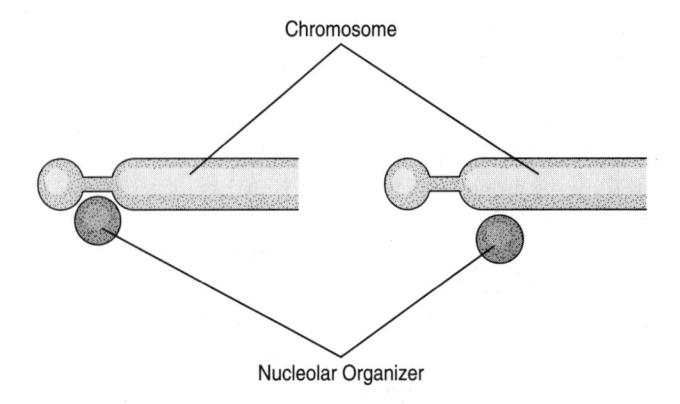

**Figure 3-3** The nucleolar organizer can nestle against a chromosome at two possible positions.

## Plant Mitosis

Thus far a clear distinction has been made between nuclear division (karyokinesis) and cellular division (cytokinesis); and although the intent at this point is to describe mitosis, it is fitting that cell division be drawn into the discussion.

Plant mitosis is in many ways similar to animal mitosis. In animal cells, the cell membrane becomes constricted, being pinched in from the outside. This appears to be accomplished by microfilaments that behave in a manner similar to a purse string, pinching the cytoplasm into two parts. In plant cells, the first evidence of cell division is seen in the center of the cell. Here, a cell plate forms and grows progressively outward to meet the cell membrane. The cell plate arises from a line of vesicles produced by the Golgi bodies.

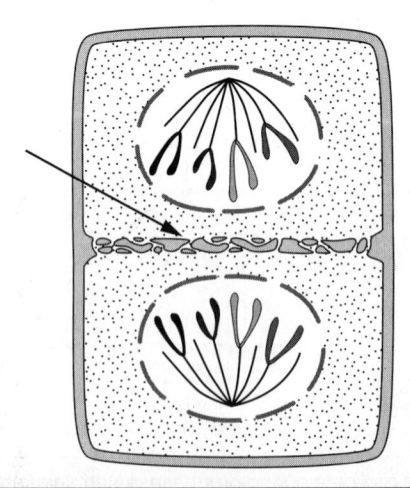

**Figure 3-4** When a plant cell divides, the new cell membrane that divides the two cells begins near the center and grows in an outward direction, as shown at the arrow.

An earlier statement regarding anaphase made reference to an impression that the spindle fibers contract to pull the chromosomes toward the centrioles at the poles. This, however, does not apply to a description of plant mitosis. For one thing, centrioles cannot be detected in most plant cells. In such plant cells, the chromosomes move toward the poles but not toward the centrioles. Even in those plant cells having centrioles, when the spindle fibers are cut and the centrioles removed by microdissection methods, the events of mitosis go on in a normal manner. If spindle fiber contraction were responsible for pulling chromosomes, cutting the fibers would bring chromosome movement to a halt.

Figure 3-5 comprises a series of drawings of plant mitosis. Because all body cells derive from mitotic divisions of the fertilized egg, an assumption can be made that all cells have the same genetic potential as does the egg—that any cell, so far as the nucleus is concerned, should be able to produce a complete and normal individual. This assumption can be tested by placing finely divided pieces of carrot in a solution that causes the intercellular cement to dissolve, thus allowing the cells to be freed from their neighbors. An individual cell can then be picked up and placed in a nutrient solution. Here, it may, by virtue of being freed from its neighbors, regain the ability to divide. Coconut milk is a favored medium for growing cells in nutrient culture. Carrot cells placed in coconut milk do recover the ability to divide, and complete carrot plants with stem, root, leaves, flowers, and seeds have been

❋ **Notes** ❋

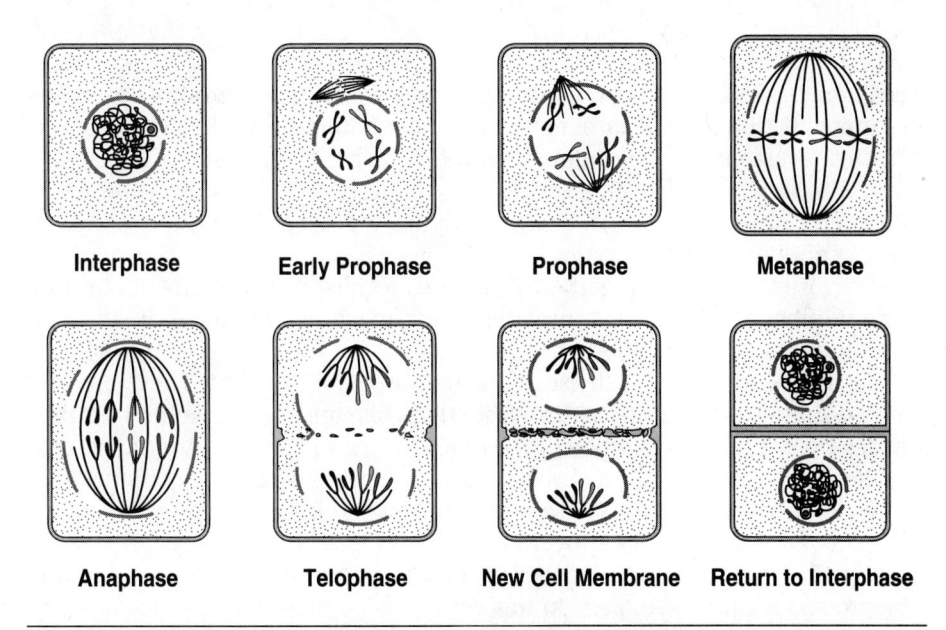

**Figure 3-5**  Plant cell mitosis. The steps are the same as in animal mitosis, but centrioles are not observed: interphase, early prophase, prophase, metaphase, anaphase, telophase, new cell wall beginning to form between daughter nuclei, and return to interphase.

**✣ Notes ✣**

produced from isolated carrot root cells. A proper word for this is **totipotency**. Any cell of a body has the capacity to produce a complete individual. This has been accomplished in the animal kingdom as well as in the plant kingdom. Frog intestine cells have been used successfully to produce frogs. This is done by enucleating an egg cell and then placing a nucleus from an intestine cell into the enucleated egg cell.

While the most common chromosome number of nonreproductive cells in plants is 2N, sometimes 3N, 4N, and 5N do occur in such cells. The condition is called **polyploidy** and is further described later in this chapter.

Although one would expect the rate of mitosis to be a constant, experiments concerning frequency of mitoses in onion root tips suggest otherwise. Onions of uniform size were selected and provided with conditions that favored root growth, including constant light and temperature. Each hour of the day, a number of root tips were examined and the number of mitoses recorded. The results were as follows:

| Time of day: | 8 | 9 | 10 | 11 | 12 | 1 | 2 | 3 | 4 |
|---|---|---|---|---|---|---|---|---|---|
| Number of mitoses: | 26 | 49 | 48 | 137 | 192 | 307 | 322 | 317 | 212 |

The preceding data show a maximum number of mitotic divisions at 2:00 P.M., when twelve times as many mitotic divisions occurred in an hour than were observed at 8:00 A.M. The frequency of mitoses, then, appears to be governed by **circadian** rhythms, or having a biological clock. (This topic is examined further in chapter 30.)

## The Basics of Meiosis

Recall that in fertilization a male gamete having a haploid chromosome number unites with a female gamete also having a haploid number. The union of gametes reestablishes the diploid condition. Each chromosome of an egg cell pairs with a like, or **homologous**, chromosome in the sperm cell; the union of gametes brings homologous chromosomes together. The chromosomes of the 2N condition occur in pairs.

Because the coming together of gametes results in a doubling of chromosome number, it follows that sometime there must be a reduction in chromosome number. This is meiosis.

There are two nuclear divisions in the process of **meiosis**. These are designated simply as Division I and Division II. While mitosis multiplies by 2 then divides by 2, meiosis multiplies by 2 then divides *twice* by 2, producing cells that have half the number of chromosomes as does the parent cell (figure 3-6).

In the early part of prophase I, the chromosomes appear as single rather than double strands. This appearance is deceptive. The chromosomes soon come to lie together in pairs. Pairing occurs between homologous strands. The process is called **synapsis**. In this coupling is a paternal chromosome and a maternal chromosome. Coupling begins at one or more points and proceeds in a zipperlike fashion along the entire length. This intimate pairing may give the impression that the chromosome number has been reduced;

except for the fact that there are two centromeres, only a single chromosome seems present. Such pairs of intimately associated chromosomes are called **bivalents**. Next, the strands replicate and are now called **chromatids**. At this point there are four parallel and closely associated strands called **tetrads**.

The four-stranded tetrads next begin to separate, one bivalent (two strands) moving toward one pole, and the other bivalent moving toward the opposite pole. The bivalents are still connected at the centromeres. This is anaphase I. Cell

❊ **Notes** ❊

**Meiosis**

**Prophase I**

**Reduction**

**Diakinesis**
recondensation
*(chromosomes reach their greatest density)*

*(homologs are separated from their partner, and the two are moved to opposite poles)*

**Diplotene**
coiling
*(homologous pairs become so widely separated that sites of cross-overs are evident)*

**Pachytene**
recombination
*(new gene alignments take place in the process of crossing-over)*

*(sister chromatids are split resulting in four gametes, each with a haploid number of chromosomes, all in the unduplicated state)*

**Zygotene**
pairing
*(each chromosome pairs with its homolog)*

**Leptotene**
condensation
*(chromosomes become visible as sister chromatids)*

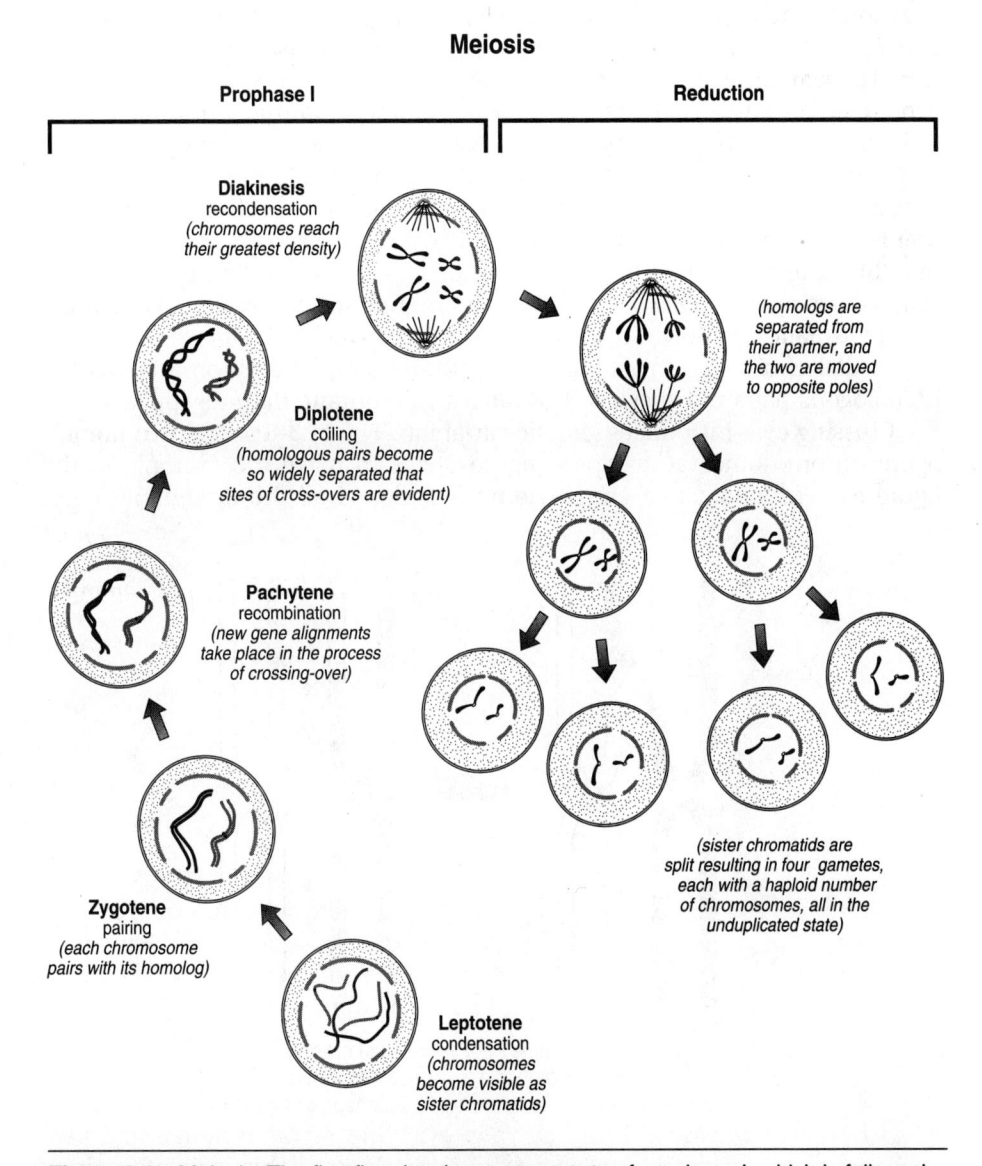

**Figure 3-6** Meiosis. The first five drawings are aspects of prophase I, which is followed by two cell divisions and a reduction in chromosome number.

## ❈ Notes ❈

division occurs, the bivalents remaining connected. Events then proceed in a manner similar to that of mitosis. The bivalents soon become disjoined. At this time the chromatids are again called chromosomes. The second nuclear division then occurs without any replication, and the chromosome number is thus reduced.

In figure 3-6, the first five cell stages shown are all part of prophase I, which takes more time than does the prophase of mitosis. The first five cell stages are **leptotene**, **zygotene**, **pachytene**, **diplotene**, and **diakinesis**, respectively. In leptotene, the chromosomes appear to be longitudinally single; in zygotene, homologous chromosomes become paired; and in pachytene, the pairing is completed, giving the appearance of chromosome number reduction. Halfway through pachytene, the bivalents appear to split longitudinally, producing tetrads. In diplotene, the paternal and maternal chromosomes in the four-stranded tetrad start to separate, the centromeres being the first parts to migrate. During this process, two of the four strands of chromatids may overlap, constituting an overlapping of male and female chromatids. These points of overlapping are called **chiasmata**. At the chiasmata, the chromatids may break and rejoin in new combinations, as shown in figure 3-7. This is called **crossing over**. Only two of the four chromatids are involved. While only two chiasmata are shown in figure 3-7, there may be several more.

In diakinesis, the chiasmata slip to the ends of the chromatids as the homologous pairs move apart. The nuclear membrane then disappears.

Crossing over introduces genetic variability. Figure 3-8 shows two homologous chromosomes at synapsis (figure 3-8a). To make the example in the figure easy to follow, one chromosome is assigned all the dominant genes,

Site of crossover or chiasma, where chromatids become physically joined

Result of crossover between two chromatids showing exchange of genes

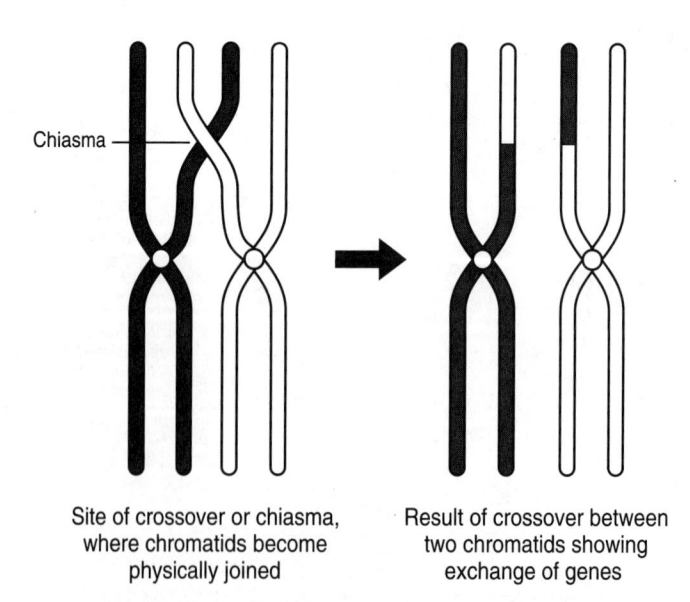

**Figure 3-7**  Chiasma formation followed by crossing over results in new combinations of genes.

A through G, and its homologue is assigned all the recessive genes, a through g. As seen in figure 3-8, the chromosomes replicate, producing a tetrad of four chromatids (figure 3-8b). In figure 3-8c, a chiasma occurs between the genes D and E, and d and e. If nothing further happens—that is, if no break occurs or if a break does occur but mends without an exchange

※ **Notes** ※

**(a)** DNA replication producing sister chromatids

**(b)** Synapsis of homologous chromosomes producing a tetrad of chromatids

**(c)** Single chiasma

**(d)** Result of crossover between two chromatids

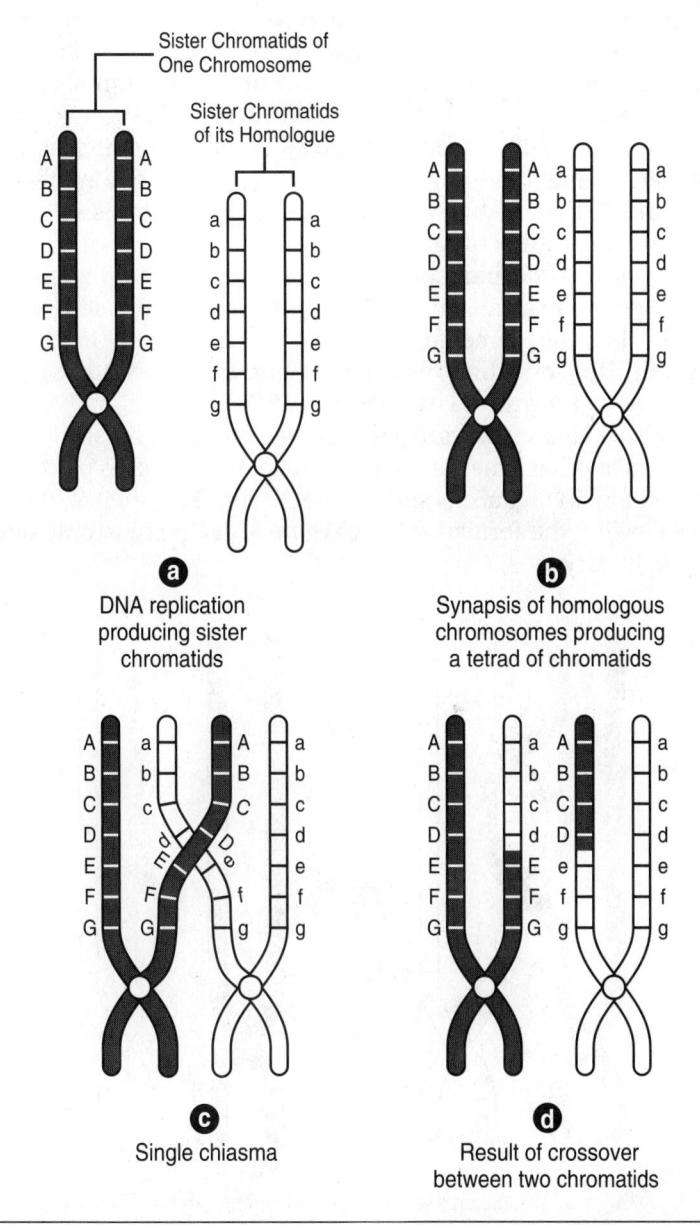

**Figure 3-8** Another example of crossing over: (a) synapsis; (b) replication produces a tetrad of chromatids: (c) a single chiasma; and (d) crossover has taken place between two chromatids.

### ✻ Notes ✻

of parts—things will progress the same as if no chiasma had taken place. If, on the other hand, a crossover occurs at the chiasma, a new combination of genes will result. One of each of the pairs of chromatids will remain unchanged, as do the outer ones in figure 3-8d. The inner two chromatids, however, will exchange parts and, thus, produce new combinations.

Two crossovers can also occur between two genes, as shown in figure 3-9. In this way the genes, here A and G, end up still linked together.

Thus, genes located on the same chromosome are not unalterably linked. They can be separated by crossing over. Further, if chiasmata and crossing over take place as readily at one position on the chromatid as at another (the chiasmata being random), crossing over may occur more frequently between A and G than between A and B. This is the basis of constructing a map of the positions of genes along the length of a chromosome. A. H. Sturtevant did the pioneering work on gene mapping, based on the concept that genes are arranged in a linear series along the chromosome and the principle that the percentage of crossover is related to the distances between genes. He worked not with plants but, rather, with the famous fruit fly, *Drosophila melanogaster*. The year was 1913.

In animals, meiosis always occurs at the time of gamete formation. In plants, it occurs at other times. In higher plants, meiosis takes place at the time of spore formation. In organisms such as the green alga Spirogyra, meiosis is the first event following the formation of a zygote; thus, there is only one diploid cell in the life history.

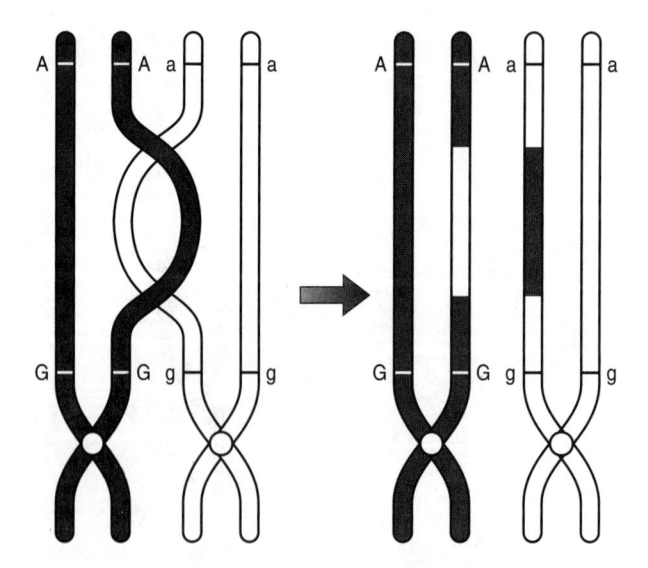

Double crossover between two chromatids

**Figure 3-9**  If two crossovers should occur between two genes, they effectively cancel each other, and the genes continue to be linked.

## Aberrations in Mitosis and Cellular Composition

Aberrations in mitosis can be induced by altering the environment. Alternating hot and cold as well as treatment with certain chemicals such as mustard gas, naphthalene acetic acid, and colchicine can cause changes in mitosis. The normal events at anaphase result in chromatids separating and moving to opposite poles. However, a process called **nondisjunction** can occur. In nondisjunction, both members of a pair go to the same daughter cell. This results in **aneuploidy**: one daughter cell will have 2N+1 chromosomes, and the other daughter cell will have 2N−1 chromosomes. This kind of aberration can occur in meiosis as well.

Some families of plants have species that naturally differ in the number of chromosome sets per cell nucleus. As mentioned earlier, this cellular condition is called polyploidy. A certain rose, for instance, has races or varieties having N, 2N, 3N, 4N, and on to 16N. A similar situation occurs in wheat, where the basic number is 7N. The wheat used for making flour is 6N (6 × 7 = 42 chromosomes). The cells of roots are frequently polyploid. In corn (*Zea mays*), most of the cells of the root have two to four times as much DNA as do normal diploid cells. Certain cells associated with the conduction of water have been found to be 32N, or 16 times the diploid number.

Aside from changes in the number of chromosomes, the structure of individual chromosomes can also be altered. Examples of such alterations are **deletion**, **duplication**, **inversion**, and **translocation**. In deletion, part of the chromosome breaks away and is therefore unable to go through mitosis. The normal pairing of homologous chromosomes produces two parallel strands, but deletion produces another pattern of pairing.

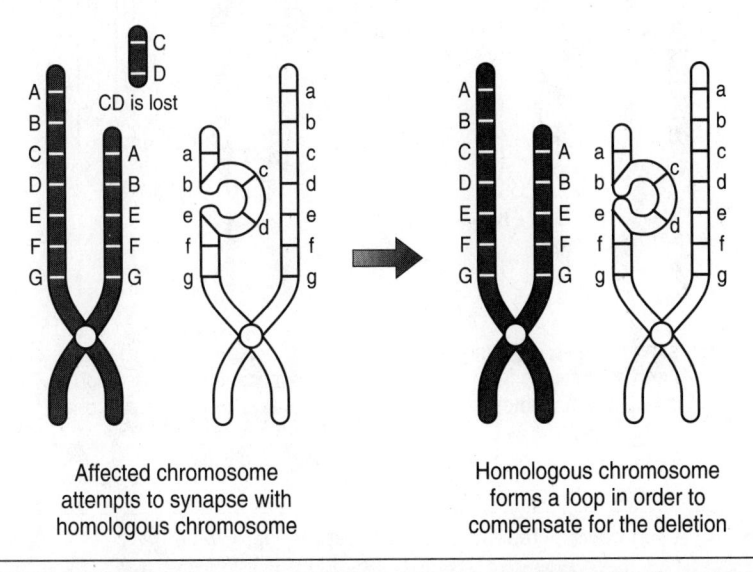

**Figure 3-10** A deletion. Genes C and D have been knocked out of the chromosome. Consequently, when the chromosome with the deletion seeks to synapse with a normal chromosome, a portion of the normal chromosome has nothing with which to pair and is represented as a loop.

## 🌸 Notes 🌸

An opposite kind of change happens in duplication. As shown in figure 3-11, the given genes A and B are represented twice. A chromosome may also break and then mend. If it mends in the same place, no alteration occurs; but if a break occurs and the broken part instead inverts before mending, a new sequencing of the genes results. This is inversion (figure 3-12).

Reciprocal translocation is a frequent chromosomal aberration and its occurrence in the Evening Primrose *Oenothera* has been much studied. At the left side of figure 3-13 are two pairs of chromosomes. One pair has the genes A through H, and the other pair has the genes I through P. A break occurs in one of each of the chromosomal pairs. In mending, the segments

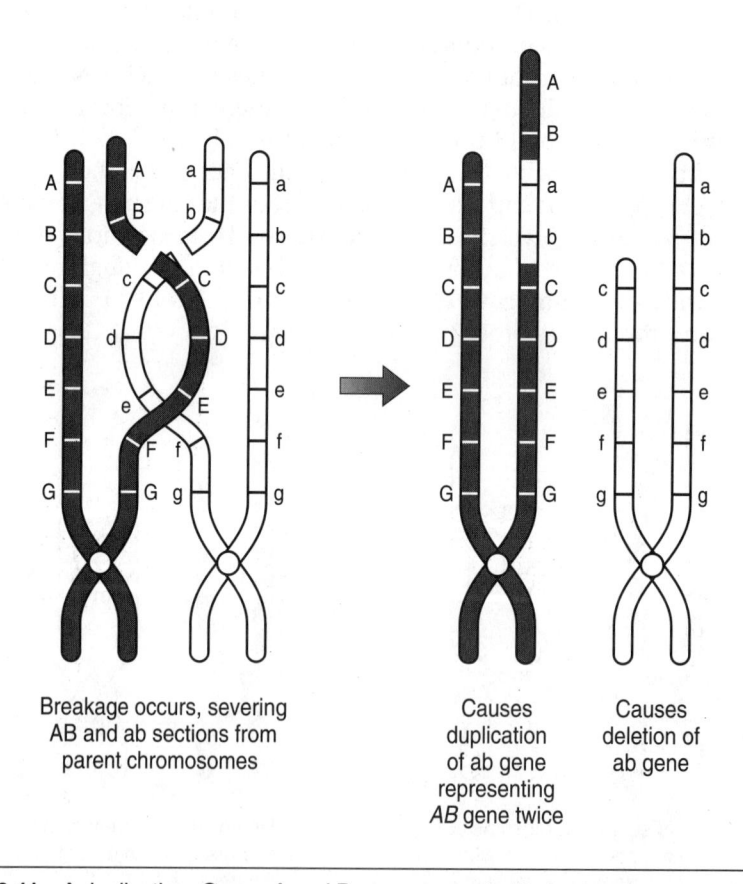

Breakage occurs, severing AB and ab sections from parent chromosomes

Causes duplication of ab gene representing *AB* gene twice

Causes deletion of ab gene

**Figure 3-11** A duplication. Genes A and B are represented twice.

trade places. When synapsis takes place, only homologous parts are able to pair. As shown in the figure, the chromosomes can synapse only along half of their lengths, and the interchanged segments cannot synapse normally because they are not homologous. Synapsis is achievable, however, as represented in figure 3-14. Two normal chromosomes position as shown at left in the figure and the altered chromosomes can then synapse as shown at right. When the chromosomes later separate during anaphase, they take the form of a circle. In *Oenothera biennis* (7N), translocations have taken place in a manner to create a circle of fourteen chromosomes during anaphase.

❉ **Notes** ❉

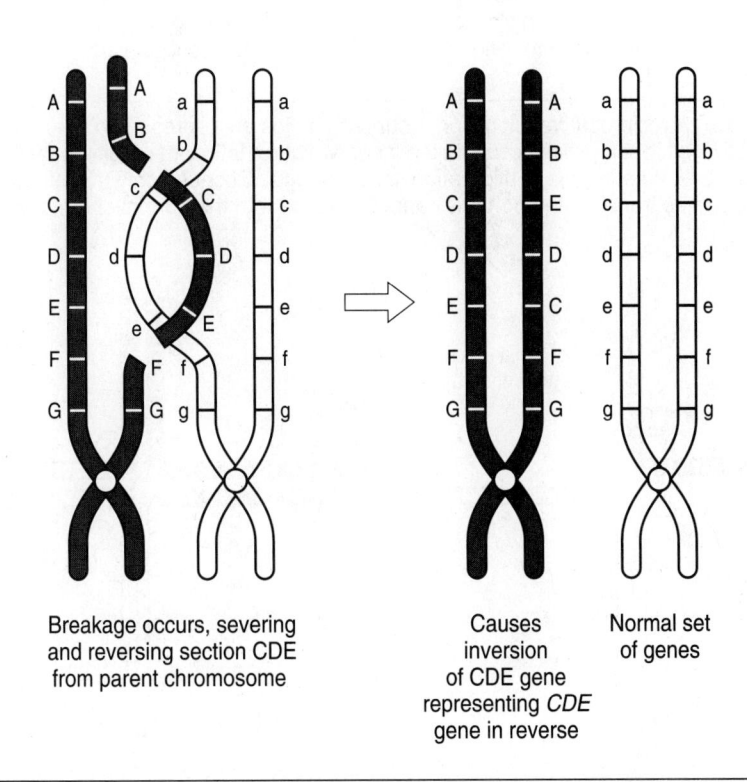

Breakage occurs, severing and reversing section CDE from parent chromosome

Causes inversion of CDE gene representing *CDE* gene in reverse

Normal set of genes

**Figure 3-12**   An inversion. Genes C, D, and E are reversed.

## ❊ Notes ❊

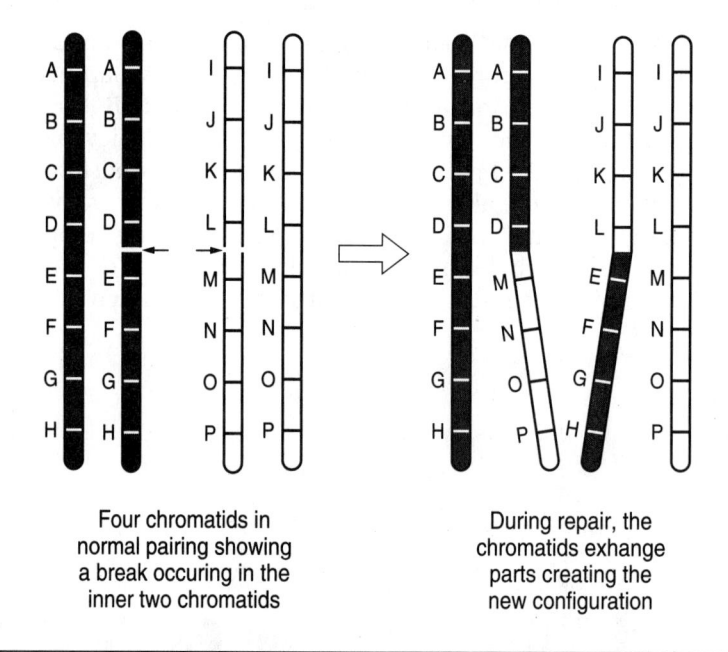

**Figure 3-13** A reciprocal translocation. Four chromatids are represented in a normal pairing at the left, but a break occurs in the inner two chromatids. In repairing, they exchange parts, creating the configuration shown at right. Because only homologous parts can synapse, the translocated portions cannot synapse with the normal chromatids.

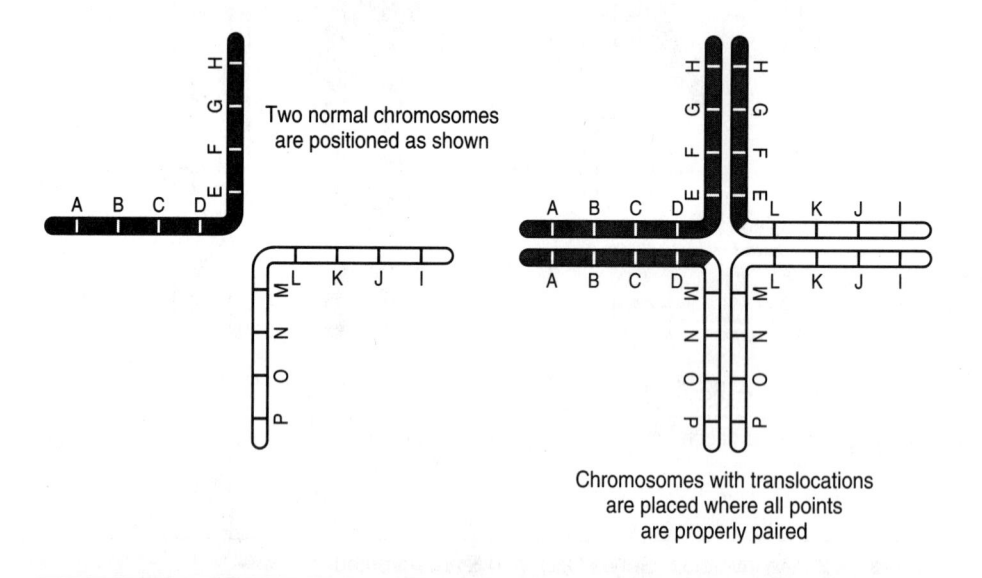

**Figure 3-14** Synapsis of chromosomes where a reciprocal translocation has taken place. Two normal chromosomes are positioned as shown at left. The chromosomes with translocations can then be effectively placed so that all points are properly paired, as shown at right.

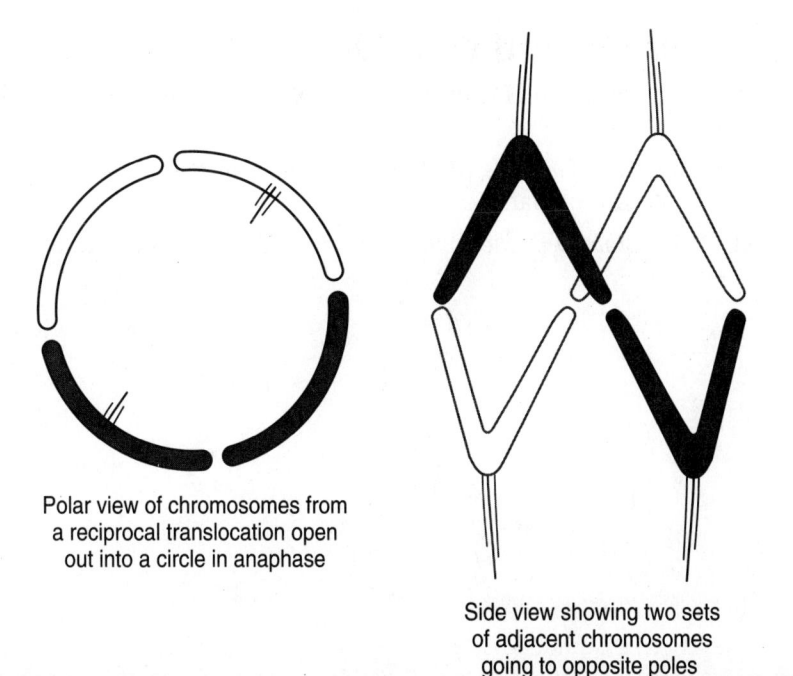

**Figure 3-15** When the chromosomes from a reciprocal translocation go into anaphase, they open out into a circle, left. Adjacent chromosomes go to opposite poles, as shown in the side view at right.

Polar view of chromosomes from a reciprocal translocation open out into a circle in anaphase

Side view showing two sets of adjacent chromosomes going to opposite poles

## Questions for Review

1. What is *crossing over* and when does it occur?
2. At what phase do chromosomes align on the equatorial plate, and at what phase are they migrating toward the poles?
3. What is the significance of meiosis in relation to sexual reproduction?
4. How is mitosis different from fission?
5. Distinguish between mitosis and cell division.
6. Define the terms *cytokinesis* and *karyokinesis*.
7. What are homologous chromosomes?
8. What is happening during the interphase of mitosis?
9. What phase of mitosis consumes the most time?
10. What organelle resides within the nucleus?

✻ **Notes** ✻

## Suggestions for Further Reading

Raven, P.H., R.F. Evert, and H. Curtis. 1981. *Biology of Plants*. New York: Worth Publishers.

Suzuki, D.T., and A.J.F. Griffiths. 1976. *An Introduction to Genetic Analysis*. San Francisco: W.H. Freeman Co.

Wolfe, S.L. 1972. *Biology of the Cell*. Belmont, CA: Wadsworth Publishing Co.

# 4

# Chemistry

Many students enter botany courses without having studied chemistry. Because much of what is presented in a botany course requires some understanding of chemistry, a basic discussion of this subject—and **organic** chemistry, in particular—is warranted.

## Empirical and Structural Formulas

Organic chemistry can be simply defined as the chemistry of carbon compounds. Inorganic chemistry, then, is all the rest. In inorganic chemistry, empirical formulas for compounds utilize the appropriate atomic symbols. For example, ammonia has one atom of nitrogen and three atoms of hydrogen in its **molecule**. Its empirical formula, therefore, is $NH_3$. In organic chemistry, however, structural formulas are necessary. In organic chemistry, ammonia (an inorganic compound) would thus be represented as in figure 4-1. The lines represent single bonds. The structural formula for carbon dioxide, $CO_2$, is shown in figure 4-2.

❀ Notes ❀

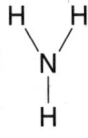

**Figure 4-1** Structural formula for ammonia.

**Figure 4-2** Structural formula for carbon dioxide.

❈ **Notes** ❈

Figure 4-2 shows that the oxygen atoms in carbon dioxide are attached to the carbon atoms. Further, two bonds are involved in the linkage. We may say, then, that oxygen has two bonds and carbon has four bonds.

Now consider the structural formula for methane (figure 4-3). This formula shows that the hydrogen atoms are attached to the carbon atoms, a fact that would not be evident if the empirical formula $CH_4$ were used. Such is the case with all compounds. Using methane as a starter, a whole series of compounds can be represented, as shown in figure 4-4.

The 4-carbon compound butane further shows the necessity of structural formulas. Both compounds shown in figure 4-5 can be represented by the empirical formula $C_4H_{10}$. They are, however, quite different compounds. The first is normal butane (n-butane), and the second is iso-butane, a substance with significantly different properties.

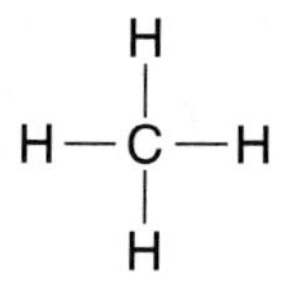

$$\begin{array}{c} H \\ | \\ H-C-H \\ | \\ H \end{array}$$

**Figure 4-3**  Structural formula for methane.

methane   ethane   propane   butane

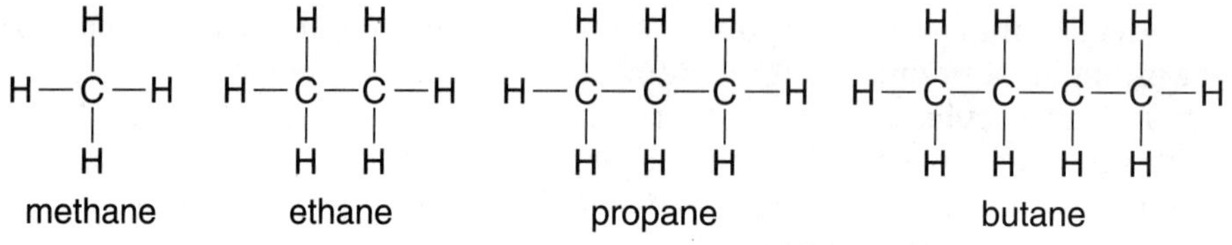

**Figure 4-4**  Structural formulas for methane, ethane, propane, and butane.

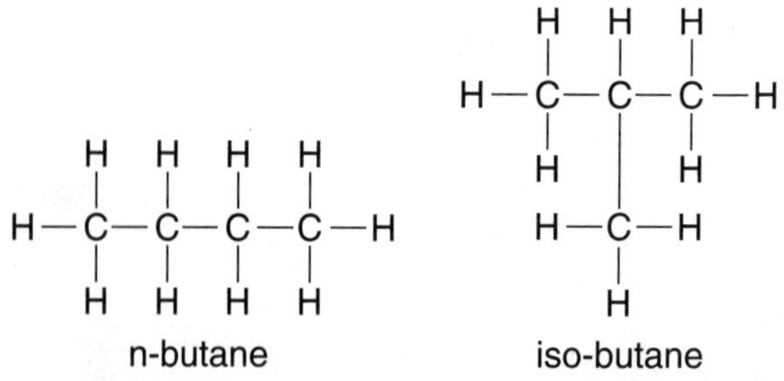

n-butane   iso-butane

**Figure 4-5**  Structural formulas for n-butane and iso-butane.

## Alcohols

If one of the hydrogen atoms of methane is replaced by an −OH group, the compound shown in figure 4-6 results. This is methyl alcohol. From this compound a whole series of alcohols can be created, as shown in figure 4-7.

## Organic Acids

If the terminal carbon atom of such a series bonds with an oxygen atom and an −OH group, as shown in figure 4-8, an organic acid is created. The group −COOH is called a **carboxyl** group (figure 4-9). Note that in all cases the carbon atom always has four bonds, the hydrogen atom always one bond, and the oxygen atom always two bonds.

Two organic compounds of this type can react together to form a longer chain, as shown in figure 4-10.

Because hydrogen is removed to create these compounds, this is referred to as a **dehydrogenation** reaction.

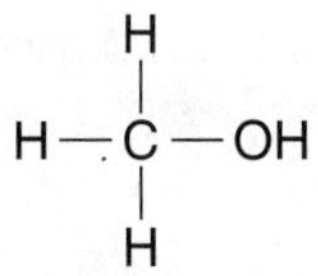

**Figure 4-6** Structural formula for methyl alcohol.

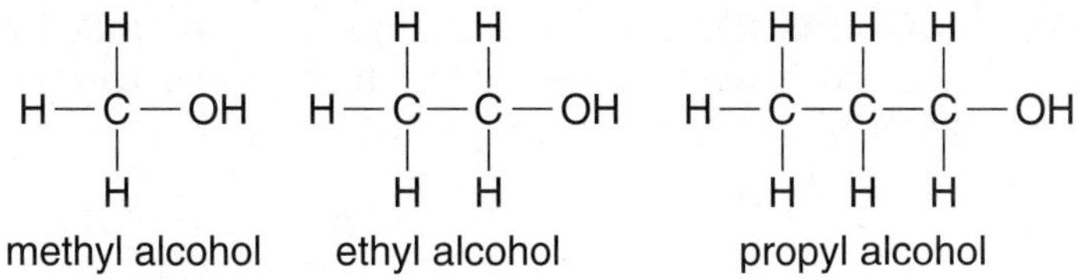

**Figure 4-7** Structural formulas for methyl alcohol, ethyl alcohol, and propyl alcohol.

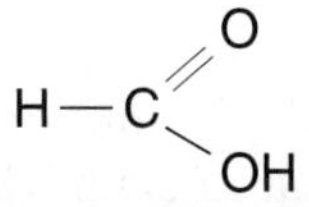

**Figure 4-8** Structural formula for formic acid.

## Notes

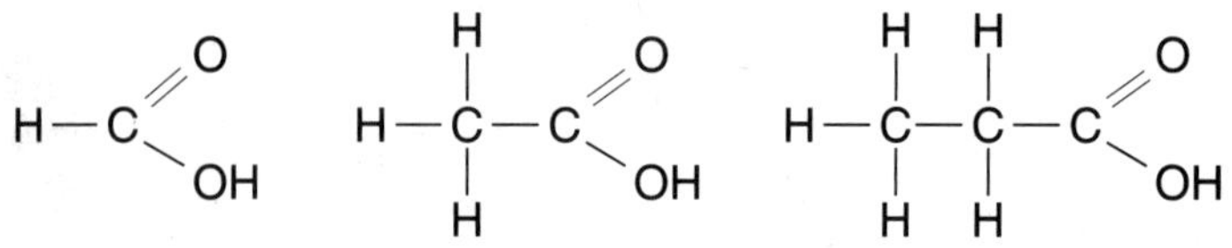

formic acid     acetic acid     propionic acid

**Figure 4-9** Structural formulas for formic acid, acetic acid, and propionic acid.

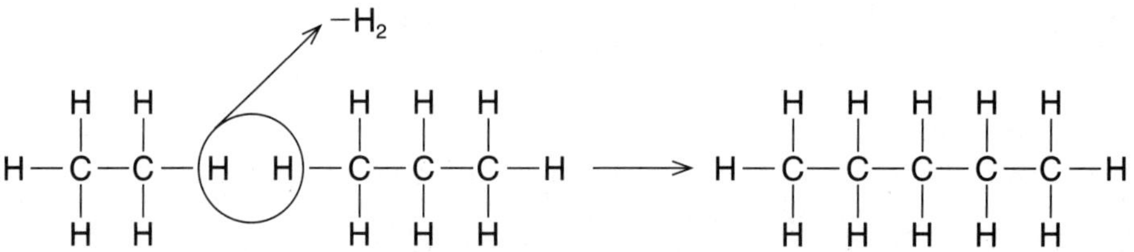

ethane + propane = pentane

**Figure 4-10** A dehydrogenation reaction causing the linkage of ethane and propane to produce pentane.

## Amino Acids

If a hydrogen atom that is attached to a carbon atom adjacent to a carboxyl group is replaced with $-NH_2$ (an amino group), the compound becomes an amino acid, as shown in figure 4-11. While this could be called an amino acetic acid, its more common name is glycine. If propionic acid is converted to an amino acid, alanine results (figure 4-12).

## Polymers

Amino acids are linked together to produce proteins. Thus, when proteins are digested, they are broken down into amino acids. The link between amino acid units is called a **peptide bond**. Uniting two amino acids is achieved by removing one molecule of water (known as a **dehydrolysis** reaction). The coupling takes place between the terminal carbon atom of one amino acid and the nitrogen atom of the other amino acid (figure 4-13). Adding a third amino acid to a union of two amino acids yields a **tripeptide**. To break such a compound into separate amino acids, one molecule of water must be added to the reaction (a **hydrolysis** reaction).

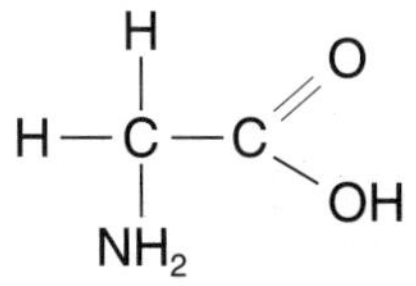

**Figure 4-11** Structural formula for glycine, an amino acid.

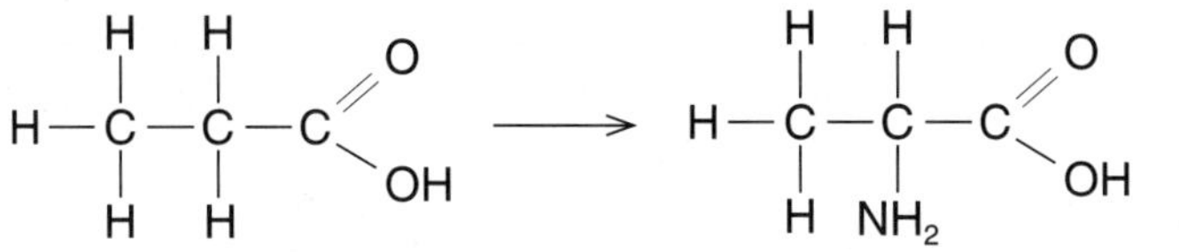

**Figure 4-12** Alanine results when propionic acid is converted to an amino acid.

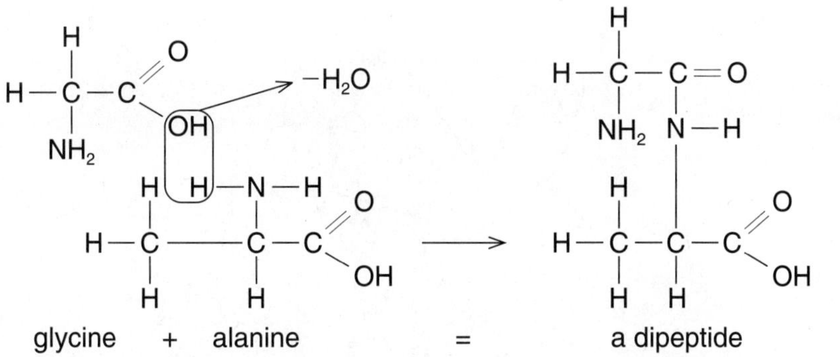

glycine + alanine = a dipeptide

**Figure 4-13** Two amino acids linked by a dehydrolysis reaction and by a peptide linkage produce a dipeptide.

## Proteins

Proteins are composed of long chains of amino acid units, which may be sequenced in any order. This makes possible a great variety of proteins. There are, however, a limited number of amino acids, some twenty of which are of interest to biologists. Long chains of repeating units of such amino acids are called **polymers**.

A given type of protein has a given sequence of amino acids, as shown in figure 4-14. The characteristics (specifically, the biological properties) of a protein can be attributed to both the amino acid sequencing and the bends and folds that occur along the chain of amino acid units (which are also constant for any particular protein) (figure 4-15).

※ **Notes** ※

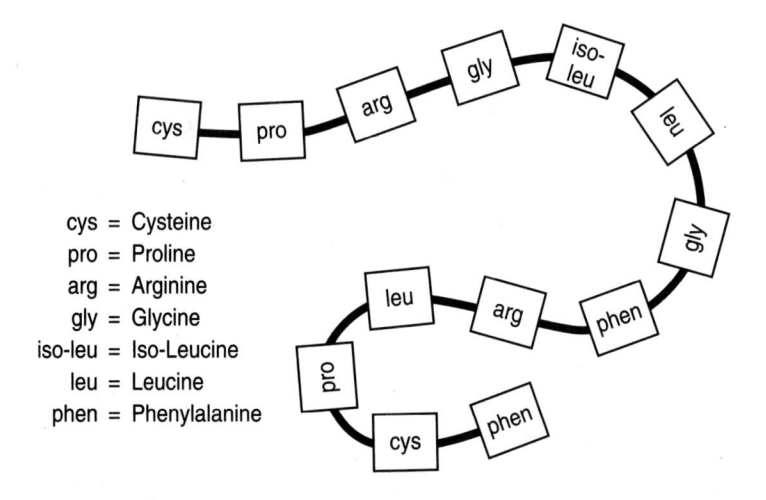

cys = Cysteine
pro = Proline
arg = Arginine
gly = Glycine
iso-leu = Iso-Leucine
leu = Leucine
phen = Phenylalanine

**Figure 4-14** A chain of amino acids linked in a particular sequence produces a protein.

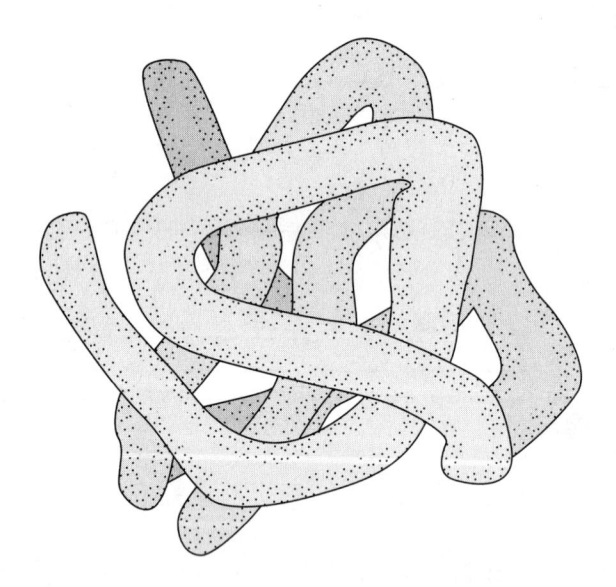

**Figure 4-15** The configuration of a protein shows bends at specific sites.

## Carbohydrates

Sugars polymerize to make starch and cellulose. Figure 4-16 shows two formulas for **glucose**.

It is necessary to show both formulas because both forms of glucose exist. The formula at left is a common way of representing glucose. It shows the

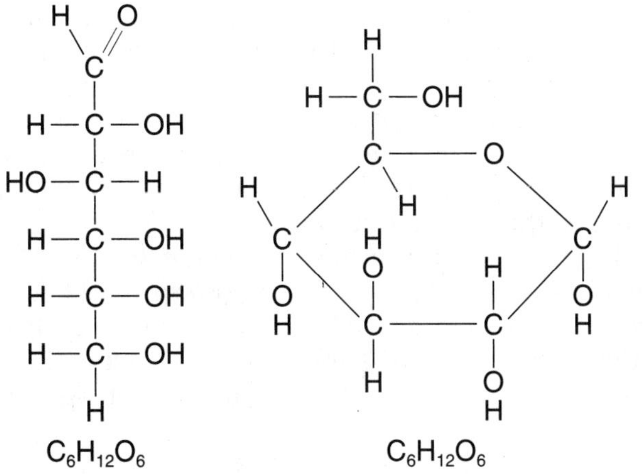

**Figure 4-16** When glucose is dissolved in water, a linkage develops between carbon-1 and carbon-5, as shown at right.

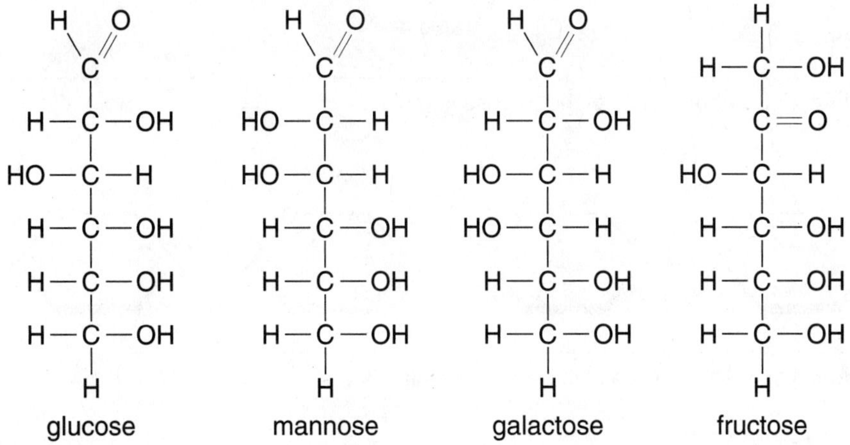

**Figure 4-17** The four sugars glucose, mannose, galactose, and fructose all have the same empirical formula, $C_6H_{12}O_6$.

oxygen atom at the end of the molecule: $-C=O$. At right, the oxygen atom is shown sharing its bonds with carbon atoms one and five. This is the way glucose occurs when dissolved in water. When starch is digested, it is broken down by **enzymes** into molecules of glucose, a **monosaccharide**, or **simple sugar** (so-called because it is the smallest form that results from digestion). Figure 4-17 shows structural formulas for several simple sugars. Note that

❈ **Notes** ❈

while each one is a different sugar, all have the same empirical formula: $C_6H_{12}O_6$. This again emphasizes the need for structural formulas in organic and biological chemistry. All the sugars shown in figure 4-17 are 6-carbon sugars and are thus called **hexoses**. Sucrose, common table sugar, is a 12-carbon sugar. When acted on by the enzyme invertase, sucrose is broken down into two molecules of 6-carbon sugar; one of glucose and one of fructose. Sucrose is, thus, a **disaccharide**.

The formula for glucose as shown at the right in figure 4-16 can be abbreviated (see figure 4-18). Starch, then, can be shown as in figure 4-19.

Both starch and cellulose are polymers of glucose; yet they are different compounds. How does this occur? Although it is difficult to show on a flat surface, figure 4-20 illustrates the two forms of glucose: alpha-glucose (α-glucose)

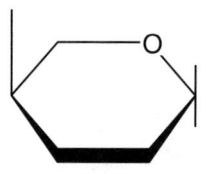

**Figure 4-18**  Abbreviation of the glucose formula.

**Figure 4-19**  Abbreviation of the starch formula.

α - glucose          β - glucose

**Figure 4-20**  Alpha-glucose and beta-glucose.

and beta-glucose (β-glucose). The difference between α-glucose and β-glucose is found in the orientation of atoms. Atom orientation affects the linkages when the sugar is polymerized into starch or cellulose. Whereas starch is composed of α-glucose units, cellulose is composed of β-glucose units. Different kinds of enzymes are involved in their digestion.

## Lipids

Fats are complex molecules composed of long chains of carbon atoms (fatty acids) linked to a 3-carbon compound called glycerol. The structure of glycerol is quite simple, as shown in figure 4-21. As previously shown, organic acids each have a carboxyl group (−COOH) at the end of a chain of carbon atoms. A fatty acid is shown in figure 4-22.

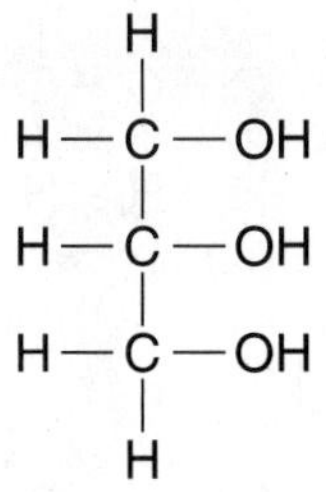

**Figure 4-21** Structural formula for glycerol.

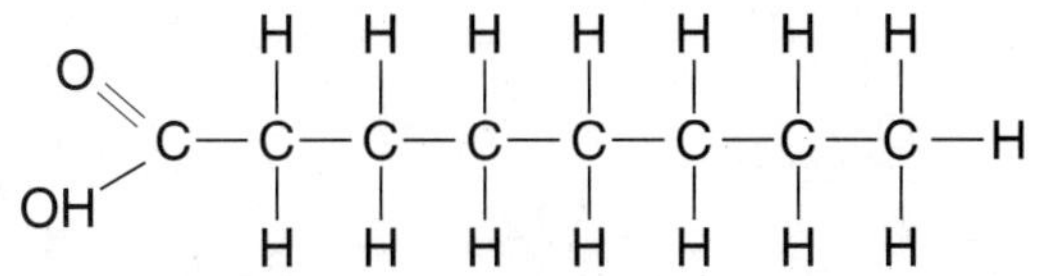

**Figure 4-22** Structural formula for a fatty acid.

46 ◆ Chapter 4

※ **Notes** ※

When fatty acids react with glycerol to produce molecules of fat, a dehydrolysis reaction takes place. Three molecules of water are removed (or created) for each molecule of fat manufactured, as shown in figure 4-23. As illustrated in the figure, the carbon atoms are linked to each other by single bonds, the remaining bonds being attached to hydrogen atoms. This is called

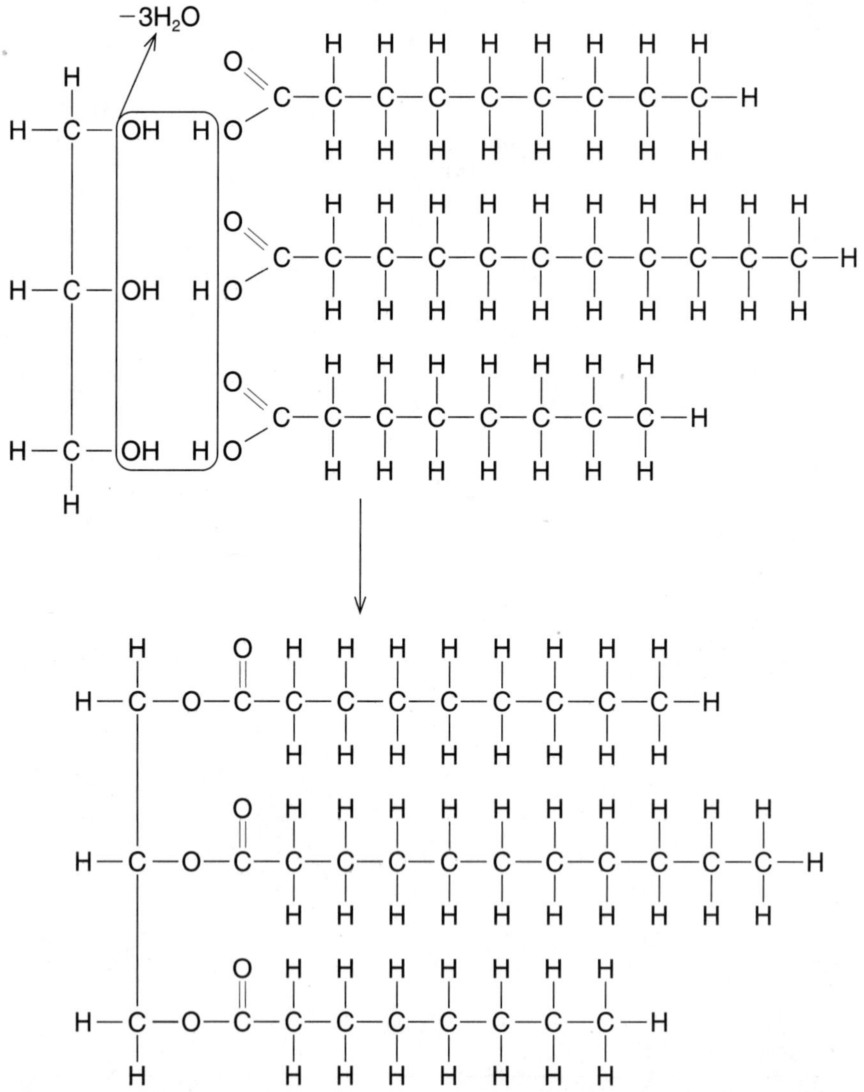

**Figure 4-23** A dehydrolysis reaction uniting glycerol and three fatty acid molecules to form a molecule of fat.

a saturated fat. An unsaturated fat has fewer hydrogen atoms and, hence, some double bonds between adjacent carbon atoms, as shown in figure 4-24. An unsaturated fat can be changed to a saturated fat by **hydrogenation**, the process of adding hydrogen atoms to a compound.

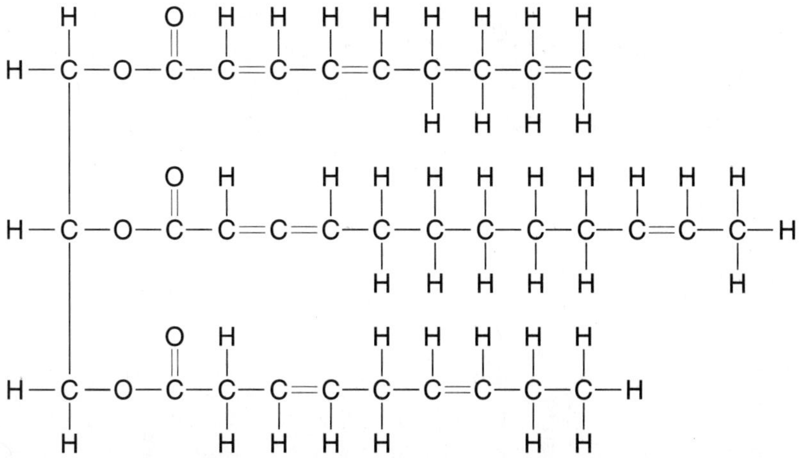

**Figure 4-24** Structural formula for an unsaturated fat.

## Questions for Review

1. Characterize carbohydrates, lipids, and proteins.
2. In what way do structural formulas reveal molecular structure that empirical formulas do not?
3. Carbon always has _____ bonds, oxygen always _____ bonds, and hydrogen always _____ bonds.
4. What group distinguishes an alcohol? Write the molecular structure of an alcohol and give its name.
5. Two organic molecules may be caused to unite into one molecule by a _____ reaction or by a _____ reaction.
6. Provide the structural formula of an amino acid and give its name.
7. Two amino acids may be united by a _____ reaction to form a _____ by the establishment of a _____ bond.
8. Distinguish between a monosaccharide, a disaccharide, and a polysaccharide. Give examples of each.

❊ **Notes** ❊

9. Provide the structural formula for glycerol and show how glycerol is involved in the formation of a lipid.

10. What is a hydrogenation reaction? What is a dehydrogenation reaction?

## Suggestions for Further Reading

Baker, J.J., and G.E. Allen. 1979. *Matter, Energy and Life: An Introduction for Biology Students*. Reading, MA: Addison-Wesley Publishing Company, Inc.

Sackheim, G.L. 1990. *Chemistry for Biology Students*. Reading, MA: Benjamin/Cummings Publishing Company, Inc.

# 5

# Mendelian Genetics

A discussion of genetics always focuses on Gregor Mendel (1822–1884). Before making his acquaintance, however, we should first consider some theorists and theories that preceded him.

※ Notes ※

## Pre-Mendelian Theorists and Theories

Jean Lamarck (that is, Jean Baptiste Pierre Antoine de Monet Lamarck, 1744–1829) asserted that species were not unalterable and that complex organisms derived from simpler ones. He proposed four laws:

1. Life tends to increase in the volume of every body up to a limit.
2. New organs come into existence as a result of new wants, new desires. (He surely was not thinking of plants!)
3. The action of organs develops in relation to how much they are used.
4. All that is acquired by individuals is conserved and passed on to successive generations.

Lamarck's four laws are known collectively as the *inheritance of acquired characteristics*, which suggests that the hands of a newborn baby being the son of a laboring man should have calluses; that because the arms of a blacksmith grow large and strong from hammering, the children born of blacksmiths should have large arms; and that the children of a prominent pianist should learn to play well with little practice. The classic example of the inheritance of acquired characteristics relates to the giraffe, the suggestion being that its long neck resulted from succeeding generations of giraffes stretching their necks to reach the leaves high in trees. This proposition had a long tenure.

August Weismann (1834–1914) challenged the whole idea of the inheritance of acquired characteristics. He is often remembered for cutting the tails off numerous generations of mice and noting that successive generations of such mice were still born with tails. Weismann was not a silly man, however,

※ Notes ※

and it would be unfair to remember him for a silly experiment. Cut-off tails are not acquired characteristics, of course; they are imposed characteristics. Weismann should instead be remembered for his theory (or observation) regarding the *continuity of germ plasm.* The body of an individual is made up of perishable **somatoplasm. Germ plasm** cannot be made anew, but rather is formed from preexisting germ plasm. An organism is a means of carrying germ plasm through time. Weismann realized that when egg and sperm unite, there is an increase in hereditary substance. He declared that somewhere along the line there must be a compensatory reduction in hereditary substance.

## Gregor Mendel

It is Gregor Mendel, however, who is considered to be the father of modern genetics, the study of heredity. He was an Austrian monk who discovered the first laws of heredity and laid the foundation for the science of genetics.

**Figure 5-1** Gregor Mendel (1822–1884), whose experiments with peas demonstrated that inheritance is related to particles. (Illustration by Donna Mariano)

Surprisingly, he failed biology and was unsuccessful in gaining a teacher's certificate primarily because of this weakness.

He joined the Augustinian Monastery in 1843, at which time he took the name Gregor. (His birth name was Johann.) He taught himself a certain amount of science, and in 1856 began some experiments on plant hybridization in the monastery garden. In time, his great weight (nearly 300 pounds) hindered his work. He had difficulty leaning over and sought to reverse his weight problem by aggressively smoking a number of cigars every day. It did not work.

Mendel experimented with garden peas, keeping statistical records regarding flower color, seed color, and whether the peas were wrinkled or smooth, the plants tall or dwarf. Two key features of his work should be emphasized: the records were statistical, and the traits he studied were either-or types of traits. He did not, for instance, track leaf size, which ranges from small to large, with all gradations in between.

Mendel became a member of the Natural Science Society, which was founded at Brunn in 1862. In 1865 and again in 1866, he presented papers before the society. After each meeting, there was gentle applause; the papers, however, were placed on a shelf to gather dust. No one seemed to have appreciated the significance of Mendel's work.

In 1900, thirty-four years after his findings were reported to the scientific community, three investigators rediscovered Mendel's work, all around the same time. Karl Erich Correns, E. Tschermak von Seysenegg, and Hugo DeVries independently obtained results similar to Mendel's, and each independently rediscovered his publications. In 1900 the scientific community was better able to listen. Fame came to Gregor Mendel only after his death. Hugo DeVries is probably best known for introducing experimental methods to the study of evolution. He originated the term **mutation** in connection with his studies of *Oenothera lamarckiana*.

## Mendel's Laws

Gregor Mendel established the fundamental truth that heredity is particulate —that is, heredity is related to substance that lies in the chromosomes. Although Mendel did not know anything about chromosomes, four fundamental laws of heredity evolved from his work: the law of **dominance**, the law of **segregation**, the law of **independent assortment**, and the law of **unit characters**. The law of dominance states that one gene of a pair may mask or inhibit the expression of the other gene of a pair. The law of segregation states that the genes that make up the pairs are separated from each other in an unaltered condition when reduction division takes place. The law of independent assortment states that the distribution of one pair of factors is independent of the distribution of another pair. The law of unit characters states that each characteristic is transmitted as an unchanging unit.

※ Notes ※

## Mendel's Experiments

Following is a discussion of various crosses performed by Gregor Mendel. Recall that Mendel worked with garden peas. In one series of experiments, he crossed round-seeded peas with wrinkled-seeded peas by placing the pollen of the wrinkled-seeded pea plant on the pistil of the round-seeded pea plant. Mendel used an uppercase *R* to represent the round-seeded form and a lowercase *r* to represent the wrinkled-seeded form. Before continuing, note that the protocol in constructing a table summarizing such experiments is to list vertically the kinds of gametes the female produces and to list horizontally the kinds of gametes the male (that is, the pollen) produces. Furthermore, the female parent of the cross is always presented first.

Thus, the diploid cell of the round-seeded form is represented by *RR*, the diploid of the wrinkled-seeded form by *rr*. When reduction division occurs, the *R*'s separate into different cells, and the haploid condition results. The gametes for the round-seeded form are all *R*, and the gametes for the wrinkled-seeded form are all *r*. The cross is often represented as shown in figure 5-2.

Figure 5-2 illustrates the law of unit characters by showing that the gametes each possess a single unit (a gene) for a trait. Mendel noted that all offspring of such a cross had round seeds. This shows the law of dominance. The trait for round-seededness is expressed at the entire suppression of the trait for wrinkled-seededness. There is no evidence of wrinkled-seededness in any of the progeny.

In another series of experiments, Mendel performed the same cross only in the reverse manner. The round-seeded form was used as the pollen source, the wrinkled-seeded form for eggs (that is, as the reciprocal cross). The result is illustrated in figure 5-3.

Again, as shown in figure 5-3, all the progeny were round seeded. Such crosses are called *hybrids* because they are produced by dissimilar parents.

It was earlier stated that in making a table, only the *kinds* of gametes produced need be represented. The round-seeded form produces only one kind of gamete, as does the wrinkled-seeded form. The progeny of such crosses are called the $F_1$ generation, or the first filial generation. If such progeny are allowed to self pollinate, the result is an $F_2$ generation.

♀ ♂
RR × rr
R    r

|   | r  |
|---|----|
| R | Rr |

rr × RR
r    R

|   | R        |
|---|----------|
| r | rR or Rr |

**Figure 5-2** A homozygous round-seeded form crossed with a homozygous wrinkled-seeded form.

**Figure 5-3** The wrinkled-seeded form is used as the egg source, and the round-seeded form is used as the pollen source, a reciprocal of the cross in figure 5-2.

The previously mentioned protocol has again been followed in figure 5-4; that is, the eggs are listed vertically, the pollen horizontally. To fill in the squares for the $F_2$ generation, the gametes from the vertical and horizontal columns are brought together. Note that while three-fourths of the progeny are round seeded, one-fourth are wrinkled seeded. The traits that occur in pairs in the diploid cells, then, do not lose their integrity. The gene for wrinkled seededness still resides in the round-seeded forms and is expressed in the $F_2$ generation. This illustrates the concepts of dominance and recessiveness. The **recessive** trait (that is, wrinkled seededness) is obscured in the presence of the dominant trait (that is, round seededness) but is expressed when it is the only trait present.

Some new terms should be introduced at this point. **Phenotype** refers to what can be seen, whereas **genotype** refers to actual genetic makeup. Consequently, *RR* and *Rr* are phenotypically the same because round seededness is a dominant trait, but they are genotypically different. The *RR* form is said to be **homozygous** (composed of two of the same type of genes) and the *Rr* form **heterozygous** (composed of two different kinds of genes).

In the $F_2$ generation in figure 5-4, the ratio of phenotypes is 3:1, the ratio of genotypes 1:2:1. One-fourth of the progeny is homozygous dominant, two-fourths is heterozygous dominant, and one-fourth is homozygous recessive. You should note that the division by fourths does not mean that there are four individuals in the progeny. There may be several hundred. The cross shows ratios rather than actual number of progeny. It is also a good time to emphasize the procedure of using an uppercase letter to represent a dominant trait and the lowercase version of the same letter to represent the corresponding recessive trait. Following this practice helps prevent confusion.

Now consider two characteristics at the same time. Using the same peas, also consider color. Mendel crossed a pure-breeding (homozygous) round-seeded yellow variety with a pure-breeding wrinkled-seeded green variety. (Note that previous experiments had already determined yellow (*Y*) to be dominant and green (*y*) to be recessive.) The results of such a cross are shown in figure 5-5.

It is immediately apparent from figure 5-5 that all offspring of this cross (that is, the $F_1$ generation) display all the traits of one parent and no traits of the other parent. This is because one parent possesses both dominant traits.

※ **Notes** ※

Rr × Rr

|   | R  | r  |
|---|----|----|
| R | RR | Rr |
| r | Rr | rr |

**Figure 5-4** An $F_2$ generation is produced by allowing the $F_1$ generation to self pollinate. Here, each parent can produce two kinds of factors.

RRYY × rryy
RY     ry  – gametes

|    | ry   |
|----|------|
| RY | RrYy |

**Figure 5-5** Homozygous, round, yellow pea seeds are crossed with homozygous, wrinkled, green pea seeds.

✿ **Notes** ✿

If, however, the heterozygous offspring of the F₁ generation are allowed to self pollinate (see figure 5-6), the traits of the recessive parent will re-emerge, illustrating the law of independent assortment. Two genes are represented in each gamete, and these genes appear with equal frequency in all combinations.

In counting the phenotypes shown in figure 5-6, you will see that nine-sixteenths possess both dominant traits (that is, round-yellow); three-sixteenths possess dominant seed coat and recessive color (that is, round-green); three-sixteenths possess recessive seed coat and dominant color (that is, wrinkled-yellow); and one-sixteenth possess both recessive traits (that is, wrinkled-green). The ratio of phenotypes, then, is 9:3:3:1. This is the basis of Mendel's law of independent assortment; the distribution of one pair of genes is independent of the distribution of the other. In this

RrYy × RrYy

|    | RY   | Ry   | rY   | ry   |
|----|------|------|------|------|
| RY | RRYY | RRYy | RrYY | RrYy |
| Ry | RRYy | RRyy | RrYy | Rryy |
| rY | RrYY | RrYy | rrYY | rrYy |
| ry | RrYy | Rryy | rrYy | rryy |

**Figure 5-6** If the progeny of the cross in figure 5-5, the F₁ generation, is allowed to self pollinate, an F₂ generation is produced.

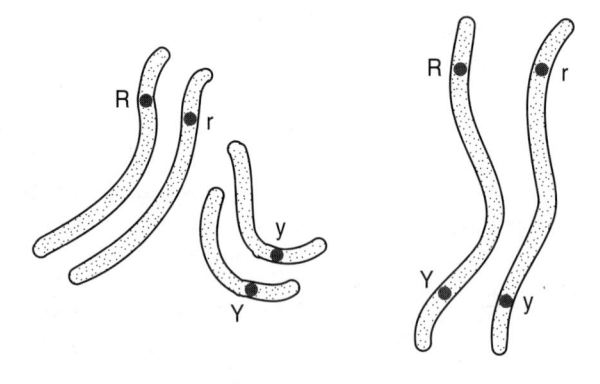

Independent Assortment
Can Take Place

Independent Assortment
Cannot Take Place

**Figure 5-7** At left, genes R and Y are on separate chromosomes, and independent assortment can occur. At right, the genes are on the same chromosome (linked), and independent assortment therefore cannot take place.

matter, Mendel was simply lucky to have chosen two traits that are on different chromosomes (see figure 5-7, left). If the genes for seed coat and seed color had been on the same chromosome (figure 5-7, right), independent assortment could not have occurred.

## Applying Genetics

Now consider a problem. We have some round-seeded peas. The phenotype is visible, but the genotype is unknown. The peas could be homozygous (*RR*), or they could be heterozygous (*Rr*). How can we determine the genotype? This is done by a **test cross**, also known as a **back cross**. A test cross is crossing an unknown genotype with a homozygous recessive. If all progeny of such a cross display the dominant traits, the unknown genotype is homozygous. If, however, the recessive trait should appear among even only a few of the progeny, the unknown genotype is heterozygous.

Consider a back cross between turnips. In turnips, purple (*P*) is dominant and red (*p*) is recessive, and cylindrical (*C*) is dominant and spherical (*c*) is recessive. When a purple-cylindrical turnip is crossed with a red-spherical turnip, the $F_1$, as expected, has both dominant traits (see figure 5-9). Now, use the members of the $F_1$ in a back cross and follow the procedure of recording only the *kinds* of gametes that can be produced. Here, the female parent can produce four kinds of gametes, the male only one. At the top of the column

$$RR \times rr \qquad Rr \times rr$$

|   | r  |
|---|----|
| R | Rr |

|   | r  |
|---|----|
| R | Rr |
| r | rr |

**Figure 5-8** A test cross (or back cross) can be used to determine genotype. Here, two round-seeded forms are crossed with the homozygous recessive to reveal the genotype. Round-seeded offspring indicate that the round-seeded parent was homozygous; any wrinkled-seeded offspring indicate that the round-seeded parent was heterozygous.

$$PPCC \times ppcc$$

|    | pc   |
|----|------|
| PC | PpCc |

**Figure 5-9** A homozygous, purple, cylindrical turnip is crossed with a homozygous, red, spherical turnip.

## Notes

in figure 5-10, the progeny show both dominant traits: purple and cylindrical. The second in the column shows dominant color and recessive shape: purple and spherical. The third in the column shows recessive color and dominant shape: red and cylindrical. The fourth in the column shows both recessive traits: red and spherical. According to the principle of independent assortment, the expected ratio would be 1:1:1:1. Yet, a ratio of 7:1:1:7 was actually observed. How could this be? The answer relates to the traits being on the same chromosomes and, thus, being linked. Based on this knowledge, one might expect there to be only two kinds of gametes: *PC* and *pc*. If linkage were unalterable, this would be the case. But as described in chapter 3 on mitosis and meiosis, crossing over allows linkage alterations; crossing over, then, is what makes possible the combinations *Pc* and *pC*.

Having examined several examples illustrating the law of dominance, where a gene finds full expression and its homologue is suppressed, consider now a situation wherein both genes of a pair are equally effective in their influence. *Mirabilis jalapa* is an annual plant often planted in flower gardens. (Its common name is "Four O'clock" because its flowers are said to open at approximately that time of day.) If a pure red variety (*RR*) were crossed with a pure white form (*rr*), the heterozygous offspring would have pink flowers, the influence of both sets of genes being expressed. This is an example of incomplete dominance. If such pink flowers are self pollinated or crossed with other pink forms, the offspring will be in the ratio 1:2:1: one-fourth red, two-fourths pink, and one-fourth white.

Dominance can be affected by environmental conditions. If Chinese primroses are grown at 68°F, for example, they will produce red flowers; but if grown at 80°F, white flowers will result.

Following is an example of multiple genes affecting a single trait. There are four genes (two pairs) that influence grain color in wheat. When red- and white-grained varieties are crossed, medium-red offspring result. When these offspring are interbred, the grain of the next generation ranges in color from red to white. This is called **polygenic inheritance**. (See figure 5-11).

Earlier, Hugo DeVries' work with *Oenothera* was mentioned. DeVries observed that from time to time, a progeny displayed a trait that seemed to

PpCc × ppcc

|    | pc    |
|----|-------|
| PC | PpCc  |
| Pc | Ppcc  |
| pC | ppCc  |
| pc | ppcc  |

**Figure 5-10** The progeny of the cross in figure 5-9 are subjected to a test cross, being crossed with the pure recessive for both traits.

$$R_1R_1R_2R_2 \times r_1r_1r_2r_2$$
$$F_1 \longrightarrow R_1r_1R_2r_2$$

|  | $R_1R_2$ | $R_1r_2$ | $r_1R_2$ | $r_1r_2$ |
|---|---|---|---|---|
| $R_1R_2$ | $R_1R_1R_2R_2$ | $R_1R_1R_2r_2$ | $R_1r_1R_2R_2$ | $R_1r_1R_2r_2$ |
| $R_1r_2$ | $R_1R_1R_2r_2$ | $R_1R_1r_2r_2$ | $R_1r_1R_2r_2$ | $R_1r_1r_2r_2$ |
| $r_1R_2$ | $R_1r_1R_2R_2$ | $R_1r_1R_2r_2$ | $r_1r_1R_2R_2$ | $r_1r_1R_2r_2$ |
| $r_1r_2$ | $R_1r_1R_2r_2$ | $R_1r_1r_2r_2$ | $r_1r_1R_2r_2$ | $r_1r_1r_2r_2$ |

**Figure 5-11** Polygenic inheritance. If four *R*s are present in the progeny, dark grains will be noted; if three *R*s are present, the grain color will be medium-dark red. If two *R*s are present, grain color will be medium red; and if no *R*s are present (that is, only small *r*s are present), the grain color will be white.

come from neither parent. He proposed that such a condition resulted from an unexpected change in a gene. He called this *mutation*. Navel oranges, nectarines, numerous types of potatoes, and variegated shrubs all result from mutation. Asexual means of propagation of such forms makes it possible to retain desired features.

Many changes occur for reasons other than mutation. (In fact, approximately only one in one thousand changes can be traced to mutations.) Other factors which affect phenotype are alterations in chromosomal configuration such as deletion of parts of chromosomes or duplication of chromosomal segments. Changes in the number of chromosomes also alter the phenotype. Most mutations produce no benefit for organisms and even have harmful consequences, but a few are beneficial. Tens of thousands of plants may be treated in the hope of causing a desired change.

*Colchicum autumnale,* the autumn crocus, produces the alkaloid colchicine, which can produce tetraploidy (4N) by arresting some part of the mitosis process. Extract from this plant is used in extreme dilution. When dividing plant cells are placed in contact with a colchicine solution, mitosis is arrested in late prophase, after chromosome replication has taken place. The cells can then be removed from the solution, at which time they recover and start the process of mitosis again from the beginning. The result is that the chromosomes replicate twice, doubling the chromosome number (tetraploidy). There are clear benefits to using the autumn crocus for this purpose. Polyploid plants are usually more vigorous than are diploid forms, and they sometimes produce larger fruits. Most cultivated irises and many lilies are tetraploids. Baldwin apples, pink beauty tulips, and Japanese flowering cherries are triploids.

Finally, although we have focused on situations wherein one gene affects one trait, there are situations wherein one gene affects more than one trait. This is called **pleiotropy.** For example, the gene involved in the synthesis of lignin may also cause numerous other effects.

**Notes**

## ❋ Notes ❋

## Questions for Review

1. Define each of the following terms: *hybrid, genotype, phenotype, homozygous,* and *heterozygous.*
2. List the four laws cited by Gregor Mendel.
3. Show the progeny of a cross between AABB and AaBb.
4. What phenotypes result from a cross between aaBb and AABb?
5. Relate some of the so-called laws proposed by Jean Lamarck. For what concept is he remembered?
6. Define the terms *somaplasm* and *germ plasm.*
7. Gregor Mendel is said to have been lucky in the choosing of traits to study for inheritance. Explain why.
8. Illustrate a cross between a heterozygous round-seeded pea and a homozygous round-seeded pea.
9. Define the term *mutation.* How may it be significant? Who introduced the concept?
10. Define the terms *pleiotropy, linkage,* and *back cross.*

## Suggestions for Further Reading

Anderson, E. 1968. *Plants, Man and Life.* Berkeley, CA: University of California Press.

Beadle, G., and M. Beadle. 1966. *The Language of Life.* New York: Doubleday and Company, Inc.

Merrell, D. 1975. *Introductory Genetics.* New York: W.W. Norton and Company.

Snyder, L.H. 1951. *The Principles of Heredity.* Boston: Heath and Company.

Srb, Adrian S., Kay D. Owens, and R.S. Edgar. 1965. *General Genetics.* San Francisco: W.H. Freeman Company Publishers.

Stebbins, G.L. 1965. From gene to character in higher plants. *American Scientist* 53:104–126.

Whitehouse, H.L.K. 1973. *Towards an Understanding of the Mechanism of Heredity.* New York: St. Martins.

# 6

# DNA

When the statistical observations of Gregor Mendel were compared with what could be seen under the microscope, it became clear that the hereditary material resided in chromosomes. Mendel observed that "factors" (that is, genes) occur in pairs, one member from each parent. The microscope revealed that chromosomes occur in pairs, one member from each parent. Mendel observed that when reproductive cells are produced, the factors separate and are distributed as units (the laws of segregation and unit characters). The microscope showed that in meiosis, the homologous chromosomes separate, with only one of each pair going to a gamete. Mendel noted that in a dihybrid, the distribution of one pair of factors is independent of the distribution of the other (the law of independent assortment). The microscope showed that in meiosis, the distribution of maternal and paternal chromosomes is random. The evidence that the hereditary material resides in chromosomes was undeniable.

## The Search for the Substance of Heredity

There are several types of molecules in chromosomes. One such molecule is protein. For a long time, protein molecules were thought to be the bearers of heredity, the reason being that heredity appeared very complicated, and proteins are certainly also complicated. It seemed logical to suppose that protein bore the "secret" of heredity. While logic is a laudable process, it can serve only as a basis for directing investigation in science.

In 1868 Friedrich Miescher (1844–1895) separated nuclear material from the rest of the cell and discovered that after the protein portion of nuclei had been hydrolyzed by pepsin, a residue remained. The residue was determined to be an acid and was thus called **nucleic acid**.

During the 1920s, some aspects of the chemical structure of nucleic acids were determined. The essential structure involves a series of 5-carbon sugars

❋ Notes ❋

### ❋ Notes ❋

(pentoses) linked in a chain and connected by phosphate groups. Attached to each sugar molecule are one or another of several types of bases. There are two types of nucleic acid, although both types have the same fundamental structure. In one type of nucleic acid, the sugar is ribose; in the other type, the sugar is deoxyribose. Deoxyribose differs from ribose in having one less oxygen atom in its molecule. The two types of nucleic acid are thus called deoxyribonucleic acid (DNA) and ribonucleic acid (RNA). The bases attached to the sugars in DNA are adenine, thymine, cytosine, and guanine; in RNA, the bases are adenine, uracil, cytosine, and guanine, uracil being in the place of DNA's thymine. These bases belong to two groups: adenine and guanine are purines, and cytosine and thymine are pyrimidines. Nucleic acids are composed of long chains of repeating units called **nucleotides**. A nucleotide is represented in figure 6-2.

The four types of nucleotides in DNA are as follows:

>    adenine-deoxyribose-phosphate
>    thymine-deoxyribose-phosphate
>    cytosine-deoxyribose-phosphate
>    guanine-deoxyribose-phosphate

In RNA, the nucleotides are as follows:

>    adenine-ribose-phosphate
>    uracil-ribose-phosphate
>    cytosine-ribose-phosphate
>    guanine-ribose-phosphate

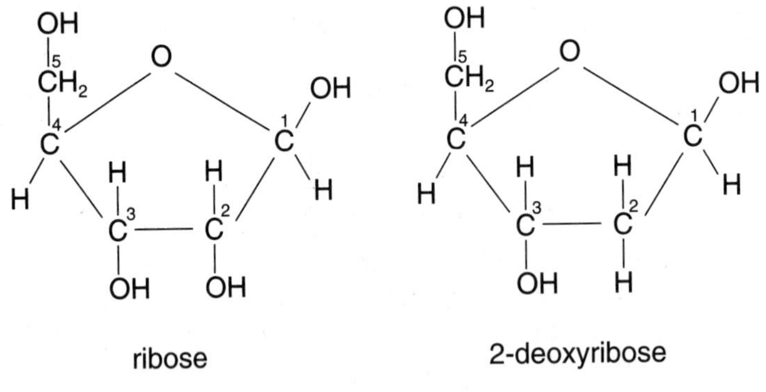

**Figure 6-1** Deoxyribose differs from ribose in having one less oxygen atom.

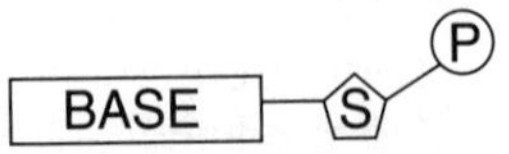

**Figure 6-2** A nucleotide is composed of a base, a sugar, and a phosphate.

Whereas DNA is primarily (but not entirely) confined to the nucleus, RNA is found in both the nucleus and the cytoplasm. Although the existence of nucleic acids was known for approximately one hundred years, the focus throughout this time period continued to be on protein as the possible bearer of hereditary material. In the 1930s, however, DNA came to be suspected of being the most genetically important part of the chromosome. Still, it was perplexing that a molecule composed of so few different parts could make such a claim. Further, the differences between the structural formulas of the bases seemed so slight.

In 1928 a bacteriologist named Fred Griffith published a paper describing some work he had been doing with pneumococcus, the bacteria that causes pneumonia. Two strains of the organism were known: a virulent strain, which could cause disease, and an **avirulent** strain, which could not cause disease. When grown in culture, the virulent strain possessed a glistening outer smooth coat; it was thus designated with the letter S for "smooth." Under prolonged

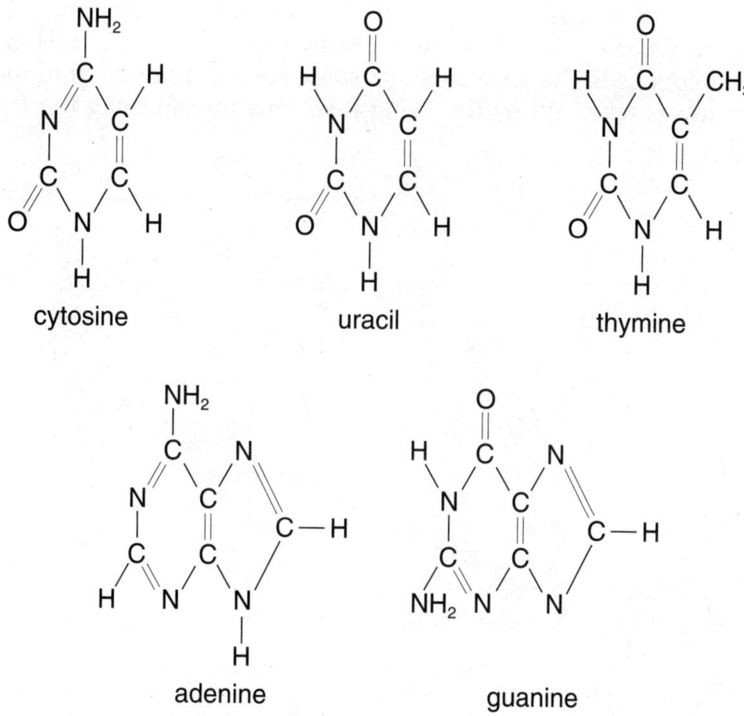

✱ **Notes** ✱

**Figure 6-3** Five kinds of bases occur in nucleic acids: adenine, thymine, cytosine, and guanine for DNA, and adenine, cytosine, guanine, and uracil for RNA. Adenine and guanine are purines; cytosine, thymine, and uracil are pyrimidines. In pairing, a purine always links with a pyrimidine.

❋ **Notes** ❋

culture in the laboratory, the virulent strain lost its outer coat and became avirulent. Because the nonvirulent strain appeared rather rough under the microscope, it was designated with the letter *R* for "rough." It was assumed that the outer capsule was responsible for causing the disease.

There are four smooth varieties of pneumococcus, each of which varies somewhat in its capsule. The four smooth varieties are designated as follows: S-I, S-II, S-III, and S-IV. Under prolonged culture, each of the smooth varieties loses its capsule and becomes incapable of causing disease; thus, there are four rough varieties, each based on the kind of capsule it previously possessed. The four rough varieties are designated as follows: R-I, R-II, R-III, and R-IV.

In Griffith's experiments, when a suspension of R-II, for example, was injected into a mouse, the mouse did not get sick because the R-II form is avirulent. When a smooth variety—say S-III—was injected into a mouse, however, the mouse sickened and died. And when a suspension of S-III was first killed and then injected into a mouse, the mouse lived. So far, there is nothing surprising in Griffith's findings. A surprising result did occur, however, when an R-type (avirulent) strain was mixed with a dead S-III (also avirulent) strain, and the mixture was injected into a mouse; the mouse sickened and died. Neither the R-type strain alone nor the killed S-type strain alone could cause disease; but a mixture of the two strains resulted in death. How could this have happened? The answer is that some genetic material from the dead S-III strain was taken up by the living R-II, thus transforming the R-II into an S-III!

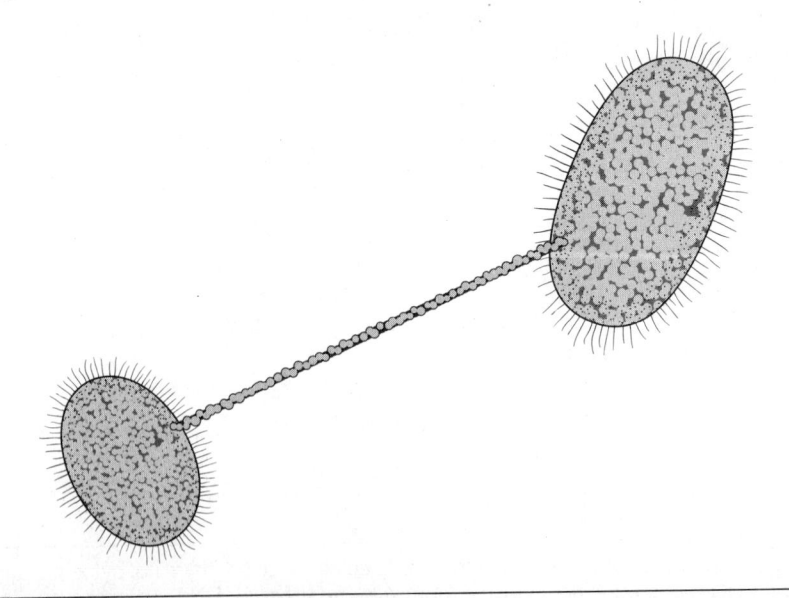

**Figure 6-4** Two bacteria connected by a conjugation tube. Genetic material from one cell can migrate through the tube to the other cell.

It was later shown that such bacterial transformation did not require the presence of the mouse. In the 1940s O.T. Avery, working at the Rockefeller Institute, fragmented the cells of pneumococcus and used centrifugation methods to separate the cell constituents into fractions. Thus, he had a fraction containing only protein and another fraction containing only DNA. When he put the protein fraction of a killed S-III strain with a living R-II strain, no change resulted. When he placed the DNA fraction of the killed S-III strain with the living R-II strain, however, the R-II took on the capacity of virulence. This provided clear, convincing evidence that DNA is the genetic material.

In 1952 Alfred Hershey and Margaret Chase's work with the T-2 bacteriophage strongly reinforced the contention that DNA is the genetic material. (Their work is described in chapter 7 on viruses.)

In time, DNA gained general acceptance as the substance of heredity; yet the structure of the DNA molecule remained very much a mystery. While it was clear that DNA consisted of a long chain of nucleotides linked together, the details of the linkage were elusive.

## The Structure of DNA

In Cambridge, England in 1953 James D. Watson and F.H.C. Crick went to work with paper, paste, tin snips, and metal, seeking to construct a three-dimensional model of DNA. They took into account all information that had been gathered by other researchers. Of particular importance was Maurice Wilkins' x-ray diffraction studies of DNA, which indicated a large spiral. The model constructed by Watson and Crick consisted of a double **helix:** two strands of alternating sugars and phosphates twisted around each other, the strands bonded to each other by pairs of bases. Recall that there are four kinds of bases in DNA—adenine, thymine, cytosine, and guanine—and that these bases are of two sorts—purines and pyrimidines. Chemical analysis had revealed that the amounts of adenine and thymine were always equal, and the amounts of guanine and cytosine were also always equal. Furthermore, the pairs of bases that linked the two helices together always measured the same distance. It was also known, however, that the sizes of purines and pyrimidines differed. Thus, if two purines were linked with each other and two pyrimidines were linked with each other, equal distances between the helices could not be achieved; if, however, a purine were always linked to a pyrimidine, the lengths of the bonding pairs would always be the same. The Watson-Crick model provided for this.

It later came to light that adenine, a purine, links only with thymine, a pyrimidine, and that guanine, a purine, links only with cytosine, a pyrimidine. These nitrogenous bases make their connections by weak hydrogen bonds.

The secret of heredity, then, had been unveiled. Learning the structure of this "most golden of all molecules" (James Watson's words) and the role it plays in heredity is considered one of the greatest discoveries in chemistry. For their efforts, Watson and Crick were awarded the Nobel prize in 1962.

※ **Notes** ※

### ❧ Notes ❧

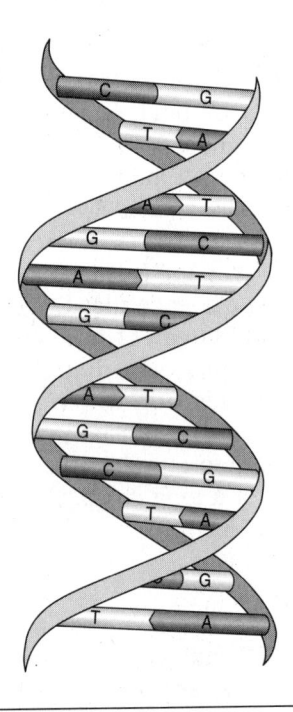

**Figure 6-5** The double helix of the DNA molecule: two spirals of alternating sugars and phosphates are bonded together by pairs of bases.

The DNA in all the cells of a given organism as well as all kinds of organisms appears to be the same: two helices of alternating sugars and phosphates bonded by pairs of bases, the pairs being of two sorts only (that is, adenine with thymine or cytosine with guanine). It is somewhat bewildering that this apparently simple structure controls the heredity of such diverse forms as plants and human beings. Hundreds of thousands of nucleotides may be joined to form a single nucleic acid molecule. The length of DNA in a plant cell is hundreds of thousands of times greater than the diameter of the cell, and a single cell may contain in its genes an amount of information that would cover several hundred thousand pages in a printed form.

The only way that any DNA molecule differs from any other is in the sequencing of the bases along the helix (although the pairing of bases—adenine always with thymine, and cytosine always with guanine—can be reversed, that is, thymine with adenine and guanine with cytosine). If the number of base pairs along the line is very high, the number of combinations of adenine, cytosine, guanine, and thymine (ACGT) can be in the hundreds of millions. There are approximately five thousand base pairs in a simple virus; in other organisms, the number of pairs may be in the millions. The variation in cells of different plants is great. A gymnosperm, for example,

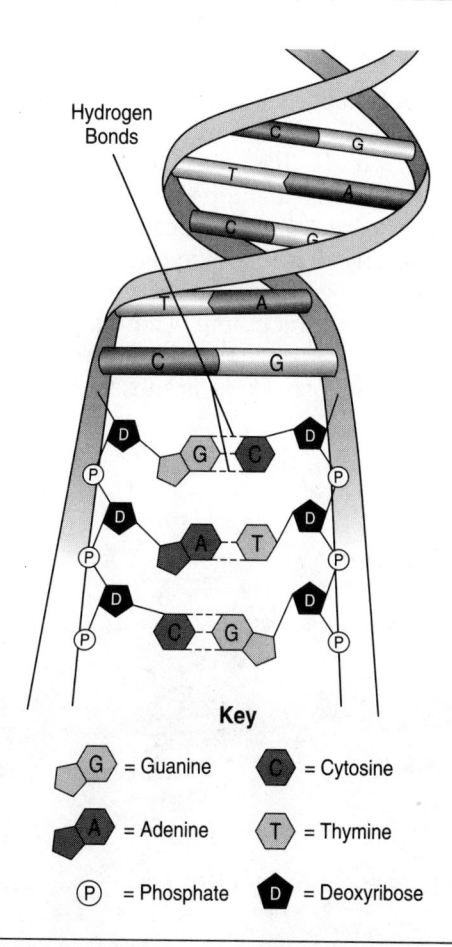

**Figure 6-6** Another representation of the double helix. Here, the five-sided figure represents the 5-carbon sugar deoxyribose; the circles represent phosphate groups; and the two chains are bonded by pairs of bases (adenine with thymine, guanine with cytosine). New nucleotides may be joined together in the process of replication.

may have ten thousandfold more base pairs than does another plant species such as *Arabidopsis thalliana*. It should be noted, however, that a high number of base pairs does not necessarily mean more genetic activity; large segments of DNA are repetitive and do not do anything.

## The Functions of DNA

What does DNA do? The answer is twofold: DNA can provide a pattern for the structuring of more DNA, which it must do when a cell divides; and DNA can provide a pattern for the making of RNA.

## 🞯 Notes 🞯   Self-Replication

To accomplish either of its functions—that is, self-replication or the patterning of m-RNA—the helices of DNA must uncoil. This is accomplished by the breaking of the hydrogen bonds that join the paired bases. In becoming uncoiled, each helical strand forms a companion strand, and the two strands again become coiled (see figure 6-7). Specifically, a freed adenine base can attract a nucleotide containing thymine; likewise, a freed cytosine base can attract a nucleotide containing guanine. As the new strand of DNA is completed, it becomes entwined with its parent. There are then two new double helices, each of which is made up of an older strand and a newly formed strand. In this manner, DNA is increased for subsequent mitoses.

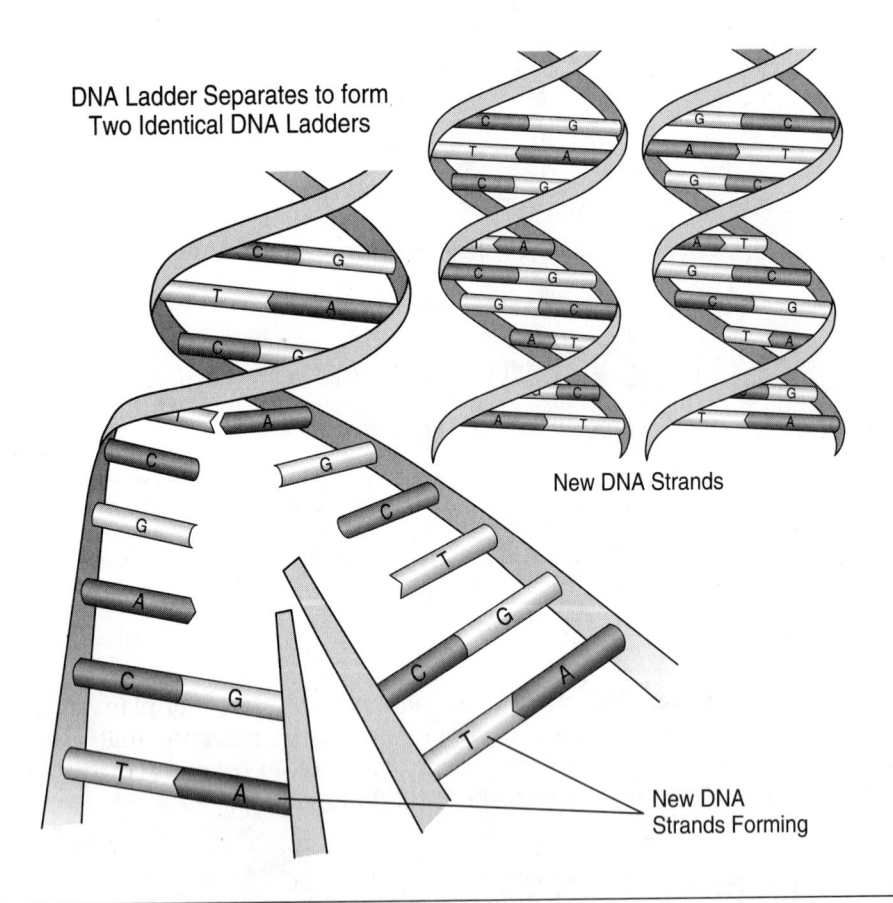

**Figure 6-7** Before DNA can replicate or act as a template for the manufacture of m-RNA, it must first uncoil. In forming a new DNA molecule, each spiral of the uncoiling helix has a newly made spiral wound around it. In this way, each new DNA molecule has a parent strand and a "newborn" strand.

## Patterning of RNA

When the structure of DNA was determined, it became clear that the bases along the helix represented coded information, that the sequencing of the bases represented a code, and that the coded information carried instructions to be used in creating proteins. DNA, however, does not leave the nucleus; it must therefore send instructions to the cytoplasm by way of a messenger. That messenger is RNA.

While most of it lies outside of the nucleus and in the cytoplasm, small amounts of RNA are present in chromosomes. RNA is manufactured in the nucleus. RNA differs from DNA in three ways. As mentioned earlier, the sugar in RNA is ribose rather than deoxyribose; and the base uracil is used in its structure in place of the thymine found in DNA. Finally RNA molecules are single- rather than double-stranded (although an RNA molecule may appear double stranded when it bends back on itself).

There are three known kinds of RNA: ribosomal RNA, messenger RNA (**m-RNA**), and transfer RNA (**t-RNA**). DNA controls what goes on in a cell by sending a messenger; the messenger molecule is m-RNA. One of the functions of DNA is to provide a template for the formation of m-RNA.

In performing this function—that is, the patterning of m-RNA—the helices of DNA again uncoil; but only one DNA strand plays a role in the fabrication of the messenger molecule. Recall that in m-RNA, uracil is used in the place of the thymine found in DNA. The new m-RNA molecule, which has been fabricated on the DNA template, moves out of the nucleus through the micropores of the nuclear membrane. It then becomes linked with the ribosomes that lie along the course of the endoplasmic reticulum. An electron microscope reveals the ribosomes to be like two beads: one smaller and the other larger. The m-RNA moves through the ribosome (or the ribosome moves along the RNA molecule), and, in doing so, the bases pass through in their previously determined sequence. As shown in figure 6-8, the first three to meet the ribosome are uracil, guanine, and cytosine (UGC). Messenger RNA is a coded molecule having code words of three letters, and the ribosome is the manufacturing center for protein. The UGC triplet (or **codon**) of the m-RNA calls on the ribosome to place the amino acid cysteine on line. When the next codon, for example AGU, passes through the ribosome, the amino acid serine comes on line and is linked to the earlier placed cysteine. The amino acids are united by peptide bonds.

# Amino Acids

Marshall Nirenberg and Heinrich Matthei, at the National Institute of Health, sought to create an artificially made m-RNA, and, in 1961, learned that UUU was the codon for phenylalanine. Nirenberg and Matthei produced an m-RNA made entirely of uracil (polyuradilic acid [U]), and, in 1964, were successful in getting ribosomes to accept this artificially produced m-RNA.

### ✤ Notes ✤

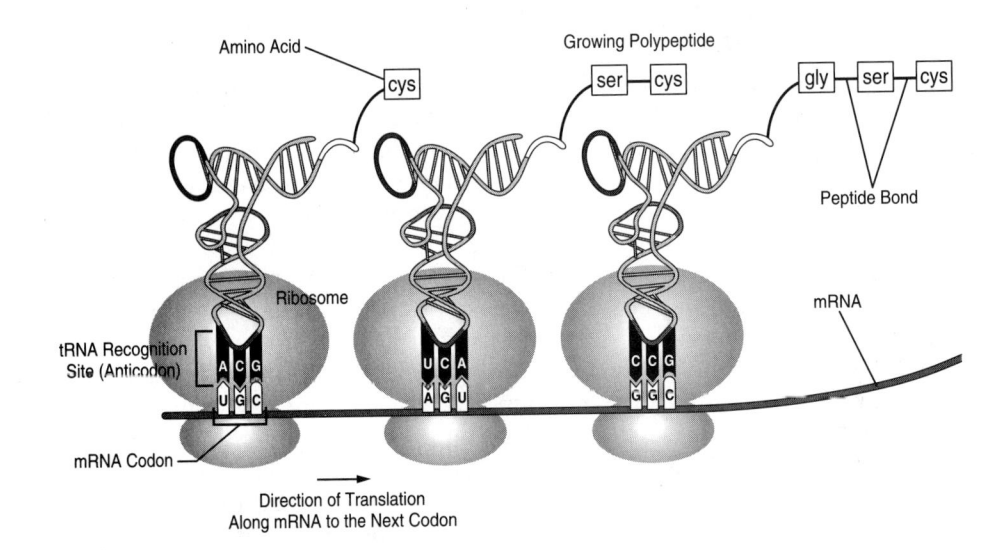

**Figure 6-8** When m-RNA molecules leave the nucleus, they go to the ribosomes, which function as protein assembly plants. The m-RNA molecules then impart to the ribosomes information implanted on the m-RNA molecules by contact with DNA; specifically, the m-RNA molecules tell the ribosomes which particular amino acid to place on line in the assembly of protein. Here, the codon UGC has caused the ribosome to place the amino acid cysteine on line; and the codon AGU, having passed through the center ribosome, has caused serine to be connected to the previously placed cysteine.

The m-RNA created by Nirenberg and Matthei was structured as shown in figure 6-9. When this m-RNA was run through the ribosomes, the resulting amino acids were always linked in the same manner, creating the polypeptide phenylalanine-phenylalanine-phenylalanine. With the discovery that the **triplet** UUU represented phenylalanine, a great deal of work followed to determine the codons for all twenty amino acids.

If the sequences of bases (that is, AGCT) on the DNA helix represent code words, how many bases (how many letters) are required to make the code word? The four bases AGCT can be put together in threes sixty-four different ways, more than is needed to yield twenty amino acids. There is also a certain amount of redundancy. For example, UCU, UCG, UCA, AGU, and AGC are all codons for the amino acid serine. Interestingly, several codons do not call for any amino acid but, rather, signal to the ribosome that the end of the molecule has been achieved. These termination codons are UAA, UAG, and UGA. Figure 6-10 shows all the combinations of bases (that is, of nucleotides) and the amino acids they represent.

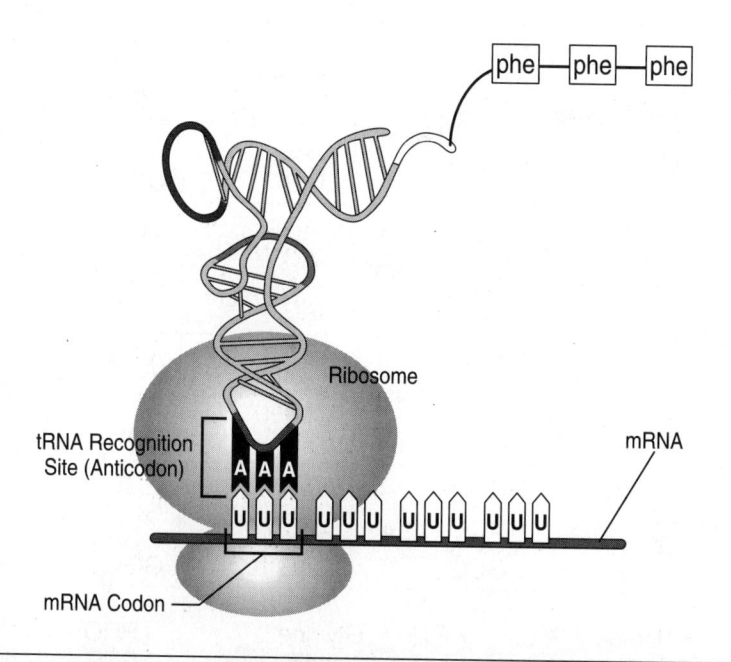

**Figure 6-9** An artificially made m-RNA molecule consisting of only uracil. UUU (uracil-uracil-uracil) is a codon for the incorporation of phenylalanine into a protein. Such a chain of uracils was successfully taken up by a cell and passed through the ribosomes, thus producing a long chain of phenylalanines.

## Transfer RNA

In order for amino acids to be available to go on line they must be brought to the ribosomes. While m-RNA seems to move without a helper, there is another kind of RNA that acts as a carrier for amino acids and brings them to the place of assembly: t-RNA. There is a different kind of t-RNA molecule for each kind of amino acid. An amino acid needed in the assembly of a polypeptide at the ribosome is escorted by a t-RNA molecule.

Transfer RNA molecules are much smaller than m-RNA molecules. Whereas an m-RNA molecule may have thousands of nucleotides, a t-RNA molecule has only seventy-three to ninety-three nucleotides. The amino acids required in a specified sequence are brought to the ribosomes one at a time by specific carrier molecules. Each molecule of t-RNA is three dimensional in appearance. When the crumpled aspect of the t-RNA molecule is flattened out, and the molecule is represented in a semi-diagrammatic fashion, it may appear as shown in figure 6-11.

## ✻ Notes ✻

|  | SECOND LETTER |  |  |  |  |
|---|---|---|---|---|---|
|  | **U** | **C** | **A** | **G** |  |
| **U** | UUU ⎤ PHE<br>UUC ⎦<br>UUA ⎤ LEU<br>UUG ⎦ | UCU ⎤<br>UCC ⎥ SER<br>UCA ⎥<br>UCG ⎦ | UAU ⎤ TYR<br>UAC ⎦<br>UAA  ?<br>UAG  ? | UGU ⎤ CYS<br>UGC ⎦<br>UGA  ?<br>UGG  TRP | U<br>C<br>A<br>G |
| **C** | CUU ⎤<br>CUC ⎥ LEU<br>CUA ⎥<br>CUG ⎦ | CCU ⎤<br>CCC ⎥ PRO<br>CCA ⎥<br>CCG ⎦ | CAU ⎤ HIS<br>CAC ⎦<br>CAA ⎤ GLN<br>CAG ⎦ | CGU ⎤<br>CGC ⎥ ARG<br>CGA ⎥<br>CGG ⎦ | U<br>C<br>A<br>G |
| **A** | AUU ⎤ ILE<br>AUC ⎥<br>AUA ⎦<br>AUG  MET | ACU ⎤<br>ACC ⎥ THR<br>ACA ⎥<br>ACG ⎦ | AAU ⎤ ASN<br>AAC ⎦<br>AAA ⎤ LYS<br>AAG ⎦ | AGU ⎤ SER<br>AGC ⎦<br>AGA ⎤ ARG<br>AGG ⎦ | U<br>C<br>A<br>G |
| **G** | GUU ⎤<br>GUC ⎥ VAL<br>GUA ⎥<br>GUG ⎦ | GCU ⎤<br>GCC ⎥ ALA<br>GCA ⎥<br>GCG ⎦ | GAU ⎤ ASP<br>GAC ⎦<br>GAA ⎤ GLU<br>GAG ⎦ | GGU ⎤<br>GGC ⎥ GLY<br>GGA ⎥<br>GGG ⎦ | U<br>C<br>A<br>G |

(FIRST LETTER on left; THIRD LETTER on right)

ALA = Alanine  
ARG = Arginine  
ASA = Aspartic acid  
ASP = Asparagine  
CYS = Cysteine  
GLN = Glutamine  
GLU = Glutamic acid  
GLY = Glycine  
HIS = Histidine  
ILE = Isoleucine  
LEU = Leucine  
LYS = Lysine  
MET = Methionine  
PHE = Phenylalanine  
PRO = Proline  
SER = Serine  
THR = Threonine  
TRP = Tryptophan  
TYR = Tyrosine  
VAL = Valine  

**Figure 6-10** Matrix showing all the ways that the four bases adenine, guanine, cytosine, and uracil can be assembled in threes to make codons, the RNA triplets of m-RNA. The first letters of the triplets are listed in the vertical column at left; the second letters of the triplets are listed horizontally across the top; and the third letters of the triplets are listed in the vertical column at right. The bases indicated by the triplets are abbreviated in the matrix; full names are given below the matrix.

One portion of the t-RNA molecule is called the **attachment site**. This is where the amino acid that is to be transported hooks on. The attachment site is at the end of the molecule, and here the sequence of three bases is always the same: CCA. Another portion of the t-RNA molecule is called the **recognition site**. It tells what kind of amino acid is to be picked up. Here, a triplet of bases constitutes an anticodon, which can match up to the codon of the m-RNA molecule. When the t-RNA molecule arrives at the ribosome, it releases the amino acid that had been attached to the free-end, CCA position. The amino acid can then join to the growing peptide chain.

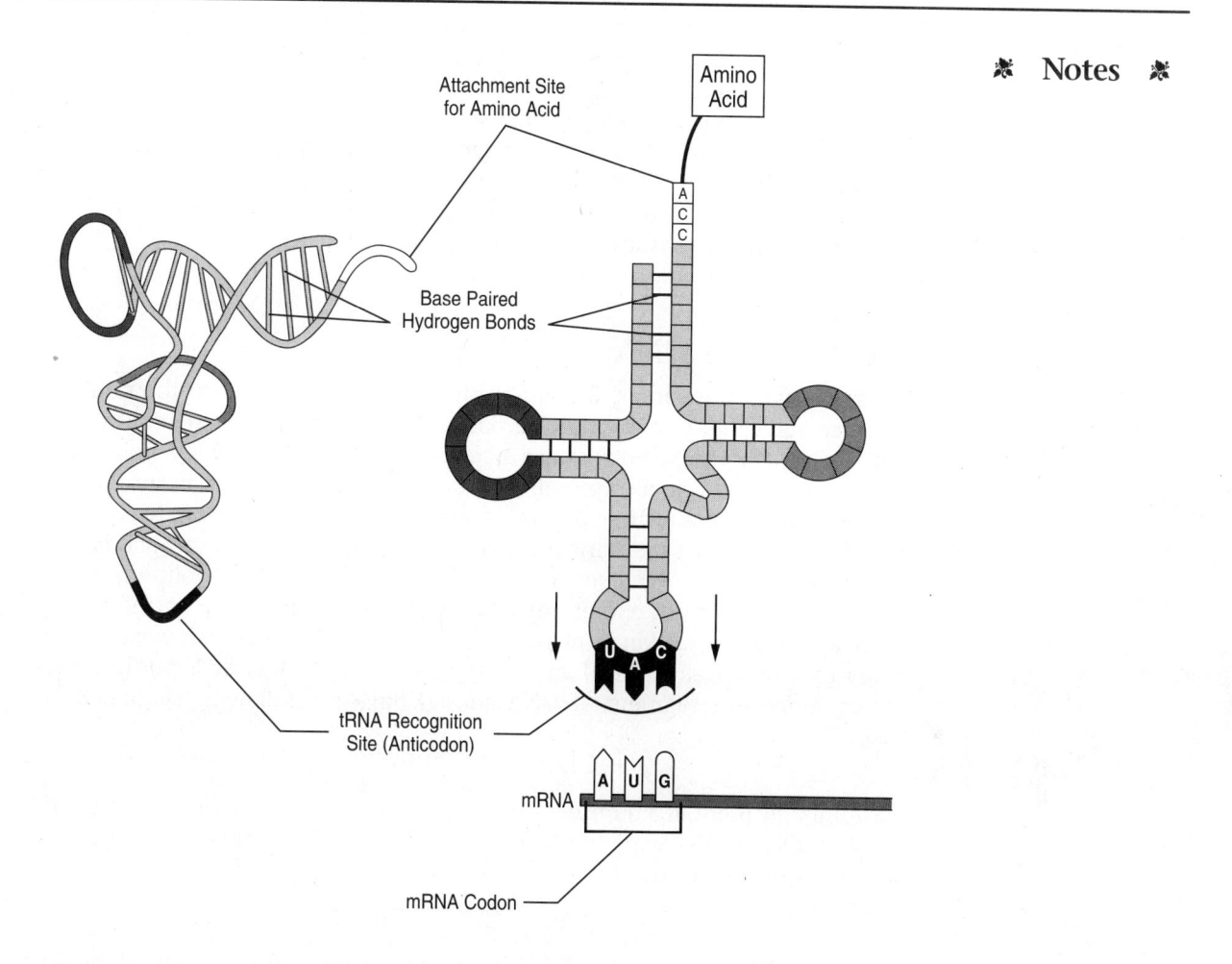

**Figure 6-11** A t-RNA molecule is much shorter than other nucleic acid molecules. Transfer RNA is a carrier molecule that brings the needed amino acid to the site of assembly at the ribosomes. There is a different kind of t-RNA for each amino acid. The amino acid to be carried is attached to the t-RNA molecule at the attachment site. Another part of the molecule is called the recognition site.

## Enzymes

DNA appears to be solely concerned with the formation of protein. A gene may be defined as that part of a DNA molecule involved in controlling the production of proteins. In general, there are two kinds of proteins: structural proteins and enzymes. Other constituents of cells, such as cellulose, lignin, suberin, **hormones**, and starch, are manufactured by the aegis of enzymes. Although not all proteins are enzymes, all enzymes are proteins; and many

### Notes

different kinds of enzymes are involved in protein synthesis. Specific enzymes, for example, are required to uncoil DNA. Other kinds of enzymes are called on to replicate DNA, while still others are needed to form m-RNA. Certain enzymes are required to connect amino acids to form polypeptides. All life processes are controlled by enzymes, and a single cell may have several thousands of kinds of enzymes. Enzymes are highly specific, a particular kind of enzyme typically catalyzing a particular reaction.

## Mutations in DNA

Consider again the earlier discussion of DNA replication. When the double-stranded molecule uncoils and the paired bases disjoin, the liberated bases can then attach to new nucleotides. Recall also that adenine always pairs with thymine, and guanine always pairs with cytosine. Well, almost always. If this pairing were completely free of error, there would be no evolution. Everything would always remain the same. But errors in replication do occur. Such errors are called *mutations*. While errors in replication may explain evolution, most such mutations do not confer any advantage. In fact, most are harmful.

There are several kinds of mutations. One type of error that can have major consequences is called an *insertion*; an extra nucleotide is put into the chain. Suppose a segment of DNA should have the following sequence of bases:

GTAACCCGGTTTGCA

Although there are no breaks or spaces between the triplets (codons) in a DNA molecule, inserting breaks on paper facilitates visualization. Thus, the preceding sequence may be represented as follows:

GTA  ACC  CGG  TTT  GCA

The m-RNA that would be fabricated on this sequence would be as follows:

CAU  UGG  GCC  AAA  CGU

representing the codons for:

histidine-tryptophane-alanine-lysine-arginine.

If a guanine were to be inserted between the second and third bases of the DNA chain (at the arrow), changes would occur all along the line. The triplets would then be as follows:

GTG  AAC  CCG  GTT  TGC  A—
 ↑

The m-RNA would then be codified as follows:

CAC  UUG  GGC  CAA  ACG

The following sequence of amino acids would then result:

    histidine-leucine-glycine-glutamine-threonine

The first amino acid in the series, histidine, is not changed, because both CAU and CAC represent the same amino acid. The others, however, are changed.

*Deletion* is another kind of mutation. Assume the same sequence of DNA bases as noted earlier:

    GTA   ACC   CGG   TTT
     ↑

If the third base in the chain (see arrow) were to be knocked out (deleted), the sequence would then be as follows:

    GTA   CCC   GGT   TT-

The deletion does not alter the first triplet, because a neighboring adenine takes the place of the adenine deleted from the first triplet. The other codons, however, are altered, and the m-RNA sequence becomes:

    CAU   GGC   CCA   AA-

Thus, the amino acid sequence is changed from histidine-tryptophane-alanine-lysine to histidine-lysine-proline. Obviously, a small alteration of DNA can result in great changes in protein structure.

Another type of DNA alteration called *substitution* appears to have lesser consequences. If the original segment of DNA were to be changed by replacing a nucleotide with one of a different kind, only one amino acid in the chain would be changed. For example, consider a substitution made at the third position in our original chain:

    GT*A*   ACC   CGG   TTA

    becomes

    GT*C*   ACC   CGG   TTA

The m-RNA in the first part, then, changes from CAU to CAG; the amino acid histidine is replaced with glycine. (Other anomalous chromosomal alterations that can affect protein synthesis [such as inversions, crossing over, and polyploidy] are described in chapter 3 on mitosis and meiosis.)

## Gene Repression

A review of the details of mitosis reminds us that all the cells of a body have the same genetic material—the same DNA. Yet, different cells do different things, and the same cells sometimes do different things at different times. Certain genes, then, can be turned on at one time and turned off at another

### ❋ Notes ❋

time. An explanation of gene repression was first developed by Francois Jacob and Jacques Monot at the Pasteur Institute. In working with *E. coli*, Jacob and Monot determined that there is a **structural gene** and an **operator gene**, which together compose a genetic unit called an **operon**. Whether the operon is able to function or not depends on another gene called a **repressor gene**. Of course, each of these genes carries out its work by the production of enzymes. In 1965 Jacob and Monot were awarded the Nobel prize for their work in this field.

## Questions for Review

1. How does deoxyribose differ from ribose?
2. Name the two types of nucleic acids.
3. Nucleotides are made up of three parts: a _____, a _____, and a _____.
4. Four types of bases enter into the formation of DNA, namely _____, _____, _____, and _____. One of these bases is not present in RNA (thymine); in its place _____ occurs.
5. While working with pneumococcus, Fred Griffith found conclusive evidence that DNA is the hereditary material. Recall and detail these experiments.
6. Who is credited with determining the molecular structure of DNA? When was this done?
7. Describe the molecular structure of DNA.
8. How is RNA different from DNA?
9. Relate how DNA is involved in the formation of protein.

## Suggestions for Further Reading

Schief, R. 1986. *Genetics and Molecular Biology*. Reading, MA: Benjamin/Cummings Publishing Company, Inc.

Watson, J.D. 1968. *The Double Helix*. New York: Atheneum Publishers.

Watson, J.D. 1970. *Molecular Biology of the Gene*. Menlo Park, CA: W.A. Benjamin Inc.

# 7

# Viruses

How does one define a **virus**? Viruses are extremely small organisms, and their diminutive size was the basis of their original definition. In 1892 D. Ivanovski was studying tobacco mosaic disease. He passed some juice squeezed from an infected tobacco plant through a ceramic filter having small enough pores to filter out bacteria. If the tobacco mosaic disease were caused by a bacterium, he reasoned, this filtering process should effectively sterilize the filtrate. What he found, however, was that the filtered tobacco juice retained the capacity to infect other tobacco plants. The infective agent, being too small to be bacterial, thus became known as a **filterable virus**. When later experimentation uncovered viruses too large to pass through such filters, other means of definition for such organisms were needed.

In 1933 Wendell Stanley, working at the Rockefeller Institute, was able to crystalize the pure tobacco mosaic virus (TMV). Upon dissolving the crystals and touching the solution to a healthy tobacco plant, infection occurred. This helped to further define viruses, but other significant findings also contributed to forming an accurate definition of these organisms. For example, viruses cannot be grown on a nutrient medium in the laboratory; they can be grown only inside of living cells. Furthermore, viruses do not have the mechanism for self-replication; in particular, they lack an adequate energy reservoir. Because viruses can be grown only inside of living cells, the invention and improvement of tissue culture methods greatly aided the study of these organisms.

It is convenient to classify viruses into four major groups: animal viruses (including viruses affecting humans and other vertebrates), insect viruses, plant viruses, and bacterial viruses (**bacteriophages**, or **phages**). In a botany course, one would expect to focus attention first on the third group; but investigations involving bacteriophages are especially pertinent and, thus, are addressed immediately following.

✻ Notes ✻

❀ Notes ❀

## T-2 Bacteriophage

The T-2 phage particle, a virus, infects the common colon bacterium *Escherichia coli*. In experiments involving T-2 and *E. coli,* a Petri dish was inoculated with *E. coli* and incubated until the bacteria covered the entire surface of the dish. Next, a much diluted suspension of the T-2 virus was spread on this bacterial "lawn." After some time had passed, cleared spots called plaques appeared where the bacteria had been lysed (broken down); wherever a T-2 particle came into contact with the *E. coli,* the bacterium was "attacked" by the T-2. This caused the bacterium to lyse. The T-2 particles, which replicated many times over in this way, spread to adjacent bacteria, which were then also lysed. The eventual result was a plaque large enough to be seen without the aid of magnification. Viruses of this kind, then, are intracellular parasites that, when grown inside a bacterium, result in the bacterium's destruction.

When the electron microscope came into use, it became possible to see things that could not be seen under a light microscope. It was even possible to see T-2 particles adsorbed (adhered) to the surface of an *E. coli* cell.

In 1952 Hershey and Chase learned that phage particles are susceptible to **osmotic shock**. They placed a suspension of phage in a concentrated salt solution. After several minutes, they suddenly diluted the solution. This caused the parts of the phage to separate. **Ultracentrifugation** then allowed the parts to be isolated and identified. Two components were found: a protein and DNA.

Phosphorus is a part of DNA but not a part of protein. Sulphur is a part of protein but not a part of DNA. In one series of experiments, Hershey and Chase made radioactive phosphorus, $P^{32}$, a part of phage DNA to provide a traceable tag that would allow them to see where the phage DNA went when placed in contact with *E. coli*. Likewise, they used radioactive sulphur, $S^{35}$, as

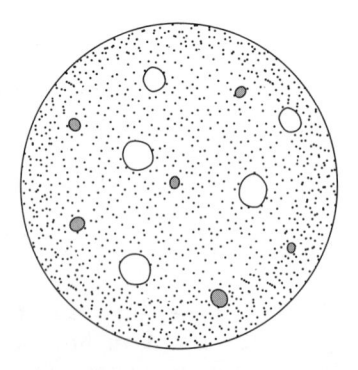

**Figure 7-1** A lawn of bacteria will show visible bare spots called plaques after scattered viruses that have touched those places have lysed and spread to neighboring cells.

**T-2 Bacteriophage**

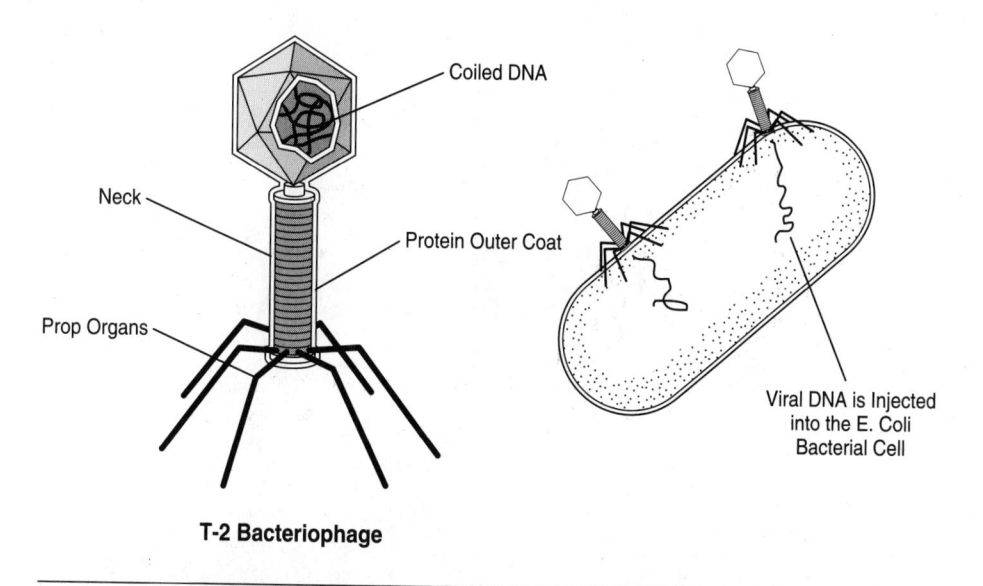

**Figure 7-2** At left, a T-2 phage particle showing the coiled DNA inside the protein outer coat; the neck; and the prop organs. At right, an *E. coli* bacterial cell to which several T-2 particles have adsorbed.

a tag to show where the phage protein went. Some T-2 phages were first cultivated in a medium containing radioactive $P^{32}$ and then were placed in *E. coli*. The T-2 phages of course grew in the *E. coli*. Although only one billionth of the phosphorus in the DNA was radioactive, this was enough. In this way, Hershey and Chase learned that most of the radioactivity was inside the *E. coli* cell; in other words, the DNA went inside the cell.

Next, Hershey and Chase cultivated T-2 phages in a medium containing radioactive $S^{35}$, a constituent of protein. When the T-2 phages were placed in *E. coli,* the protein part of the phage remained outside of the bacterial cell.

Hershey and Chase's findings coupled with electron microscope studies showed that the T-2 phage is composed of an outer protein portion and an inner DNA portion; and that the phage becomes adsorbed to the surface of *E. coli* and extrudes (forces) its DNA into the *E. coli* cell. The only part of T-2 that infiltrates the bacterial cell, then, is information. This again demonstrated that DNA is the genetic material. The electron microscope reveals the phage outer coat to take the form of a hexagonal head part, a tail portion, and some prop organs.

When the phage's DNA enters the *E. coli*, it "takes over," in a sense. By exploiting this energy source, many more phage particles are manufactured. When the phages increase to a certain number, the bacterial cell is lysed; the T-2 then spreads to adjacent *E. coli* cells. It is interesting to note that the components of the phage are not immediately assembled upon manufacture.

78 ♦ Chapter 7

❀ **Notes** ❀

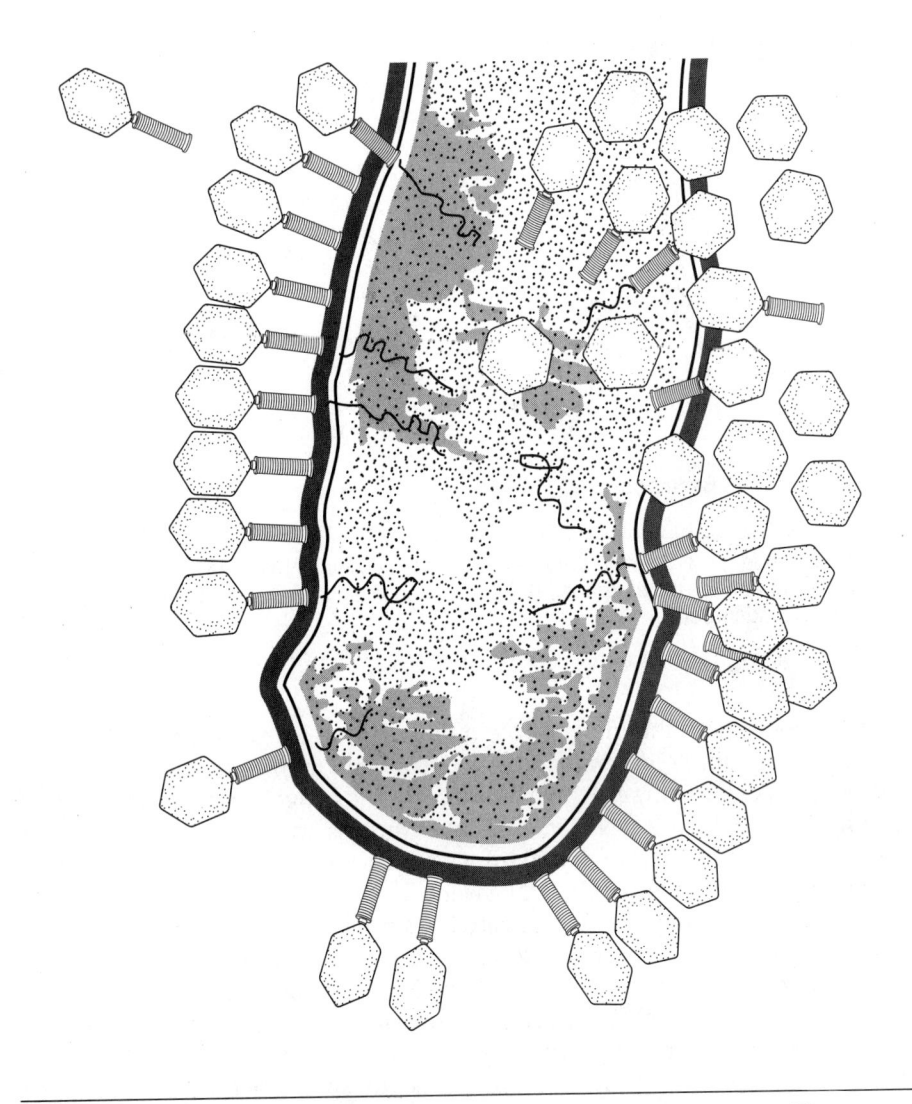

**Figure 7-3** An electron micrograph showing a large number of T-2 virus particles attached to the outer surface of the bacterium *E. coli*.

The electron microscope shows that at one point in the cycle, head parts, tail parts, and prop organs exist separately in the cell. Only at the last moment, just before lysis, are the complete phages assembled; the DNA is taken into the head, the tail and prop organs are properly attached, and the bacterial cell is lysed. These events take approximately ten and one-half minutes from the time the T-2 first contacts the outside of the bacterium.

## Plant Viruses

While the genetic material of bacteriophages is DNA, the genetic molecule of plant viruses is most often RNA. Bacteriophages and plant viruses also replicate differently. Plant viruses generally do not rupture the cells in which they form; rather, they seep out through cell membranes and to neighboring cells.

More than one thousand plant diseases can be traced to viruses: tobacco, potato, sugar beet, peas, orange, elm, apples, cucumber, bean, cauliflower, tulips, and rice. It seems that freedom from viral infection is an uncommon condition in the plant kingdom.

Viral infections generally result in necrosis, yellowing, mottling, leaf spots, streaks, and wilting. Not all plant viruses have negative effects, however. The highly esteemed Rembrandt tulip, for example, owes its colorful streaks to a viral infection. A mosaic virus that damages rice plants has quite a different effect on jute (a plant used in making rope and burlap bags); it improves the growth of jute plants.

Viral transmission between plants can result from touching an infected plant and then touching a healthy plant. In nature, viruses are most commonly carried by insects (especially aphids and leafhoppers) and nematodes. Some plant viruses multiply within the bodies of their insect carriers as well as in the plants themselves. It follows, then, that insect control is a proper means of combating viruses. Plants also may become infected by viruses through wounds.

People who raise potatoes know that attention must be paid to controlling viral diseases. It is inadvisable, for example, to use one's own potatoes as seed pieces for successive crops, as gradual degeneration will result. Rather, seed potatoes should be obtained from other regions in order to minimize the hazard of an accumulation of viral diseases. Such viral diseases are transmitted by **aphids**. Weeds also can be carriers of viral diseases, and dahlias carry a virus that infects tomato plants.

While most plant viruses have RNA and several have DNA, no plant viruses have both. The tobacco mosaic virus is approximately 300 nanometers long, and its RNA molecule has 6,000 bases. These bases make up three genes. Other plant viruses contain as many as 200 genes. A first stage in the replication of plant viruses is the shedding of the protein coat. Recall that in bacteriophage, the protein remains outside of the cell. In other viruses, the protein may be shed inside the infected cell. The viral nucleic acids then direct the production of viral enzymes. These, in turn, direct the manufacture and assemblage of protein and the production of more viral RNA. The end result may be thousands of new virus particles in the cell.

Each virus particle is commonly organized into a twenty-sided figure, a shape called an **icosahedron**. RNA viruses have either a single strand of RNA (ssRNA) or a double strand (dsRNA). (DNA viruses can be classified in the same way.) Single-stranded RNA viruses can be further subdivided into two groups: plus strand or minus strand. Plus-stranded RNA acts directly to program replication; minus-stranded RNA acts as a template for the production

### Notes

of an m-RNA that then directs virus replication. Plus-stranded viruses can be further subdivided for classification purposes based on whether or not the virus particle has an envelope. Further details on classification of viruses can be found in the suggested readings.

## Questions for Review

1. What is a phage?
2. How do viruses replicate?
3. Are viruses alive? What evidence supports your position?
4. In 1892, Ivanovski demonstrated something about viruses that helped distinguish them from bacteria. Explain.
5. Wendell Stanley did some significant work with tobacco plants and a virus. Describe this work.
6. What happens when a bacteriophage such as T-2 comes in contact with the bacterium *Escherichia coli*?
7. How was it demonstrated that the T-2 phage does not completely enter the bacterial cell?

## Suggestions for Further Reading

Gibbs, A.J., and B.D. Harrison. 1979. *Plant Virology: The Principles*. New York: John Wiley and Sons, Inc.

Luria, S.E., et al. 1978. *General Virology*. New York: John Wiley and Sons, Inc.

# 8

# Physical Properties of Protoplasm

There are great differences among living organisms, yet all organisms are made of cells. When the most minute organelles of cells are examined, even of very diverse life forms, the differences disappear. Mitochondria, Golgi bodies, ribosomes, endoplasmic reticulum, centrioles all seem to be the same wherever they are found. This is also largely true of chemical composition. All cells have proteins, fats, carbohydrates, nucleic acids, minerals, and enzymes. This living substance is called *protoplasm*, and although it may vary within a certain range from organism to organism, the likenesses are certainly more striking than are the differences. All kinds of protoplasm possess the capacity to respond (irritability), the capacity to move within a cell (motility), the capacity to grow, and the capacity to reproduce. What is life? We may attempt to define life by stating of what life is made, or by explaining what life does. Life, however, is more than the sum of the substances of which it is made. (One of the most amazing substances of life is DNA, which justified separate treatment in chapter 6.)

## Composition of Protoplasm

Water is often the most abundant substance of protoplasm. Water can account for more than 90 percent of protoplasm weight, although in seeds, the figure can be as low as 10 percent. Proteins can account for from 7 to 10 percent; fats 1 to 2 percent; vitamins and enzymes approximately 1 percent; and minerals approximately 1 percent. Another way this might be expressed is to say that for each molecule of protein in protoplasm, there may be 18,000 molecules of water, 10 molecules of fat, and 100 molecules of inorganic materials.

❧ **Notes** ❧

## ❈ Notes ❈

## Colloids

Some of the substances of protoplasm are in solution, some in the form of ions, and some are **colloids**. Colloids are either aggregates of molecules or very large molecules (such as proteins) dispersed in water. Colloidal particles, while not in true solution, remain dispersed for a long time. Each colloid has an electrical charge, either positive or negative. The fact that like charges repel helps maintain the colloidal dispersion.

The finely divided state of colloidal particles is significant to the events that occur in protoplasm because this state offers a large surface area for substance interaction. Consider a cube 1 centimeter in length on each side. Such a cube has 6 square centimeters of surface area. If the cube were cut in two, 8 square centimeters of surface area would be exposed. If cut through again, 10 square centimeters of surface area would be exposed (see figure 8-2). Without increasing the amount of substance, the surface area of the substance is increased by dividing the substance into smaller parts. If this same 1 cubic centimeter of material were divided into the colloidal state, 60,000 square centimeters of surface area would be exposed.

A colloid is constituted of two parts called phases: a **continuous phase** and a **dispersed phase**. Oil can be dispersed in water by agitating the water; but when the agitation ceases, the droplets of oil promptly coalesce. If a third substance called an emulsifying agent is added to the mixture, however, the oil will disperse and remain so in a colloidal state. The **emulsifying** agent surrounds each minute droplet of oil and prevents the droplets from coming

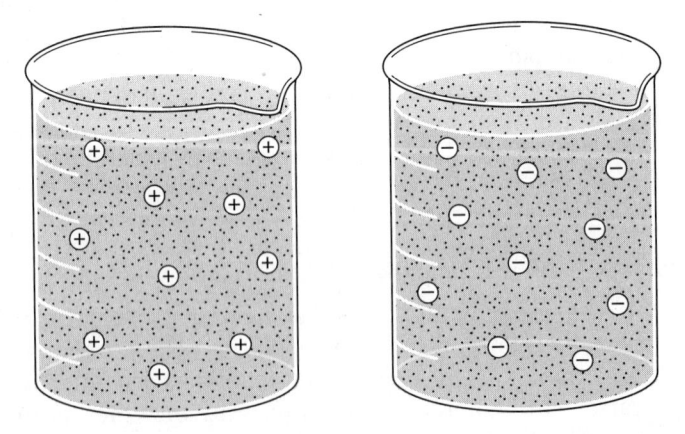

**Figure 8-1** Because like charges repel, dispersed (colloidal) particles having the same electrical charge tend to remain dispersed.

together again. In figure 8-3, oil is the dispersed phase and water the continuous phase. This could be reversed, however, by having water dispersed in oil.

A watery colloid is called a **sol**; a viscous colloid, a **gel**. Protoplasm is a mixture of sols and gels, and one may change into the other and back numerous times.

✱ **Notes** ✱

6 square cm.   8 square cm.   10 square cm.

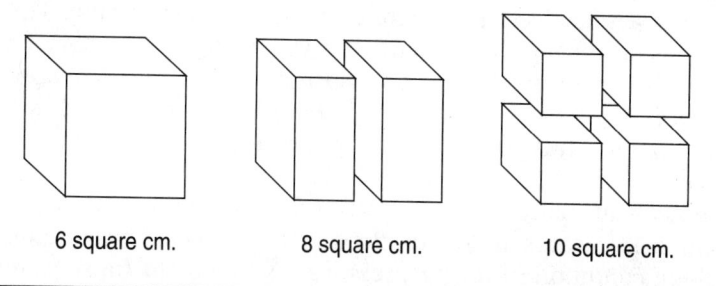

**Figure 8-2** One cubic centimeter of substance exposes six square centimeters of surface. If such a substance were cut into two parts, the surface area exposed would increase by two square centimeters without increasing the mass of the substance.

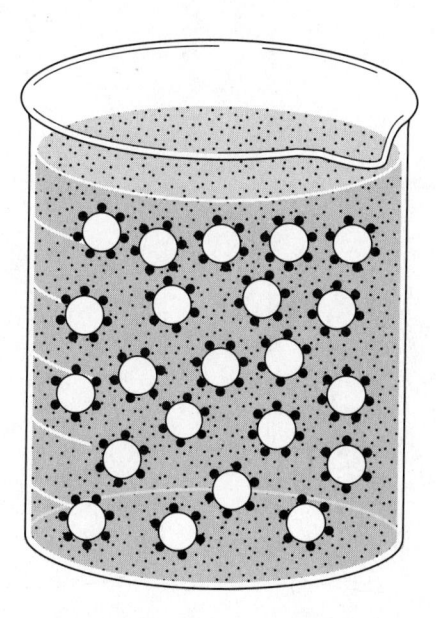

**Figure 8-3** Particles in colloidal dispersion can be kept in a dispersed state by surrounding each particle with an emulsifying agent such as soap.

※ **Notes** ※

## Diffusion

Consider some substances in true solution. Substances tend to diffuse, that is, move from a region where they are in greater concentration to a region where they are in lesser concentration. This is simply because molecules are in constant motion. If a cell membrane is involved, water can pass readily through; but dissolved material may or may not be able to pass through. Whether or not dissolved material is able to pass through a cell membrane depends in part on the size of the molecules. Remember, a cell membrane is a selectively permeable, or osmotic, membrane. The process of water passing through the membrane is called **osmosis.** Water will tend to flow from a region of greater concentration of water to a region of lesser concentration of water. Another and perhaps more common way of saying this is that water tends to flow toward a greater concentration of dissolved substances.

When cells are placed in pure water, there being a dissolved substance inside the cells, water tends to flow into the cells, producing **turgor.** The resulting condition of **turgor pressure** then tends to cause water to flow out of the cell. Soon, equilibrium is attained. The flow of water, then, does not stop; rather, there is an equal rate of flow in both directions. The pressure that builds up in the cell can sometimes cause the cell to burst, or lyse. In some plant cells, this is prevented by the presence of rigid cell walls.

**Figure 8-4** Osmosis. Water flows toward a region of greater concentration of dissolved substance. A cell placed in pure water soon becomes turgid by the inflow of water, as shown at right.

When cells are placed in a concentrated solution such as syrup, the concentration of dissolved substances (sugar) is greater outside the cells, and water tends to flow out of the cells, causing the cells to shrink.

A solution having a lesser concentration of dissolved material as is present in the cells is called a **hypotonic** solution; a solution having the same concentration of dissolved material as is present in the cells is called an **isotonic** solution; and a solution having a greater concentration of dissolved material than is present in the cells is called a **hypertonic** solution.

Plant cells having rigid cell walls are good material for demonstrating diffusion, turgor, and differences between solutions. Such cells placed in pure water (a hypotonic solution) become turgid, the cell membranes pressing against the cell walls. The cells do not lyse, and the process is reversible. Such cells placed in a hypertonic solution will shrink, water flowing out of the cells and thus causing the cell membranes to pull away from the cell walls. This process, called **plasmolysis**, can be reversed.

Figure 8-6 shows a U-tube. An osmotic membrane at the bottom of the tube forms a partition between the two sides. A 6-percent solution of sucrose is placed on one side, and a 2-percent solution of sucrose is placed on the other side. Water, tending to flow toward a greater concentration of dissolved substance, will flow from the 2-percent solution to the 6-percent solution, causing the level of the liquid on the left side to rise. This, in turn, causes a pressure buildup in the other direction, from left to right, and an equilibrium of flow is attained.

❋ **Notes** ❋

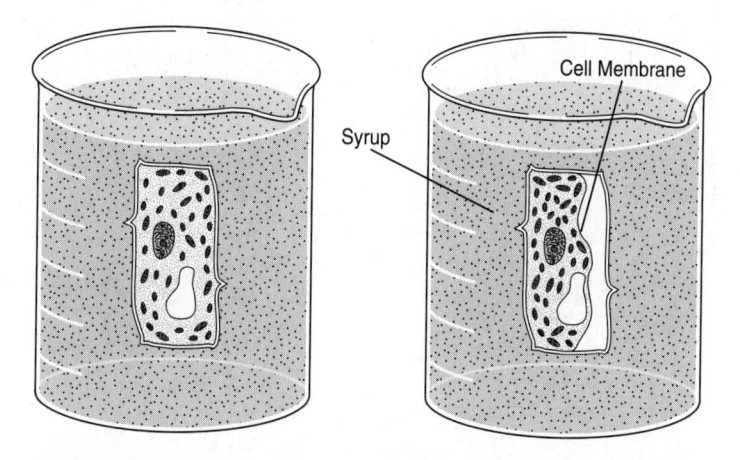

**Figure 8-5** The direction of water flow is reversed when cells are placed in syrup, water tending to flow toward a region of greater concentration of dissolved substance. The process is the same, only the direction is changed.

## ✽ Notes ✽

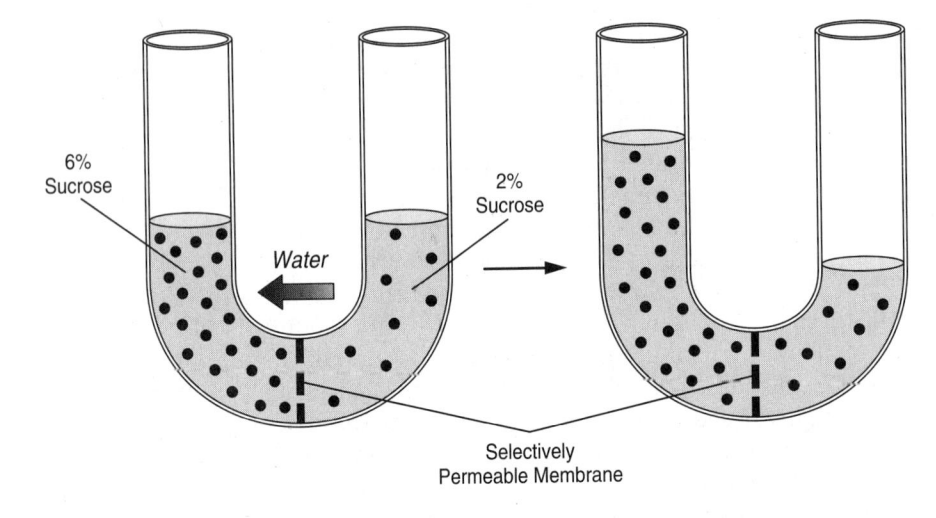

**Figure 8-6** A U-tube with an osmotic membrane centered at the bottom. Water flows toward a greater concentration of sugar.

Figure 8-7 shows a U-tube with a 6-percent solution of sucrose on the left side and a 2-percent solution of sodium chloride on the right side. Again, water will flow toward a greater concentration of dissolved substance (toward the sugar). Whereas the selectively permeable membrane does not allow the sugar

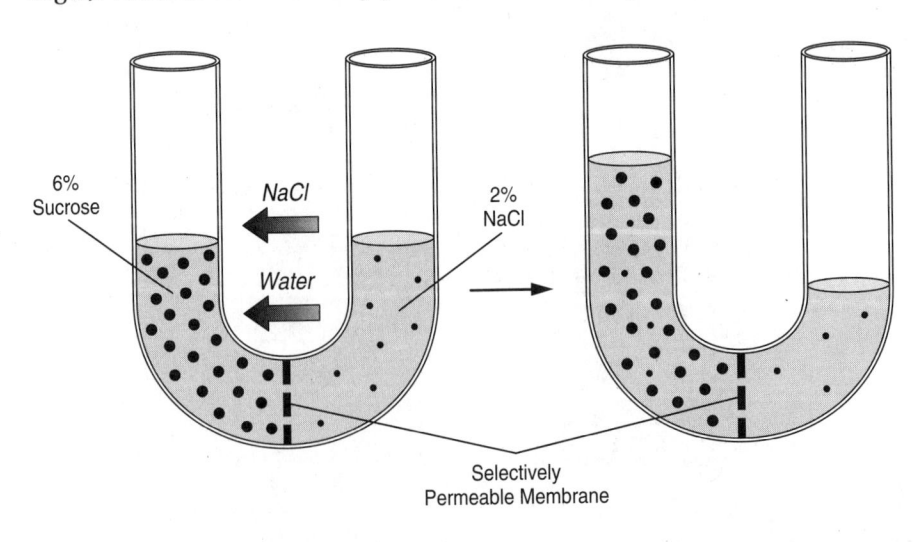

**Figure 8-7** A U-tube with a sodium chloride solution on the right side of the tube. The same tendencies would occur causing the volume on the right side of the tube to diminish.

to pass through, the sodium chloride is readily able to pass through. This causes the concentration of sodium chloride to become the same on both sides.

Figure 8-8 shows a U-tube with a 6-percent solution of sucrose and a 2-percent solution of sodium chloride (salt) on the left side, and pure water on the right side. In time, the salt concentration will equalize on both sides. Whereas osmosis is the movement of water, **dialysis** is the movement of salt.

In the U-tube examples, the force that allows the movement of water and solutes is diffusion, which is itself caused by the molecules being in constant motion. In living systems, the same force is at work; but other factors also come into play. Membranes, being alive, are able to exert additional forces. For example, solutions can be moved against a concentration gradient, that is, in a direction opposite that previously described. Such movement requires the expenditure of energy and is called **active transport**. The large, brown seaweeds offer an example of active transport. These organisms are able to concentrate iodine. Iodine is present in extremely minute quantities in sea water, but it is taken in through the living membranes and concentrated in these seaweeds. For a long time these seaweeds, called kelp, were used as a commercial source of iodine. Certain kinds of plants are able to concentrate potassium. There may be a 50 to even 1,000 times greater concentration of potassium in the cells of such plants than exists in the surrounding soil. Certain fungi concentrate metals. All such examples of active transport require the expenditure of energy.

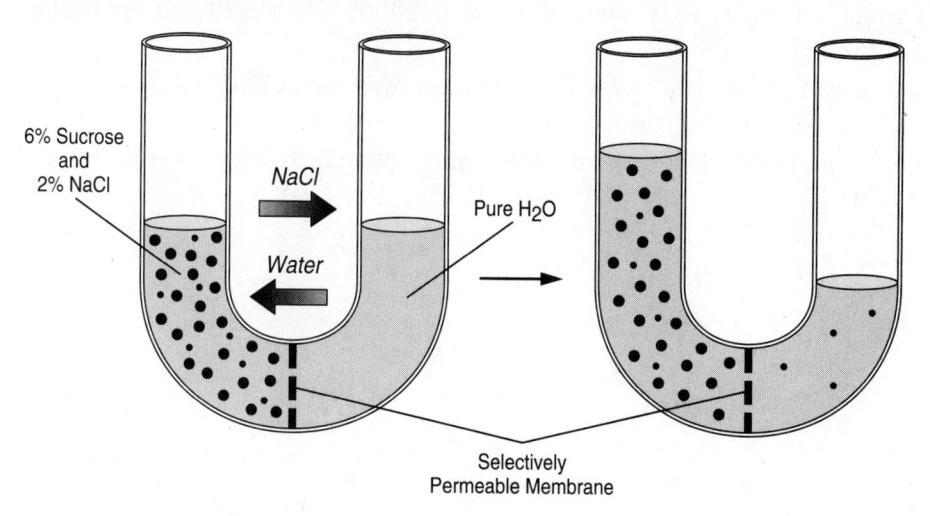

**Notes**

**Before Osmosis**     **After Osmosis**

**Figure 8-8** If both sugar and sodium chloride are placed in the same side of the tube, the rules are the same; a dissolved substance tends to flow toward a region of lesser concentration of that substance. This applies to both the sugar and the salt. But whereas the sugar confronts a barrier in the osmotic membrane, the osmotic membrane does not present a barrier to the sodium chloride. In time, the concentration of sodium chloride will be the same on both sides of the membrane. The sugar, however, will remain on the left.

## Notes

## Questions for Review

1. What are the physical properties of colloids?
2. Define the terms *diffusion* and *osmosis*.
3. Relate an experiment that illustrates osmosis.
4. What is meant by the term *active solute absorption*?
5. Recount an experiment that illustrates plasmolysis and deplasmolysis.
6. In reference to surface exposure, what is the result of breaking something into a finely divided state?
7. What is the most abundant substance in protoplasm?
8. Define the terms *emulsion, continuous phase, hypotonic,* and *dialysis.*
9. Seaweeds are able to concentrate iodine in their tissues by extracting it from sea water. This involves a flow against the _____ _____ from lesser concentration to greater concentration.

## Suggestions for Further Reading

Bonner, D.M. 1961. *Control Mechanisms in Cellular Processes.* New York: Ronald Press.

Jensen, William A. 1964. *The Plant Cell.* Belmont, CA: Wadsworth Publishing Co.

Kennedy, D. 1965. *The Living Cell: Readings from the Scientific American.* San Francisco: W.H. Freeman Co.

Robinson, David G. 1985. *Plant Membranes.* New York: John Wiley and Sons, Inc.

# 9

# Photosynthesis

Photosynthesis, the process whereby the green pigment in plants (chlorophyll) traps energy from the sun, is responsible for the maintenance of *all* life on the planet, whether plant or animal. Animals are maintained by eating either plants or other animals that are nourished by plants. Life on earth is made possible by the sun. All oxygen in the atmosphere was put there by photosynthesis. All energy stored in coal and oil was supplied by photosynthesis. Obviously, photosynthesis is a topic worthy of attention.

## Early Research

As a general rule, knowledge is gained in a gradual, stepwise, cumulative fashion. Such is the case with what we know about photosynthesis. We started to learn about it several hundred years ago, and we are still learning.

The account could begin with J.B. van Helmont (1577–1644), who gave this report:

> I took a vessel and placed in it 200 pounds of earth which had been carefully dried and then weighed. Then I placed in it a stock of willow weighing 5 pounds. I faithfully watered it always with rain water and after five years had gone by I took the tree from the soil and found that it weighed 159 pounds and 3 ounces. The earth in which the tree had grown I carefully protected from the accumulation of dust by placing over the vessel an iron plate perforated with holes. I again dried and weighed the earth in which the willow had grown and found that it weighed the same, 200 pounds minus about 2 ounces, and hence I concluded that 164 pounds of wood had been derived from the element of water alone.

This creditable experiment uncovered part of the truth about photosynthesis. Although van Helmont knew nothing of either the atmosphere or any role that the atmosphere plays in the process, he did learn something very significant: that water contributes to the growth in weight of a plant. This work indicates a careful investigator and a good thinker.

## ✲ Notes ✲

Henry Cavendish (1731–1810), an English chemist and physicist, was a pioneer in the study of gases and is credited with determining that water is a compound. He also determined that the air produced by **fermentation** in vats and by putrefaction has the same properties as the "fixed air" obtained from marble. His 1784 paper summarizing his findings was entitled *Experiments on Air*.

Joseph Black (1728–1799) discovered the existence of an atmospheric gas that differed from common air: carbon dioxide, or fixed air. He performed detailed studies on the fixed air produced in fermentation vats, in respiration, and by burning charcoal.[1]

Joseph Priestley (1733–1804) was a dissenting English clergyman and a chemist. He discovered that plants make **dephlogisticated air**, or oxygen. Following is the essence of his report:

> I took a mass of air made thoroughly noxious by mice breathing and dying and putrefying in it and divided it into two parts. To one I added a sprig of mint and the other I let stand in the same position as that portion containing the sprig of mint. This was about the beginning of August, 1771, and after eight or nine days I discovered that a mouse lived perfectly well in that part of the air in which the sprig of mint had grown, but died the moment it was put in the other part of the same quantity of air which had no plant growing in it.

His report was titled *On the Purification of Air by Plants*.

Antoine Laurent Lavoisier (1743–1794) is credited with demolishing the phlogiston theory. He gave the name *oxygen* to dephlogisticated air. Although Joseph Priestley had prepared and discovered it, Lavoisier named it. He is also remembered for claiming that water is formed of hydrogen and oxygen.

Jan Ingenhousz (1730–1799), an English physician working in Holland, described the interchange of gases between plants and the atmosphere. He found that dephlogistication (the release of oxygen) takes place only in sunlight and that "plants behave like animals" (producing carbon dioxide) at night. In 1772 Joseph Priestley observed that plants could purify air; later, Ingenhousz learned that only the green parts of plants are able to restore air and that sunlight alone has no such power. Upon his return to London in 1779, Ingenhousz published a book entitled *Experiments on Vegetables, Discovering Their Great Power in Purifying the Common Air in Sunshine, but Injuring it in the Shade or at Night*. (Many years later, in 1932, Robert Hill was able to show that intact chloroplasts freed from all other cell constituents could trap light energy and liberate oxygen. Still later, Daniel Arnon established that isolated chloroplasts could use light energy to reduce carbon.) In 1782 a theologian named Jean Senebier discovered that fixed air (carbon dioxide) was required for the process described by Ingenhousz.[2] Then in 1804, Nicholas Theodore de Saussure (1767–1845) used Lavoisier's methods of

---

[1] One of Joseph Black's friends was James Watt, who invented the steam engine.

[2] Although the concentration of carbon dioxide in the atmosphere is only .03 percent, hundreds of millions of tons of carbon are fixed in photosynthesis each year.

**Figure 9-1** Joseph Priestley (1733–1804). (Illustration by Donna Mariano)

quantitative measurement to show that equal volumes of carbon dioxide and oxygen are exchanged during photosynthesis and that the plant retains carbon. He showed that plants gain more weight during photosynthesis than can be accounted for by the carbon alone, the rest of the matter, with the exception of minerals, coming from water (figure 9-2).

When it was determined that the green substance in plants is chlorophyll and that the first product of photosynthesis is glucose, a balanced chemical equation was produced (figure 9-3). This equation indicates that carbohydrate is formed from a combination of carbon and water molecules and that the oxygen liberated comes from splitting carbon dioxide. This seemed acceptable until van Niel reported on his studies of **purple sulphur bacteria**.

$$1 \text{ volume } CO_2 + H_2O \longrightarrow \text{carbon compounds} + 1 \text{ volume } O_2$$

**Figure 9-2** During photosynthesis, the volume of carbon dioxide consumed and the volume of oxygen liberated are equal.

$$6CO_2 + 6H_2O \longrightarrow C_6H_{12}O_6 + 6O_2$$

**Figure 9-3** A balanced chemical equation showing the end products of photosynthesis.

### Notes

While at Stanford University, C.B. van Niel studied a group of bacterial organisms capable of photosynthesis. Such organisms have a chlorophyll-like pigment (unlike the pigment in flowering plants) that can catalyze photosynthesis without releasing oxygen. One such group of bacteria utilize hydrogen sulfide rather than water, and liberate free sulphur rather than oxygen (figure 9-4).

These bacteria seemed to manufacture water in the process. Van Niel, thus, proposed the generalized formula shown in figure 9-5. In this formula, *A* represents sulphur, some other substance used by photosynthetic bacteria, or water. Thus, van Neil proposed that it is the lysis of water rather than the breakdown of carbon dioxide that liberates oxygen during photosynthesis.

## Modern-day Research

A long time passed before this proposal was proven correct. Following World War II, radioactive isotopes of many elements became available for research. A radioactive isotope of oxygen known as $O^{18}$ made possible the utilization of carbon dioxide and water in photosynthesis wherein the oxygen of the carbon dioxide was different from the oxygen of the water. A carefully arranged experiment was devised wherein the oxygen of carbon dioxide was $O^{16}$ and the oxygen of the water was $O^{18}$. This difference enabled researchers to trace the source where the oxygen in each compound formed. The chemical properties of the two types of oxygen were exactly alike; only the weights differed. In this way, it was clearly established that all the oxygen liberated in photosynthesis comes from the breakdown of water. An equation could now be written as shown in figure 9-6. Unlike the water on the left side of the equation, the water on the right side is manufactured water.

$$CO_2 + H_2S \longrightarrow (CH_2O)\ \text{carbohydrate} + H_2O + S$$

**Figure 9-4** The photosynthetic reaction of purple sulphur bacteria. Hydrogen sulfide is utilized in the place of water, resulting in the liberation of elemental sulphur.

$$CO_2 + 2H_2A \longrightarrow (CH_2O) + H_2O + A$$

**Figure 9-5** A generalized equation of photosynthesis that does not designate a specific compound in the reaction and indicates that water is manufactured in the process.

$$6CO_2 + 12H_2O \longrightarrow C_6H_{12}O_6 + 6O_2 + 6H_2O$$

**Figure 9-6** An upgraded equation for photosynthesis, showing that water is manufactured.

## Chlorophyll

A number of different kinds of chlorophyll have been discovered. All flowering plants have chlorophyll a and chlorophyll b, as well as the yellow pigments carotene and xanthophyll. The German chemist Richard Willstatter (1872–1942) determined the molecular structures of chlorophylls a and b and demonstrated that the yellow pigments carotene and xanthophyll are also present in green leaves. He reported that these four pigments are present in approximately the same relative proportions in all green plants. (Willstatter is said to have taken delight in undertaking the most complex chemical problems.)

Later studies revealed the existence of five different chlorophylls, twelve carotenes, and from seventy to eighty xanthophylls. Although the various chlorophylls differ only slightly in structure, only chlorophyll a is capable of directly catalyzing photosynthesis. Pigments other than chlorophyll a do absorb light; but if it is to be effective in photosynthesis, the energy absorbed by other pigments must be transferred to chlorophyll a.

It is remarkable to note that the molecular structure of hemoglobin, the red pigment of blood, is very similar to that of chlorophyll (figure 9-7). One significant difference, however, is found at the center of the molecules. Whereas chlorophyll has an atom of magnesium, hemoglobin has an atom of iron.

## Light

Light that passes through or is reflected by a leaf cannot contribute energy to the photosynthetic process; only light that is absorbed by the leaf can contribute to photosynthesis. The light most readily absorbed by leaves is red or blue-violet light; these wavelengths are most responsible for the manufacture of carbohydrate from carbon dioxide and water.

In 1882 T.W. Engelmann devised a clever and incredibly small demonstration. He placed a filamentous green alga in a solution rich in a form of bacteria that are attracted to oxygen. He then exposed the algal filament to a tiny spectrum of colors. Figure 9-8 shows both how the experiment was set up and what was observed. One part of the filament was exposed to violet light, another part to blue, another to green, and so forth all the way to red light. The oxygen-seeking bacteria clustered mostly in the violet and red zones, and it was from these regions that the most oxygen was liberated. In other words, most photosynthesis took place at these wavelengths.

In 1905 the English plant physiologist F.F. Blackman measured the rate of photosynthesis at various light intensities. Starting with dim light and increasing light intensity, Blackman discovered that while increasing the intensity of light did increase rate of photosynthesis, no increase occurred above a certain intensity. He then determined that it was the temperature increase brought on by bright light that accelerated the rate of photosynthesis. Seeing

### ❋ Notes ❋

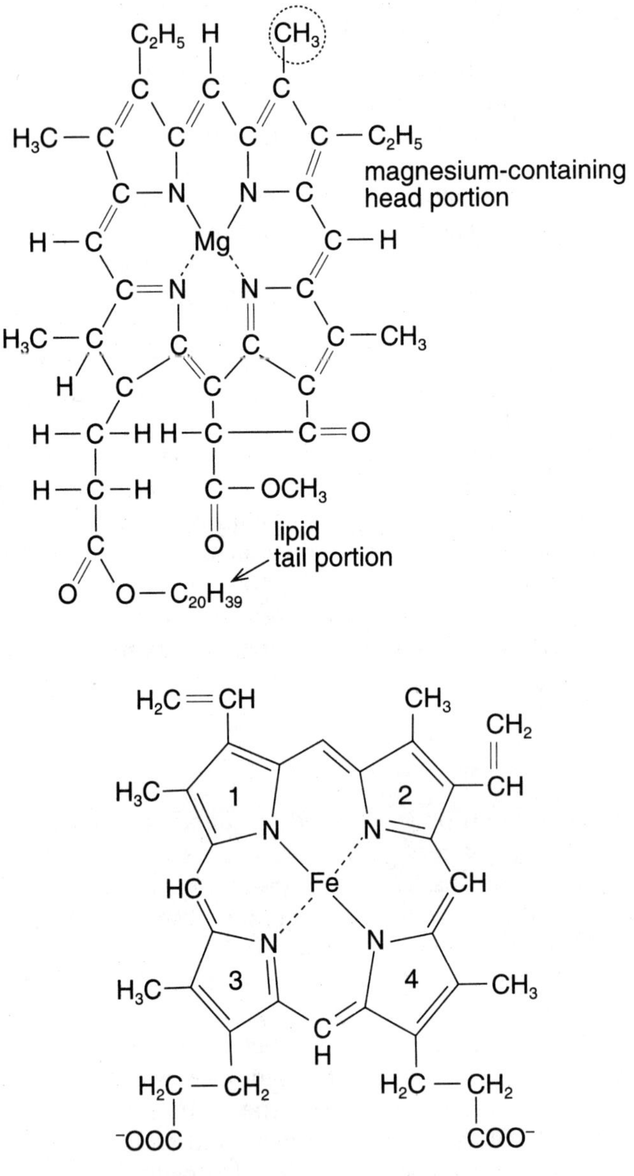

**Figure 9-7** Top: chlorophyll; bottom: hemoglobin.

that the reactions were temperature sensitive led Blackman to conclude that such reactions are governed by enzymes.

*Chlorella*, a one-celled green protist, is a favored organism in studies of photosynthesis. Laboratory experiments with Chlorella show that photosynthesis efficiency increases when light is alternately turned on and off.

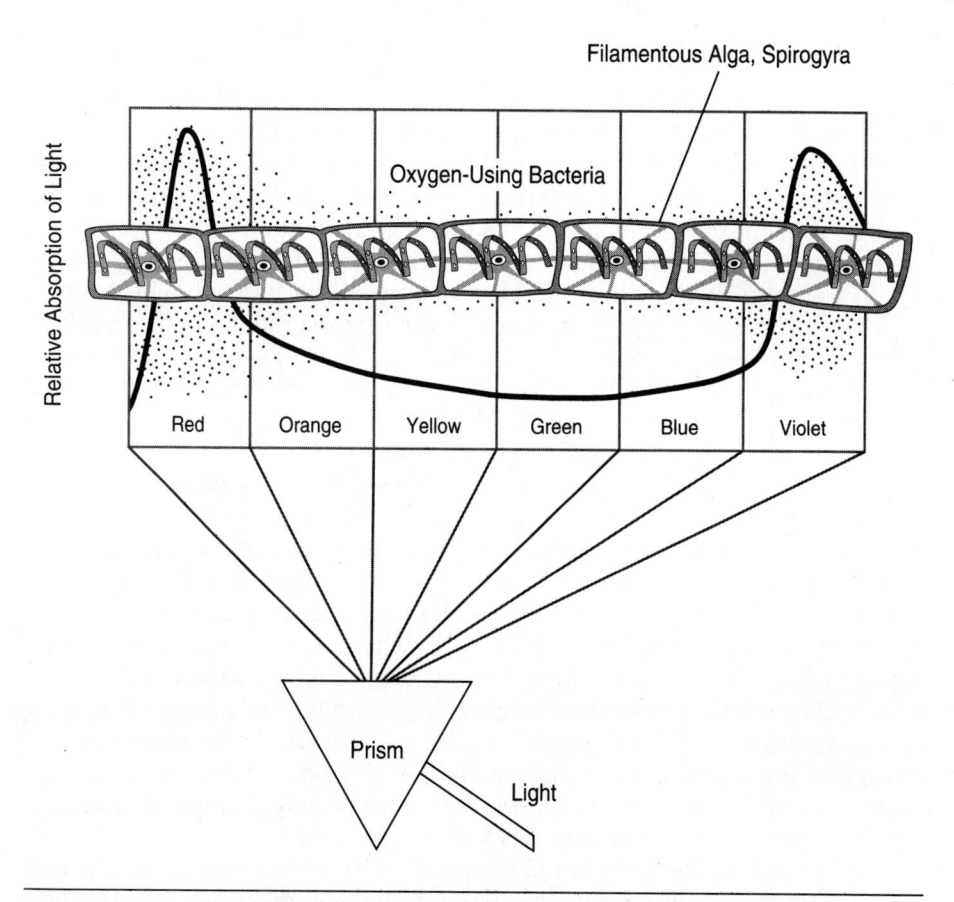

**Figure 9-8** The Engelmann experiment. A filamentous alga such as *Spirogyra* is exposed to a spectrum of light. Bacteria that thrive in an environment of higher oxygen concentration migrate to the red and blue portions of the spectrum, where increased photosynthesis releases more oxygen.

This does not mean that more photosynthesis takes place in a given unit of time; rather, more photosynthesis takes place per amount of light used.

This and other observations led to the understanding that a series of reactions occur in the course of photosynthesis, that some reactions are light dependent, and that some do not require light. These reactions have been termed *light reactions* and *dark reactions*. The name *dark reaction* is in a sense an unfortunate misnomer because it suggests requiring darkness. The so-called dark reactions take place in light; but they are not light dependent. The conversion of carbon dioxide and water into carbohydrate is accomplished by a number of intermediate compounds, and it was to the dark reactions that Blackman referred when he concluded that certain changes were controlled by enzymes.

�֍ Notes �֍

## Electron Transfer

What is the role of light in photosynthesis? When a unit of light, a photon, strikes and is absorbed by a molecule, an electron in the molecule dislodges from its orbit and moves to an orbit farther out from the atomic nucleus. That is, it moves to a higher energy level. Because this condition is unstable, the electron may promptly fall back to its original orbit (figure 9-9). In doing so, the energy that was invested is released. Although some of the energy will be lost in the form of heat, the remainder may take the form of light, causing **fluorescence**. Chlorophyll, then, acts as both a receiver and a source of light. Because some energy is lost as heat, less energy is available for light production than was invested in the molecule. The color of the light in fluorescence is therefore shifted toward the red. Fluorescence in a solution of chlorophyll can readily be demonstrated in the laboratory.

Of course, if energy is dissipated as heat or light, no manufacture occurs. In order for the energy that is trapped when light shines on chlorophyll to become available for carbohydrate manufacture, it must be harvested in certain energy-rich molecules. Two such molecules are nicotine amide dinucleotide phosphate (NADPH) and adenosine triphosphate (ATP).

In the normal events of photosynthesis, when light energy is absorbed by the chlorophyll and an electron is dislodged, it is passed on to another molecule rather than returning immediately to its original place. The molecule that captures the electron is called an *electron acceptor*; the molecule that provides the electron is called an *electron donor*. The electron acceptor molecule quickly passes the electron on to another acceptor molecule, and this process may continue through several intermediaries.

An electron dislodged from chlorophyll takes one of two pathways as it moves from station to station. One pathway has the dislodged electron first traveling to several intermediate molecules and then returning to the chlorophyll. This is called *cyclic electron transfer*. In the other pathway the dislodged

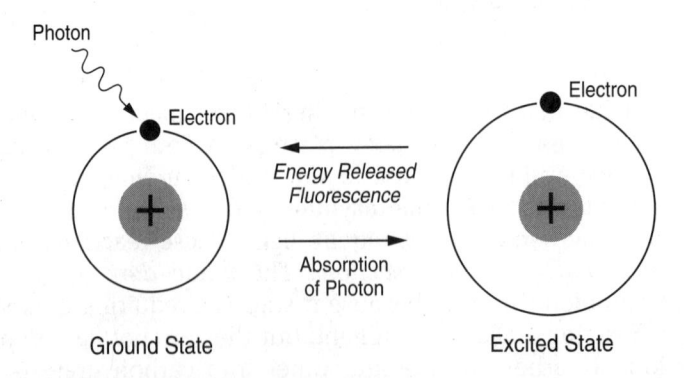

**Figure 9-9** Energy from light can cause an electron of a chlorophyll molecule to dislodge from its orbit and move to a higher orbit. The electron may then immediately fall back to its original orbit, thereby releasing energy and, in so doing, causing an emission of light (fluorescence).

electron does not return to the chlorophyll. The then positively charged chlorophyll is again neutralized (that is, regains an electron), however, by an electron derived from water. This is called *noncyclic electron transfer*.

Consider first the cyclic electron transfer pathway. Because light invests energy in the chlorophyll, an electron in the chlorophyll is excited and moves to another molecule (an electron acceptor) in a higher energy state. As it is passed from molecule to molecule, the electron loses some of its energy until, by the time it returns to the chlorophyll, all the energy with which it was invested has dissipated. The chlorophyll thus returns to its normal state.

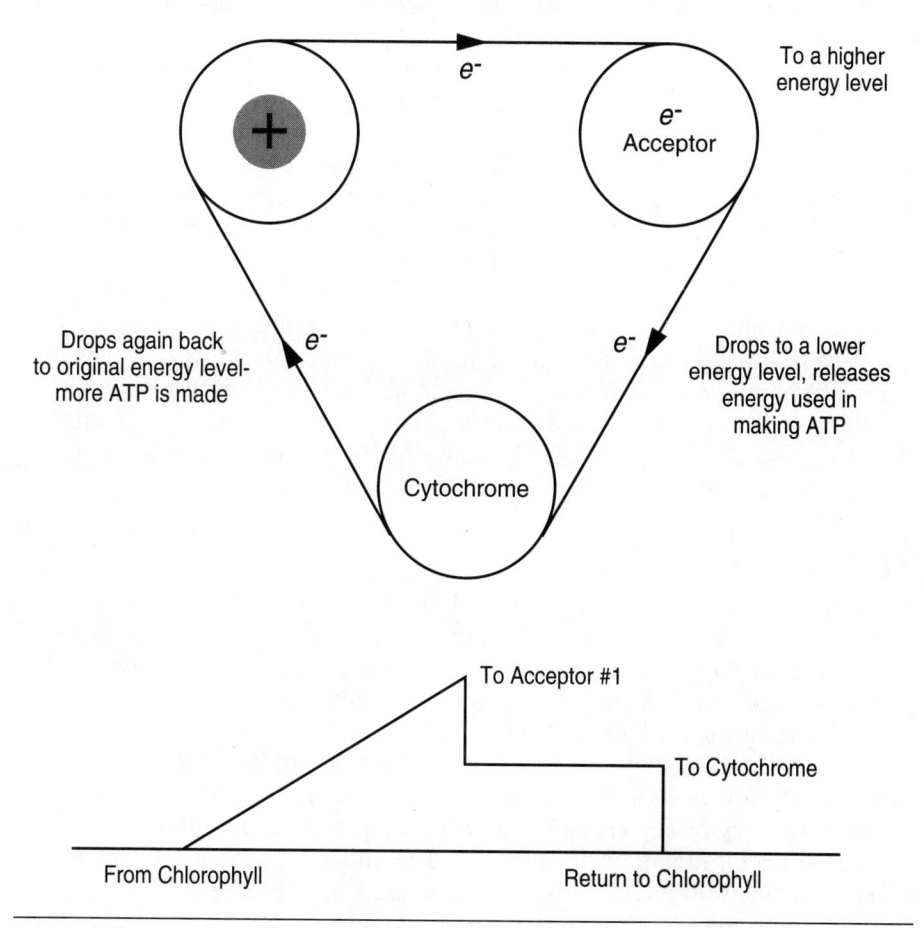

**Figure 9-10** At top, cyclic electron transfer wherein an electron moves from a chlorophyll molecule to an electron acceptor (at a higher energy level) and then drops down in energy in two steps (first to a cytochrome and then back to the chlorophyll molecule). The energy released at each step is used in the manufacture of ATP. At bottom, the same events. Light shining on a molecule of chlorophyll transfers an electron from the chlorophyll molecule to another molecule (electron acceptor 1) at a higher energy level. The electron then falls back to its original place in the chlorophyll molecule, the transfer process being accomplished in two steps. Absorbed energy is released at each step, and the released energy is used in the manufacture of two molecules of ATP.

### Notes

As energy is released from intermediate molecules, it is harvested to make ATP. Adenosine diphosphate (**ADP**) present in the cell is linked to a third phosphate group (also present in the cell), thus making ATP. Energy is stored in the bond that links the third phosphate group. The energy-rich ATP molecule can then provide energy needed by the cell for its metabolic activities. The process is called **photophosphorylation**. The discovery of ATP manufacture in chloroplasts was a significant finding because it had previously been supposed that ATP was made only in mitochondria. This finding is credited to Daniel Arnon, who did his work at the University of California at Berkeley in 1954.

Noncyclic electron transfer produces other energy-rich molecules. This pathway is more complicated than is cyclic transfer and is, therefore, presumed to have evolved later than the more primitive cyclic method. As in the cyclic pathway, an electron moves from the chlorophyll molecule to other intermediate molecules and energy is harvested in certain molecules. Nicotineamide adenine dinucleotide phosphate (NADP) acts as an electron acceptor, becomes negatively charged (is reduced), and, thus, can receive a positive charge in the form of a **proton**. The proton (which is also a hydrogen nucleus) that attaches to NADP comes from the breakdown of water. The NADP thus becomes NADPH, an energy-rich molecule. Water is split into electrons, protons, and oxygen. Two molecules of water must be lysed in order to produce one molecule of oxygen (figure 9-11).

Because ATP is also produced, noncyclic electron transfer yields two kinds of energy-rich molecules. Oxygen is liberated to the atmosphere, and electrons flow to the electron-deficient chlorophyll. These electron transfers, the formation of ATP and NADPH, the lysis of water, and the release of oxygen are facets of the light reactions of photosynthesis.

## The Calvin Cycle

Recall now what F.F. Blackman observed in 1905: in certain situations, the rate of photosynthesis is governed by temperature, indicating that the chemical changes of photosynthesis are controlled by enzymes. Such reactions are the so-called dark reactions of photosynthesis. Carbon is converted to carbohydrate during the dark reactions.

Melvin Calvin (1911–    ) used a radioactive isotope of carbon, $C^{14}$, to trace the carbon of carbon dioxide as it went through several intermediate changes. He very briefly exposed the cells of algae to carbon dioxide that had been manufactured with radioactive $C^{14}$. This made it possible to trace where this carbon went on its pathway to carbohydrate. Catching the $C^{14}$ in an intermediate compound was accomplished by careful timing. Calvin reasoned that if the algal cells were exposed for a very short time to the carbon dioxide made with $C^{14}$ and the cells were then quickly thrust into a fixing agent, it

$$2H_2O \longrightarrow 4e + 4H^+ + O_2$$

**Figure 9-11**  Two molecules of water must be broken down to liberate one molecule of oxygen.

would be possible to detect an early form of intermediate compound. Likewise, if the algal cells were exposed to the $C^{14}$ for a longer time before being thrust into the fixative, a later-formed compound could be detected.

In this way, Calvin determined that the first stable carbon compound formed in photosynthesis is the 3-carbon compound phosphoglyceric acid (**PGA**). Because only one of the three atoms of carbon in PGA was radioactive, it seemed reasonable to conclude that the carbon of carbon dioxide was added to a 2-carbon compound already present in the cell. No 2-carbon compound could be found, however. It was later learned that the carbon of carbon dioxide is added to a 5-carbon rather than a 2-carbon compound. This creates a 6-carbon compound, which is promptly cleaved into two molecules of the 3-carbon PGA. Figure 9-12 constitutes a simplistic representation of this process, showing only the carbons.

The 5-carbon compound was eventually identified as the sugar phosphate ribulose diphosphate (**RuDP**). The 6-carbon compound formed from it is another very short-lived sugar. Each molecule of PGA is phosphorylated using the energy of ATP and NADPH, producing phosphoglyceraldehyde (**PGAL**), also a 3-carbon compound. Ten of every twelve molecules of PGAL are recycled in the production of more RuDP. The other two molecules of PGAL are united to produce the 6-carbon sugar glucose. These events are called the **Calvin cycle**. The glucose is then used in making other constituents of the cell, starch, cellulose, lipids, and amino acids. Melvin Calvin was awarded the Nobel prize in 1961 for his discoveries.

## The $C_4$ Plants

Numerous plants produce the 4-carbon compound oxaloacetic acid rather than the 3-carbon compound in PGA photosynthesis. Those that produce PGA are called $C_3$ plants. Those that produce oxaloacetic acid are called $C_4$ plants.

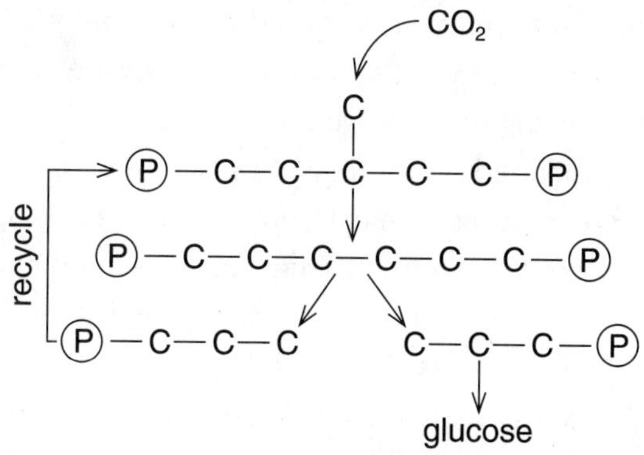

**Figure 9-12** The manufacture of glucose is achieved by a number of intermediate steps. First, the carbon of carbon dioxide is added to a 5-carbon diphosphate to produce a 6-carbon diphosphate. This 6-carbon diphosphate is then promptly cut into two molecules of 3-carbon phosphoglyceric acid (PGA), some of which is used to make glucose and some of which is recycled.

### Notes

The oxaloacetic acid molecule is formed when a 3-carbon phosphoenol pyruvate (PEP) unites with carbon dioxide with the help of the enzyme phosphoenol pyruvate carboxylase. The $C_4$ plants possess two kinds of chloroplasts: large chloroplasts with few grana (which are found in the bundle sheath cells surrounding the veins of a **leaf**) and smaller, more numerous chloroplasts (which occur in the mesophyll of the leaf). The large chloroplasts have numerous starch grains not present in the smaller chloroplasts. The $C_4$ plants thrive at much higher temperature ranges than do the $C_3$ plants, and the rate of photosynthesis in $C_4$ plants is much faster than that in $C_3$ plants.

## The CAM Carbon Pathway

There is yet another pathway taken by carbon in its flow from carbon dioxide to carbohydrate. It occurs in plants of the Crassulaceae family (mostly fleshy herbs) and in cacti. Such plants grow in regions of high light intensity (which is interesting given that the carbon pathway described following is not light dependent). The pathway is called CAM for crassulacean acid metabolism. Plants in which this pathway occurs accumulate malic and isocitric acids at night. These acids are converted back to carbon dioxide during the day. The stomates of CAM plants tend to close during the day, which prevents the entrance of carbon dioxide from the atmosphere. Plentiful carbon dioxide is available, however, from the reservoir of the malic and isocitric acids.

## Questions for Review

1. What occurs during the light reactions of photosynthesis?
2. Write a chemical equation showing what takes place during photosynthesis.
3. What is the concentration of carbon dioxide in the atmosphere?
4. What accounts for the movement of carbon dioxide into leaves?
5. What colors of light are most absorbed by chlorophyll?
6. Define the old terms *phlogiston* and *dephlogisticated air*.
7. Who is credited with giving oxygen its name?
8. All the oxygen liberated in photosynthesis comes from _____.
9. Name two pigments other than chlorophyll that are present in green leaves.
10. Chlorophyll contains an atom of what metal?

## Suggestions for Further Reading

Fogg, G.E. 1972. *Photosynthesis*. New York: American Elsevier Publishing Company, Inc.

Rabinowitch, I.E., and Govindjee. 1965. The role of chlorophyll in photosynthesis. *Scientific American* 213(16–17):74–83.

Ting, I.P., and M. Gibbs. 1982. *Crassulacean Acid Metabolism*. Baltimore: Waverly Press.

# 10

# Respiration and Fermentation

A chemical equation showing the initial and final products of **respiration** is commonly represented as the breakdown of glucose (see figure 10-1). In this process, energy is released. This energy can then be used in doing the work of the cell. This energy is not used immediately following carbohydrate breakdown however. Rather, it is transferred to a reservoir, where it can be stored and subsequently called on. The storage reservoir for energy is adenosine triphosphate (ATP). Its chemical structure is shown in figure 10-2.

✺ Notes ✺

$$C_6H_{12}O_6 + 6O_2 \longrightarrow 6CO_2 + 6H_2O$$

**Figure 10-1** A chemical equation for respiration showing the breakdown of glucose.

**Figure 10-2** The structural formula for adenosine triphosphate (ATP).

101

## ❀ Notes ❀

## The ATP Molecule

Note that the adenosine part of the ATP molecule has two parts: an adenine and the 5-carbon sugar ribose. Hence, another name for this molecule could be adenine ribose triphosphate. Adenosine triphosphate is the universal energy carrier molecule, and respiration can be considered an energy harvesting process.

In a simplified scheme, the phosphate group can be represented as —Ⓟ, and the bond that links this group to the rest of the molecule can be shown as either ___ , if a low-energy bond, or ∼ , if a high-energy bond. The ATP molecule, then, can be represented as shown in figure 10-3. ATP manufacture is a part of the respiration process in all organisms (figure 10-4).

## Respiration and Photosynthesis

Note that the equation for respiration shown in figure 10-1 is the reverse of that for photosynthesis. Whereas photosynthesis takes carbon dioxide out of the atmosphere, respiration restores carbon dioxide to the atmosphere. These two processes, then, exist in precise balance. In other words, all the world's photosynthesis is in balance with all the world's respiration, decay (which is also respiration), and burning of fuels. By these opposing processes, the concentration of carbon dioxide in the atmosphere is maintained at a nearly constant .03%. Carbon, then, is taken out of the atmosphere in one process and returned to it in another. Approximately sixteen billion tons of carbon are processed in this way every year. The cycle can be graphically represented as shown in figure 10-6.

| adenosine |—| P |∼| P |∼| P |

**Figure 10-3**  A graphical way of representing ATP showing two high-energy bonds.

$$\text{ADP} + \text{P} + \text{E} \xrightleftharpoons[\text{energy stored}]{\text{energy released}} \text{ATP}$$

**Figure 10-4**  When a phosphate is added to adenosine diphosphate to make adenosine triphosphate, energy is invested. When a phosphate is broken away from ATP (thus returning it to a diphosphate), energy is liberated.

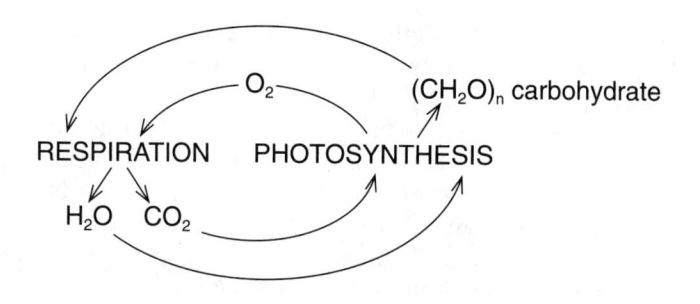

**Figure 10-5** Respiration and photosynthesis are opposite types of reactions, and carbon dioxide and oxygen go through cycles.

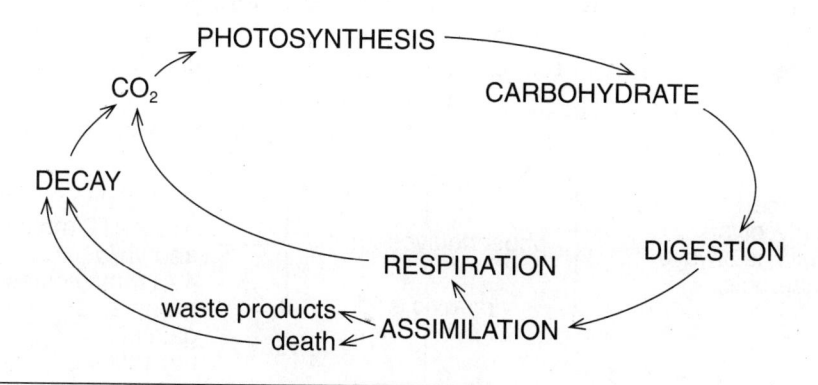

**Figure 10-6** A more detailed version of figure 10-5.

## The Anaerobic and Aerobic Pathways

Although the events of respiration may be thought of as the reverse of the events of photosynthesis, they are not altogether the same events in the reverse order. The first stage of respiration, called **glycolysis**, is an anaerobic process of the initial breakdown of sugar. The 6-carbon glucose is broken down into two 3-carbon molecules of pyruvic acid. In order for this to occur, however, the glucose must first be linked to phosphate groups (a process known as **phosphorylation**). The molecule is then cut in two to make pyruvic acid (see figure 10-7). Glycolysis, then, requires some energy; but it yields slightly more energy than it uses. There is, in fact, an energy gain of two ATP molecules. Following glycolysis, the pathways diverge according to whether the events are **anaerobic** or **aerobic**. Anaerobic processes, which can be called fermentation, produce one set of products, whereas aerobic processes produce another. Fermentation such as occurs in yeast cells yields ethyl alcohol; in other types of cells, lactic acid is the product of fermentation. Aerobic

### Notes

processes first make acetic acid. The acetic acid then enters the mitochondria, where a series of cyclic events takes place. These events are summarized in figure 10-7.

Aerobic processes are far more efficient in making ATP than are the anaerobic processes. Whereas aerobic breakdown of a **mole** (gram molecular weight equal to 180 grams) of glucose releases 684,000 calories of heat, anaerobic breakdown of the same amount of glucose releases only 28,000 calories. Thus, 24.1 times more heat is evolved in aerobic respiration than in anaerobic respiration. In fermentation, two ATP molecules are made for each molecule processed.

On the aerobic pathway, the compound formed immediately following pyruvic acid is acetic acid. The acetic acid then enters the mitochondria, where the remaining chemical changes of respiration occur. These changes take a cyclic pattern, with oxaloacetic acid being formed next and, after a total of nine intermediate steps, oxaloacetic acid being formed again. The complete cycle results in the formation of thirty ATP molecules.

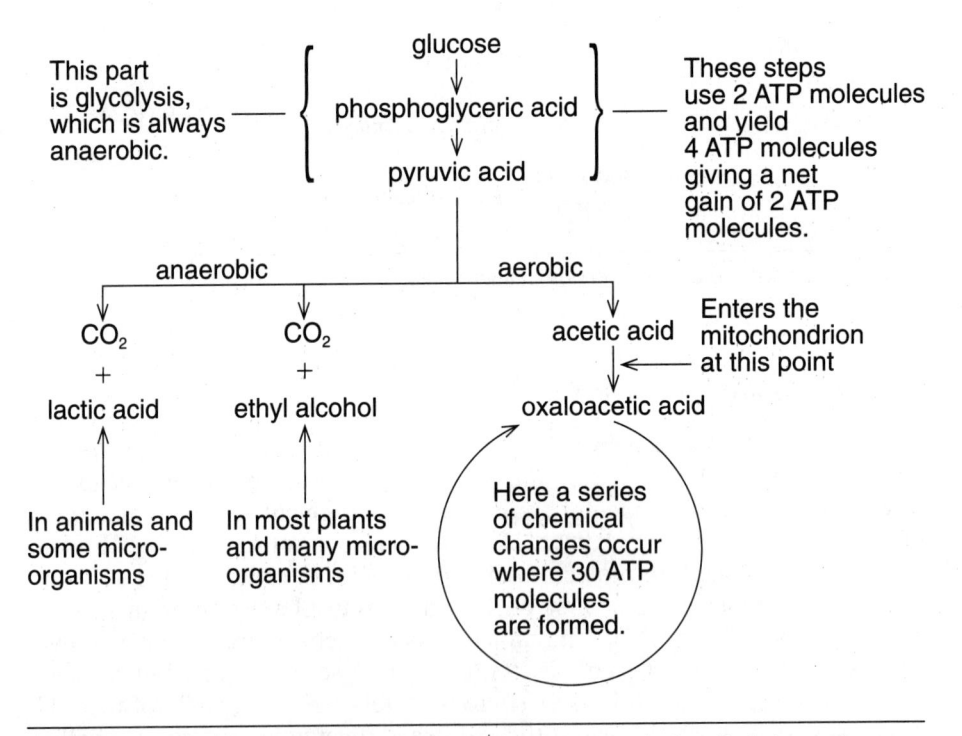

**Figure 10-7** The events of respiration. The first steps, collectively called glycolysis, are always anaerobic. The pathways then diverge, going one way if oxygen is available and another way if oxygen is unavailable. On the aerobic pathway, the chemical changes take a cyclic pattern known as the Krebs cycle.

In 1937 Sir Hans Krebs[1] uncovered the chemical changes that occur in the mitochondria. Although these changes were first collectively described as the citric acid cycle, they are now known as the *Krebs cycle*. Each step in the cycle sees the elimination of one acetic acid molecule and the formation of two carbon dioxide molecules and four pairs of hydrogen atoms. Each of the nine steps in the Krebs cycle requires a specific enzyme, and the enzymes are arranged in sequence for effective exploitation. The Krebs cycle occurs in plants, molds, bacteria, and animals.

## Hydrogenation

The production of ethyl alcohol by one pathway and acetic acid by the other cannot be attributed solely to one process being anaerobic and the other being aerobic. These chemical changes are controlled by enzymes, and the kinds of reactions that occur depend on the kinds of enzymes available. One type of reaction that occurs as carbohydrate is broken down is called *dehydrogenation* (wherein hydrogen atoms are removed from the molecule). The particular enzyme required for dehydrogenation is called **dehydrogenase**. When the hydrogen bonds are broken, energy is released. Some of this energy is used to maintain the reaction, some is lost as heat, and some is used in the formation of ATP. The removal of hydrogen atoms requires a hydrogen acceptor. In aerobic respiration, the hydrogen acceptor is oxygen. Hydrogenation in aerobic respiration, then, results in the manufacture of water. This is not a one step process; rather, it is accomplished by a series of transfers from carrier to carrier (figure 10-8). As the hydrogen passes from one carrier to the next, energy is released at each step (figure 10-9).

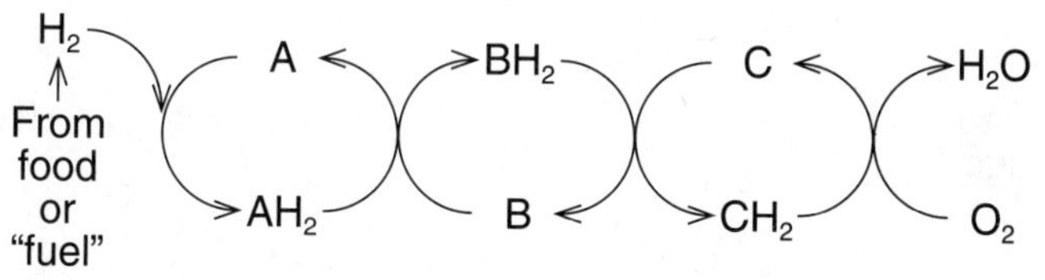

**Figure 10-8** Hydrogen obtained from "food" passes through several different steps before contributing to the formation of water.

---

[1]Sir Hans Krebs fled from Germany in 1932, at the beginning of the rise of Adolf Hitler, and was knighted in 1958.

✼ **Notes** ✼

**Figure 10-9** An elaboration of figure 10-8. As hydrogen is passed from one molecule to another, energy is released and used in the manufacture of ATP.

## The Carbon Cycle

ATP is manufactured in a similar way when carbon-to-carbon bonds are broken. The carbon cycle is an energy compacting process. The addition of a third phosphate group to adenosine diphosphate yields a high-energy bond (figure 10-10).

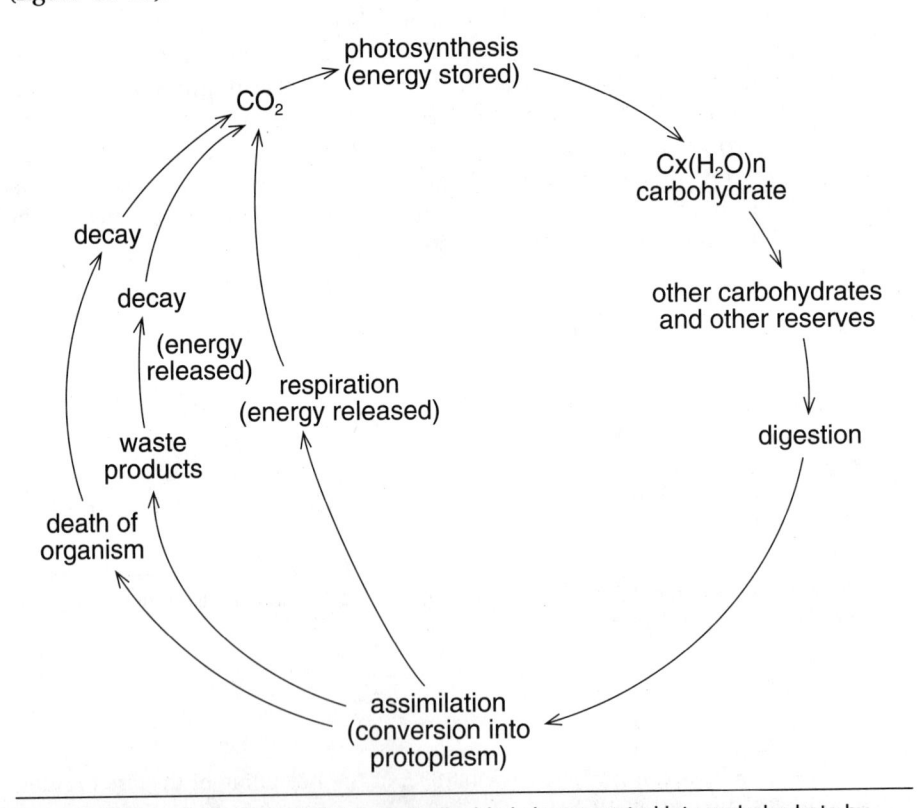

**Figure 10-10** The carbon cycle. Carbon dioxide is incorporated into carbohydrate by photosynthesis. Then, by a series of degradation processes including respiration, it is returned to the atmosphere as carbon dioxide.

## Questions for Review

1. What is glycolysis?
2. Write a chemical equation representing the events of aerobic respiration.
3. As energy is released in the process of respiration, it is transferred to the molecule _____.
4. Most of the steps of ATP manufacture take place in the _____.
5. The equation for respiration is in a large sense the reverse of the equation for _____.
6. Another name for anaerobic respiration is _____; _____ is formed in the process.

## Suggestions for Further Reading

Bassham, J.A. 1962. The paths of carbon in photosynthesis. *Scientific American* 206(6):88–100.

Salisbury, F.B., and C. Ross. 1978. *Plant Physiology*. Belmont, CA: Wadsworth Publishing Co.

Siegenthaler, U., and H. Oeschger. 1978. Predicting future atmospheric carbon dioxide levels. *Science* 199(4327):388–395.

# 11

# Bacteria

**A**nton van Leeuwenhoek (1632–1723), a dry-goods merchant who liked to manufacture and use his own microscopes, is credited with first describing **bacteria** in the year 1683. The bacteria he studied may have come from scrapings of his teeth. Van Leeuwenhoek lived and worked at a time when it was commonly thought that small creatures arose from dust or bred from corruption, and his work helped uncover the truth that all life comes from previously existing life.

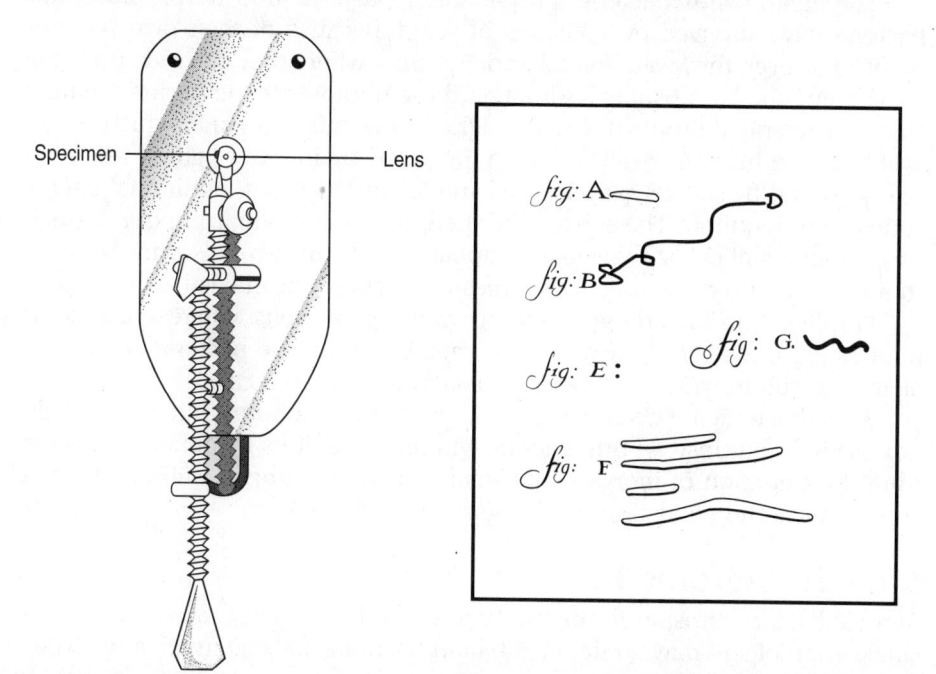

❋ **Notes** ❋

**Figure 11-1**  At left, one of Leeuwenhoek's microscopes; at right, drawings of his discoveries.

🕸 Notes 🕸

## Is Bacteria a Plant?

Because we are using a five-kingdom system of classification, how do we justify covering bacteria in a botany course? While bacteria are not technically plants, some bacteria do have a form of chlorophyll and carry on photosynthesis. They thus deserve brief consideration.

## The Original Bacteria

Bacteria belong to the kingdom Monera. All bacteria are prokaryotic. Bacteria are the oldest, smallest, and most abundant organisms. They can be found in the fossil record tracing back 3.2 billion years; and for the major portion of the time prior to the appearance of the first more complex organisms, they were the sole occupants of the world. For 2.8 billion years, life was microscopic, prokaryotic, anaerobic, and did not liberate oxygen in photosynthesis. Three kinds of bacteria dominated the scene: green sulfur bacteria, purple sulfur bacteria, and purple nonsulfur bacteria (which may be purple, red, or brown). The green sulfur bacteria have a form of chlorophyll called **chlorobium chlorophyll**; the other two forms of bacteria have **bacteriochlorophyll** (which is blue-gray in color). The first bacteria to produce oxygen in photosynthesis were the **cyanobacteria** (also called **blue-green algae**). Their respiration was anaerobic. In time, the liberation of oxygen as the waste product of photosynthesis necessitated the retreat of anaerobic forms, and aerobic respiration became more common.

The green sulfur bacteria, purple sulfur bacteria, and purple nonsulfur bacteria have survived over billions of years. Because of their own requirements, however, they were forced underground, where they can hide from the oxygen-enriched atmosphere. Although these photosynthetic bacteria require "light," bacteriochlorophyll absorbs light in the infrared zone, which is not visible to the human eye. The energy provided by infrared light is less than that provided by shorter wavelengths and is partly wasted. Additional energy sources are required. These bacteria, then, may tap sulfur compounds such as hydrogen sulfide or thiosulfate, aliphatic acids, or alcohols underground. In a laboratory setting, they sometimes directly exploit molecular hydrogen.

Purple sulfur bacteria grow in decaying organic soils, where sulfur compounds are available. By way of their metabolism, these bacteria create free elemental sulfur, which can be seen inside the bacterial cells.

Some bacteria are chemoautotrophic rather than photosynthetic. They do not use light but instead bring about synthetic reactions by obtaining energy from the oxidation of inorganic molecules such as nitrogen, sulfur, and iron compounds. They may also oxidize gaseous hydrogen.

## Modern, Aerobic Bacteria

Aerobic bacteria prosper in the modern world. Raven and Curtis[1] state that a single gram of soil may contain 2.5 billion bacteria, 400,000 fungi, and 50,000

---

[1]Raven, Peter, and Helena Curtis. 1970. *The Biology of Plants.* New York: Worth Publishing Co.

algae. The total weight (worldwide) of these organisms is greater than the combined weight of all animal organisms, and most of the respiration that takes place in the world is due to bacteria and fungi.

Bacteria are abundant in sewage, stagnant water, and the bodies of living and dead organisms. Some bacteria require oxygen, while other bacteria require that oxygen be absent. Many, however, can grow in either type of situation.

## Characteristics of Bacteria

Bacteria resemble plants in that they have rigid cell walls, although the walls are generally of nitrogen rather than cellulose. Some forms have flagella, while others are nonmotile. Bacteria have ribosomes and nucleic acids but lack mitochondria, plastids, and the central vacuole. Bacteria are distinguished from each other, in part, on the basis of their shapes. Rod-shaped bacteria are called *bacilli*, round forms are called *cocci*, and curved forms are called *spirilli*. Cocci can occur in pairs (*diplococci*), clusters (*staphylococci*), or chains (*streptococci*).

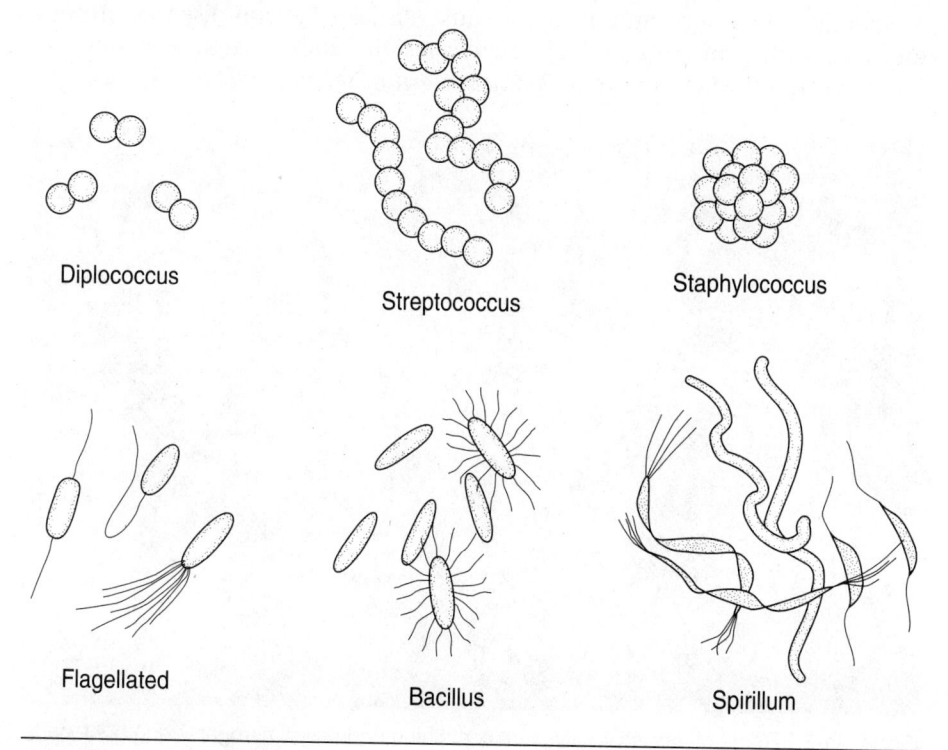

**Figure 11-2** Types of bacteria.

## Notes

## Benefits of Bacteria

In these times of interest in ecology, bacteria get high credit for their role in decay, whereby carbon dioxide is returned to the atmosphere and goes through the carbon cycle again. Without bacteria, plant and animal remnants would continually accumulate and perhaps overwhelm the planet. Bacteria are also important as sources of **antibiotics**. Furthermore, cheeses, acetic acid, vinegar, and various amino acids and enzymes used commercially are produced by bacteria.

Although many types of bacteria cause plant diseases, there is some injustice in the bad reputation of these organisms. Some bacteria, in fact, hold inherent virtues for plants. For example, the bacteria that reside in the root nodules of leguminous plants such as alfalfa, clover, soy bean, and peas can fix (that is, make useful compounds of) atmospheric nitrogen. They convert nitrogen to nitrate, thereby making it available to plants. This is an example of **symbiosis**. Two soil organisms are valuable in fixing nitrogen: *Nitrosomonas*, which converts ammonia compounds to nitrates, and *Nitrobacter*, which converts nitrite to nitrate. Bacteria are also important agents of decay, clearing away dead vegetation and animals and building up the soil.

## Hazards of Bacteria

As mentioned earlier, some bacteria cause disease. Several diseases of potatoes are attributable to bacteria: *Pectobacterium carotovorum*, *Pseudomonas solanacearum* (brown rot), and *Corynebacterium sepedonicum* (ring rot). Shot

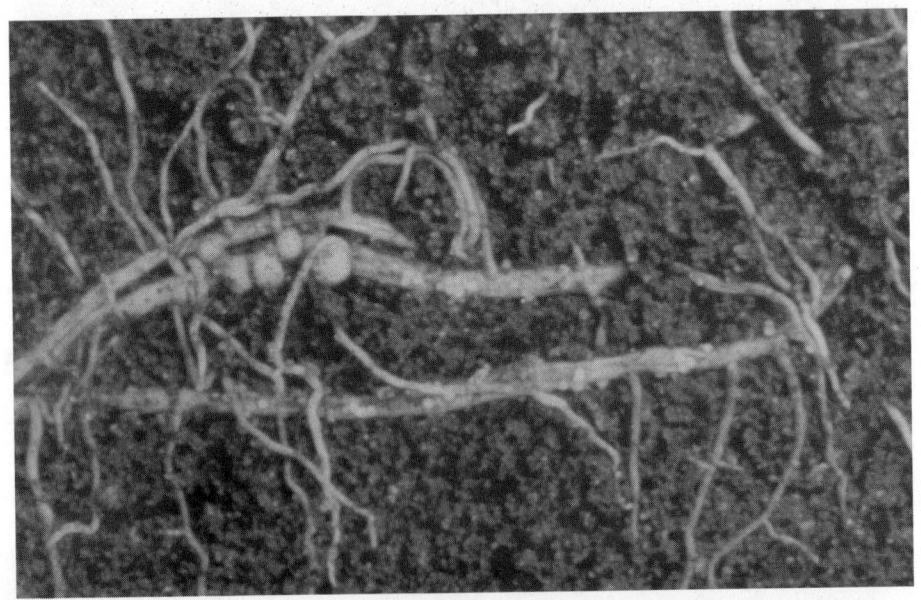

**Figure 11-3** Roots of a bean family plant containing nodules of nitrogen-fixing bacteria.

hole is a bacterial disease that produces numerous small holes in leaves. The second most common problem among cotton plants is a bacterial disease called leaf spot, or black arm. Sugar cane is prey to a bacterial infection called *Bacterium vasculorum* (gumming disease). Tobacco suffers from a similar disease.

Bacterial diseases, as well as viral and fungal diseases, are frequently carried by insects, and, thus, can often be brought under control by the use of insecticide sprays or dusts. Infected seeds can be treated with bichloride of mercury, with formaldehyde, or with hot water. Timing of the latter treatment is particularly important.

## Identifying Bacteria

Identification of bacteria may require a number of specialized tests designed to reveal a number of characteristics. One such characteristic is whether or not the bacterium causes a disease (and if yes, which one). The color of a bacterium colony as it grows on a nutrient medium is an important identifying characteristic, as is the kind of medium on which a bacterium will grow. Whether a bacterium is Gram positive or negative is also very useful in identifying the bacterium. (The Danish microbiologist Hans Christian Gram discovered that the cell walls of certain bacteria have a lipopolysaccharide that renders the bacterial cells unable to retain a dye called gentian violet. Such bacteria are labeled Gram negative. Conversely, bacterial cells lacking the lipopolysaccharide will retain the dye, and are, thus, labeled Gram positive.) Whether a bacterium forms spores and requires oxygen, and the temperature at which the bacterium best grows are also valuable in determining species. The presence of a gelatinous sheath outside the bacterial cell wall can also be useful in identification. While more than 1,600 species of bacteria have thus far been discovered and described, many more are as yet unknown.

## Bacterial Growth

Bacteria divide by **fission** and may do so every twenty minutes in favorable living conditions. This suggests that in a given day, a given bacterium could come to weigh 2,000 tons. This, of course, does not happen.

Because bacteria divide by fission, they are known as Schizophyta, or Schizomycetes (a term introduced by von Naegeli in 1857). The cytoplasmic membrane grows inwardly, the cell wall splits from the outside toward the center, and the cell divides into two daughter cells. Bacteria grow on organic matter by secreting enzymes outwardly into the surrounding medium, making possible the absorption of nutrients through the cell membrane. (Such enzymes are called **exoenzymes**.)

Some bacteria form (usually one but sometimes more) **endospores**. The protoplast rounds up in the cell by the loss of water and develops a heavy

※ Notes ※

## ❦ Notes ❦

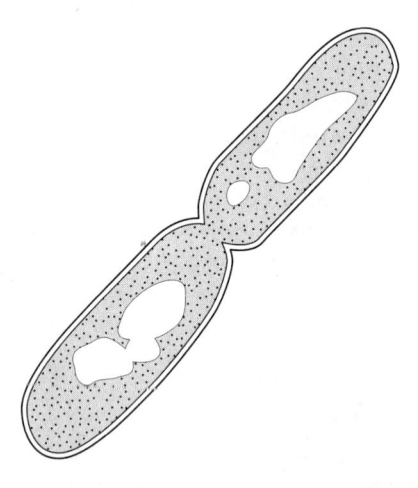

**Figure 11-4**  Bacteria divide by simple fission.

cell wall. In this way, the endospores can survive adverse conditions (even boiling water) for several hours. Endospores can be sealed in vacuum, reduced in temperature to −270°C (in liquid helium), and then restored to active growth again.

Do bacteria have nuclei? Because the definition of a nucleus requires an enclosing membrane, the answer is no. But bacteria do have nuclear bodies—sometimes two—because nuclear division generally precedes cell division. Each nuclear body consists of a circle of double-stranded DNA, a single chromonema (said to be 1,000 times longer than the cell in which it resides). Although reproduction occurs by transverse fission, a sexual process (a form of conjugation) has been observed.

Consider, now, some work done by Lederberg and Tatum involving a strain of *Escherichia coli*, the bacterium that grows in the colon. This bacterium is able to synthesize the amino acids it needs. This organism can thereby grow on a medium containing only glucose and some inorganic salts—a **minimal medium** (figure 11-5a). Lederberg and Tatum theorized that if such an *E. coli* were to lose the ability to make a needed amino acid because of a mutation (sudden, unexpected change in genetic material), it would no longer be able to thrive on a minimal medium; if, however, the amino acid that the mutated *E. coli* was unable to make were added to the medium, the bacterium would grow (figure 11-5b).

Lederberg and Tatum used ultraviolet radiation to induce mutations in *E. coli* and, thereby, successfully developed two strains, each of which had lost the ability to manufacture three amino acids. The three amino acids that strain number II could not make were not the same three as those that strain number I could not make. Let us say that strain number I could not make amino acids *a*, *b*, and *c*. It could not, therefore, grow on the minimal medium.

It could, however, grow on the same type of medium if the amino acids that it was unable to make (that is, types *a, b,* and *c*) were added to the medium (figure 11-5b). Similarly, strain number II could not make amino acids *d, e,* and *f,* and could not grow on the minimal medium. It could, however, grow on a medium that provided amino acids *d, e,* and *f* (figure 11-5c). Although strain number I could not make amino acids *a, b,* or *c,* it could make amino acids *d, e,* and *f;* likewise, although strain number II could not make amino acids *d, e,* or *f,* it could make amino acids *a, b,* and *c.* When Lederberg and Tatum mixed these two strains of bacteria and plated the mixture onto the minimal medium, a good growth of bacteria was observed (figure 11-5a). The explanation for this is shown in chapter 6, figure 6-3. The electron microscope reveals a conjugation tube extending between two cells through which genetic material passes from one cell to the other. In this manner a cell having the full capacity to grow on a minimal medium is created. The DNA takes the form of a fine thread as it passes across the tube bridge. One bacterium is a donor cell and the other is a recipient. The DNA strand is at first in the form of a circle, the ends being united (figure 11-6). The circle breaks at a specific point, and the long strand then moves across the bridge. Varying portions of the molecule may be transferred in this process called **parasexual**.

❋ **Notes** ❋

**a**
Agar, sugar salts (minimal medium)

**b**
Agar, sugar salts and three amino acids a, b and c

**c**
Agar, sugar salts and three amino acids d, e and f

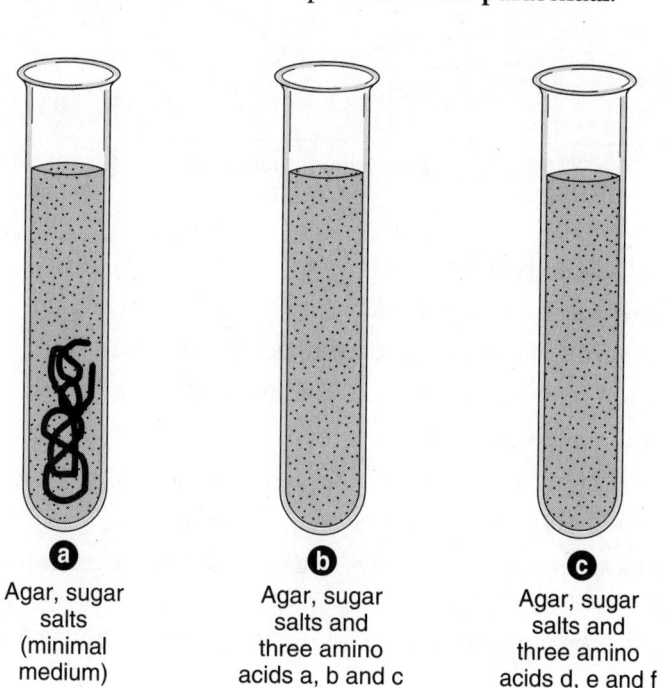

**Figure 11-5** Culture tube (a) contains only a minimal medium composed of agar plus some sugar and salts. Culture tube (b) contains the same ingredients plus three amino acids; *a, b,* and *c.* Culture tube (c) contains the same ingredients as does culture tube (a) plus the amino acids *d, e,* and *f.*

## ✽ Notes ✽

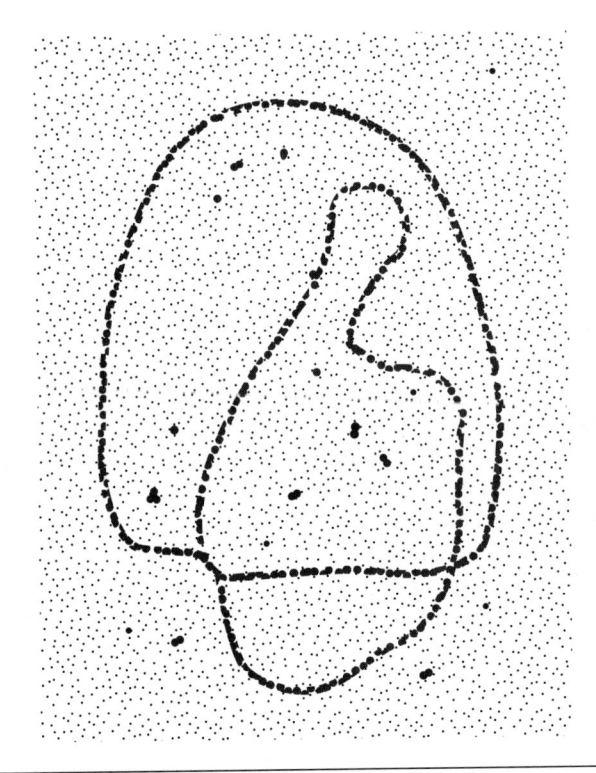

**Figure 11-6** The DNA filament in a bacterial cell takes the form of a complete circle.

## Questions for Review
1. How do bacteria transfer DNA?
2. Make clear the distinction between bacteria and viruses.
3. Distinguish between aerobic bacteria and anaerobic bacteria.
4. Who discovered bacteria?
5. How do bacteria play a role in the welfare of plants?
6. Name some plant diseases caused by bacteria.
7. How and where are bacteria involved in making nitrogen into a form that can be used by plants?
8. Some bacteria are able to carry on photosynthesis. Name a form that can do this.
9. Some bacteria are able to liberate _____, which forms crystals inside the cell.
10. Bacteria reproduce by _____, but they are also capable of a sexual union. The latter, however, is not a form of reproduction. Explain.

## Suggestions for Further Reading

Krogmann, David W. 1981. Cyanobacteria (blue-green algae)—Their evolution and relation to other photosynthetic organisms. *BioScience* 31:121–124.

Margulis, Lynn. 1982. *Early Life*. Portola Valley, CA: Science Books International Publishers.

Sonea, S., and M. Paniset. 1983. *A New Bacteriology*. Boston: Jones and Bartlett Publishers, Inc.

Woese, C.R. 1981. Archaebacteria. *Scientific American* 244(6):98–122.

## ❋ Notes ❋

# 12

# The Blue-green Algae

**Notes**

The algae are not a group of related forms; in fact, various forms of alga are assigned to different kingdoms. Because they have no nuclei, the blue-green algae are assigned to the kingdom Monera. They are prokaryotic. Members of the three phyla Euglenophyta, Chrysophyta, and Pyrrophyta are assigned to the kingdom Protista because none of the members of these phyla are multicellular. Members of the phyla Chlorophyta, Rhodophyta, and Phaeophyta are assigned to the kingdom Plantae. These assignments have both positive and negative implications; they are also tentative.

## Primordial Ooze

Fossils of blue-green algae have been found in **Precambrian** rocks known to be more than 4 billion years old. For approximately five-sixths of the time that has lapsed since these organisms appeared, only prokaryotic life existed. Then, approximately six hundred million years ago, an event of profound significance took place; eukaryotic life appeared. Dr. Lynn Margulis of Boston University theorizes that this was accomplished by invasion, wherein a smaller prokaryotic cell moved into a larger prokaryote, took up residence there, and became the nucleus. Until this time—for 3 billion years—the only life on Earth was the primordial ooze that we know as the Cyanophyta: the blue-green algae. If it is correct to assert that all life comes from pre-existing life, perhaps this **primordial** ooze represents the ancestors of all subsequent life.

## Characteristics of the Blue-green Algae

Being in the kingdom Monera, blue-green algae lack membrane-enclosed nuclei. They are, thus, **acellular**, and closely related to bacteria. In fact, they

119

❧ Notes ❧

have been given the name *cyanobacteria*. Monera lack not only nuclei, but also mitochondria, Golgi bodies, and plastids.

The prokaryotic blue-greens can be characterized by their cell walls and their pigments. As they are in bacteria, the cell walls of blue-green algae are composed of several layers of mucoprotein and other polysaccharides including lipopolysaccharides. Cellulose, which is found in the cells of flowering plants, is mostly absent.

The pigments in blue-green algae are chlorophyll a, carotenoids, **phycocyanin** (blue), and phycoerythrin. These organisms often are not blue-green in color; they are sometimes yellowish, red, purple, violet, or nearly black. *Trichodesmium erythraeum*, a reddish blue-green alga that occasionally grows in great abundance, imparts a red color to water. The Red Sea is so named for this. The pigment chlorophyll a is directly responsible for photosynthesis, the carbohydrate product of which is glycogen (the same carbohydrate produced in animal livers).

The colored portion of the cell lies around the periphery, and the central part of the cell (called the *incipient nucleus*) contains the chromatin. Vacuoles do not occur in healthy young cells. Some forms of blue-green alga produce **endospores**.

The blue-greens are well known for their capacity to thrive in adverse conditions. Some can grow on snow and ice, while others can be found in hot springs where the temperature approaches boiling. Blue-greens can also grow in desert soils when water is present. Some forms of blue-green alga grow in nodules in the roots of Cycads in the same manner as nitrogen-fixing bacteria grow on the roots of legumes. Some blue-greens serve as significant sources of food for fish, an arrangement that sometimes results in the fish themselves becoming poisonous. (Cattle have, in fact, died from drinking blue-green infested water.) Blue-green algae sometimes grow inside amoebas, **diatoms**, and other algal cells. A form of *Anabaena* (figure 12-1c) grows in the hollow leaves of *Azolla*, and *Nostoc* (figure 12-1b) grows in the thallus of *Anthoceros*, a bryophyte. Both *Anabaena* and *Nostoc* grow in the shells of turtles and snails, the intestines of some animals, and the hair follicles of the three-toed sloth.

Reproduction among the blue-greens is entirely **asexual**. They reproduce only by fission or a form of fragmentation into hormogonia. Sterols are important to sexual reproduction, and these compounds do not occur in the blue-green algae.

## Types of Blue-green Algae

There is some confusion regarding classification of algae. Some researchers have suggested that there may be as many as 7,500 species of blue-greens. This estimate may be high, however, given the tendency of alga to take on different forms in differing environments.

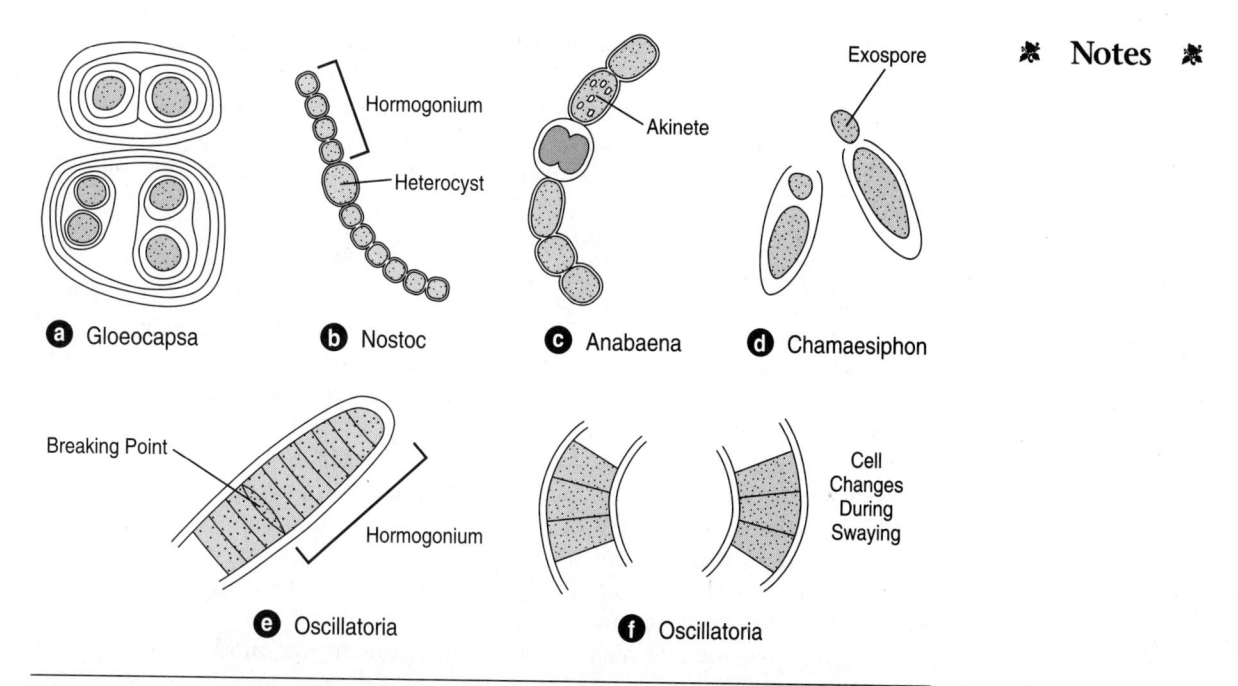

**Figure 12-1** Several blue-green algae: (a) *Gloeocapsa;* (b) *Nostoc,* possessing a heterocyst and a hormogonium; (c) *Anabaena,* showing an akinete (a form of resting spore); (d) *Chamaesiphon,* with an exospore; (e) *Oscillatoria,* showing a breaking point and a hormogonium, and how the cell shape changes, (f), in the swaying movement.

*Gloeocapsa* (figure 12-1a) is a very simple blue-green that grows as single cells or in small clusters of cells. A beginner examining *Gloeocapsa* under a microscope may think that a nucleus is seen. This deceptive appearance is caused by the matrix that is secreted by and lies outside of the cell. Some cells show a double layer of matrix. When the cell attains a certain size, the protoplast divides, and each daughter cell produces a new matrix. There are twenty-three species of this genus in the United States. They may be found growing on wet rocks and the sides of aquarium tanks, and in ponds.

*Oscillatoria* (figure 12-1e) is a common blue-green that grows in filaments. It may be found in lakes, ponds, and moist soil, where the filaments appear singly or in mats. There are approximately thirty species of *Oscillatoria* in North America. These blue-greens give water a putrescent odor. The formation of filaments is caused by the cells dividing transversely. The cells of the filament are essentially alike, and each cell can be regarded as an individual. *Oscillatoria* is capable of an undulating movement, and this is the basis of the name. The movement is so slow that patience is required to detect it. Although several theories have been offered to explain this movement, the mechanism is not definitely known.

### ✣ Notes ✣

The drawing of *Oscillatoria* in figure 12-1e shows a dead cell where the filament can break, thus producing two shorter segments. The shorter pieces are called **hormogonia**, and this is one method of reproduction. Figure 12-1 also shows how the cells change shape during the swaying movement of the filament. The left side of the cell has a greater vertical dimension when the filament bends to the right (figure 12-1f); the water then shifts in the cell to produce a greater vertical dimension on the right side when the filament bends to the left (figure 12-1g).

*Nostoc* (figure 12-1b) forms in colonies as large as a plum. In such a colony, there are thousands of filaments embedded in a jellylike matrix. Although *Nostoc* does occur in soil, it is more commonly found floating in water. Both *Nostoc* and *Anabaena* (figure 12-1c) are capable of using atmospheric nitrogen to make compounds that can be used by higher plants. It is reasoned that the fertility of rice paddies is maintained by the presence of these algae. In this sense, they resemble the bacteria that grow in nodules on the roots of bean-family plants (and that can also manufacture useful compounds from atmospheric nitrogen).

Figure 12-1b shows empty cells lying intermittently along the filament. These are called **heterocysts**. The filament breaks at these points, producing hormogonia. Nostoc colonies may be found growing symbiotically with fungi, thus forming **lichens**.

The illustration of *Anabaena* (figure 12-1c) shows an **akinete**, a large, resistant cell containing a food reserve. Its thick cell wall allows it to withstand adverse conditions. When living conditions again become favorable, it can germinate and grow a new algal filament.

Some blue-greens produce endospores. *Chamaesiphon* (figure 12-1d) is an example.

## Questions for Review

1. List two ways whereby blue-green algae are similar to bacteria and two ways they are different.

2. Define the terms *prokaryotic* and *eukaryotic*.

3. Dr. Lynn Margulis postulates a way whereby the eukaryotic condition came into being. Explain this postulation.

4. What pigments occur in blue-green algae?

5. List two species of blue-green algae.

6. Define the terms *hormogonia, heterocyst,* and *akinete*.

## Suggestions for Further Reading

Blume, J.L. 1956. The ecology of river algae. *Botanical Review* 22:291–341.

Chapman, V.J. 1957. Marine algal ecology. *Botanical Review* 23:320–350.

Chase, F.M. 1941. Useful algae. *Smithsonian Institution Annual Report* 401–452.

Dawson, E.Y. 1961. The rim of the reef. *Natural History* 70:8–16.

Lewin, J.E., and R.L. Guillard. 1963. Diatoms. *Annual Review Microbiology* 17:377–414.

Papenfuss, G.F. 1951. Phaeophyta. In Smith, G.M. (ed.) *Manual of Phycology.* Waltham, MA: Chronica Botanica.

Prescott, G.W. 1954. *How to Know the Freshwater Algae.* Dubuque, IA: William C. Brown and Company.

Round, F.E. 1965. *The Biology of the Algae.* London: Edwin Arnold, Ltd.

Smith, G.M. 1950. *Freshwater Algae of the United States.* New York: McGraw-Hill Book Co.

Tseng, C.K. 1944. Agar: A valuable seaweed product. *Scientific Monthly* 58:24–32.

## ❋ Notes ❋

# 13

# The Green Algae

There are more than 20,000 species of green algae. All are eukaryotic, possessing definite nuclei. All contain the pigments chlorophyll a, chlorophyll b, carotenes, and xanthophylls—the same pigments that occur in flowering plants. In addition to well-defined nuclei, all possess the other organelles found in eukaryotic cells: nucleoli, vacuoles, ribosomes, mitochondria, endoplasmic reticulum, and chloroplasts. Genus can sometimes be recognized by the shape of chloroplasts. The product of photosynthesis is starch similar to that found in higher plants. Both motile and nonmotile forms occur, and both sexual and asexual reproduction take place. The motile forms have whiplash-type flagella and are of equal length. Some forms grow on snow and ice, some on the trunks of trees; in protozoans, hydra, flatworms, animal fur, and sponges; or on the backs of turtles. Green algae grow as symbionts in the formation of lichens. Most green algae are found in fresh water.

A generally accepted way of classifying green algae is to group them according to three presumed lines of evolution. The first is the **Volvacine** line, encompassing flagellated, unicellular or colonial forms; the second is the **Tetrasporine** line, encompassing nonmotile cells in filaments or sheets; and the third is the **Siphonous** line, characterized by a plurinucleated plant body and generally absent cross walls.

While these lines of evolution are based on considered thought, it is important to note that they are theoretical and, thus, not cast in stone. Revisions to this theory, then, may be forthcoming.

❋ Notes ❋

## The Volvacine Line
### *Chlamydomonas*
*Chlamydomonas* represents an ancient algal form. *Chlamydomonas*-type cells appear in the fossil record tracing back 1 billion years.

## ✾ Notes ✾

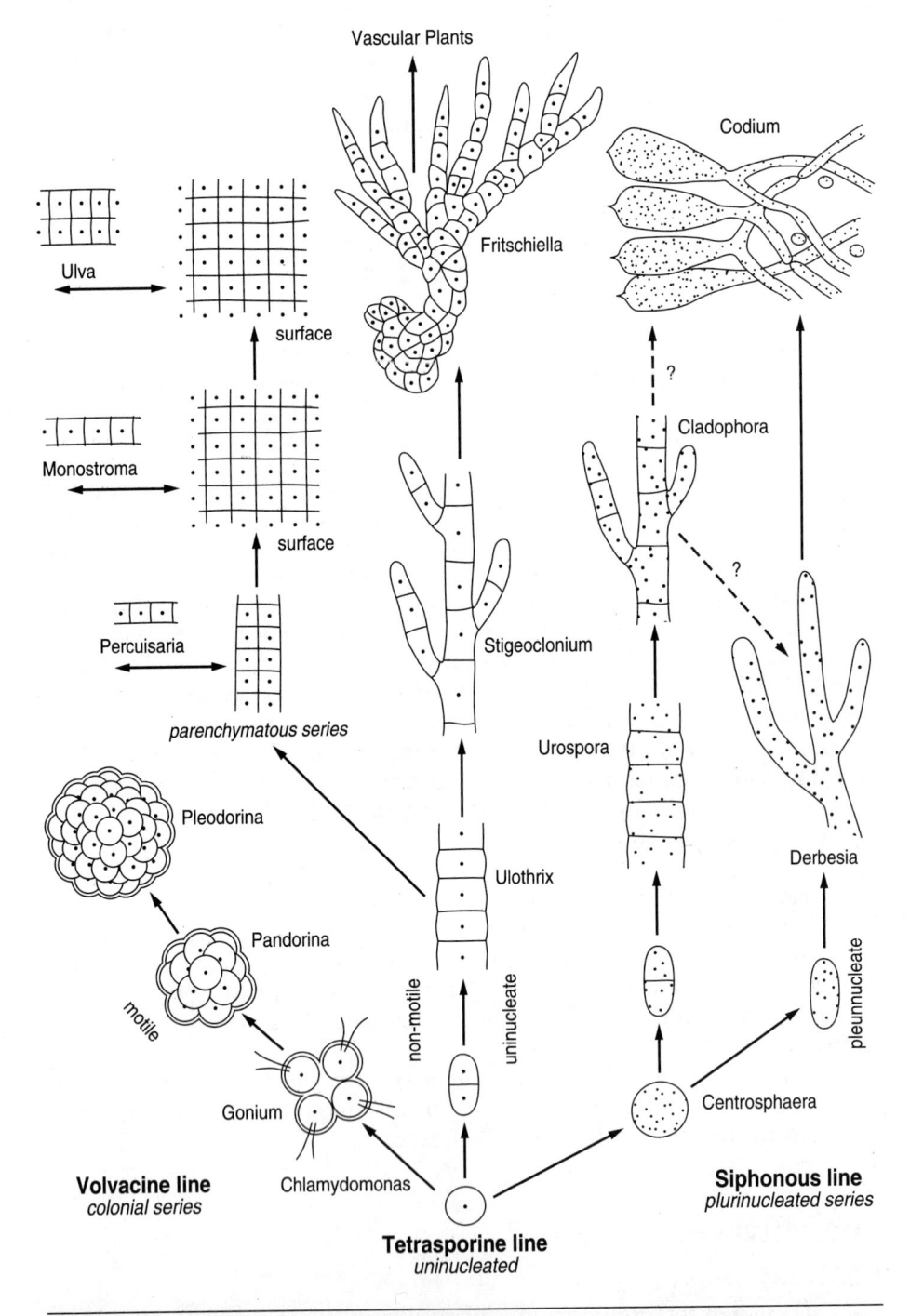

**Figure 13-1** Three postulated lines of evolution of the green algae: the Volvacine line, encompassing flagellated unicellular or colonial forms; the Tetrasporine line, encompassing nonmotile, uninucleated forms; and the Siphonous line, encompassing plurinucleated forms.

*Chlamydomonas* is a minute (approximately ten-thousandth of an inch long), single-celled, flagellated green alga. It is round or pear shaped and possesses a rigid cellulose cell wall, contractile vacuoles, a cup-shaped chloroplast, and a pair of whiplash flagella at the anterior end. The flagella pass through tiny holes in the cell wall. Each, of course, has a cell membrane immediately inside the cell wall. *Chlamydomonas* also each have a red-pigmented **eye spot**, or **stigma**, considered to be light sensitive, and a pyrenoid body that functions as a center for the formation of starch. The starch eventually changes into an oil. Although pyrenoids are widely distributed in algae, they do not occur in most land plants. While *Chlamydomonas* are green, photosynthetic cells, they may get some of their nutrients ready-made from the surrounding water. The word *chlamys* is Greek for "cloak." The name *Chlamydomonas* derives from the fact that the chloroplast appears to form a cloak around the nucleus.

※ **Notes** ※

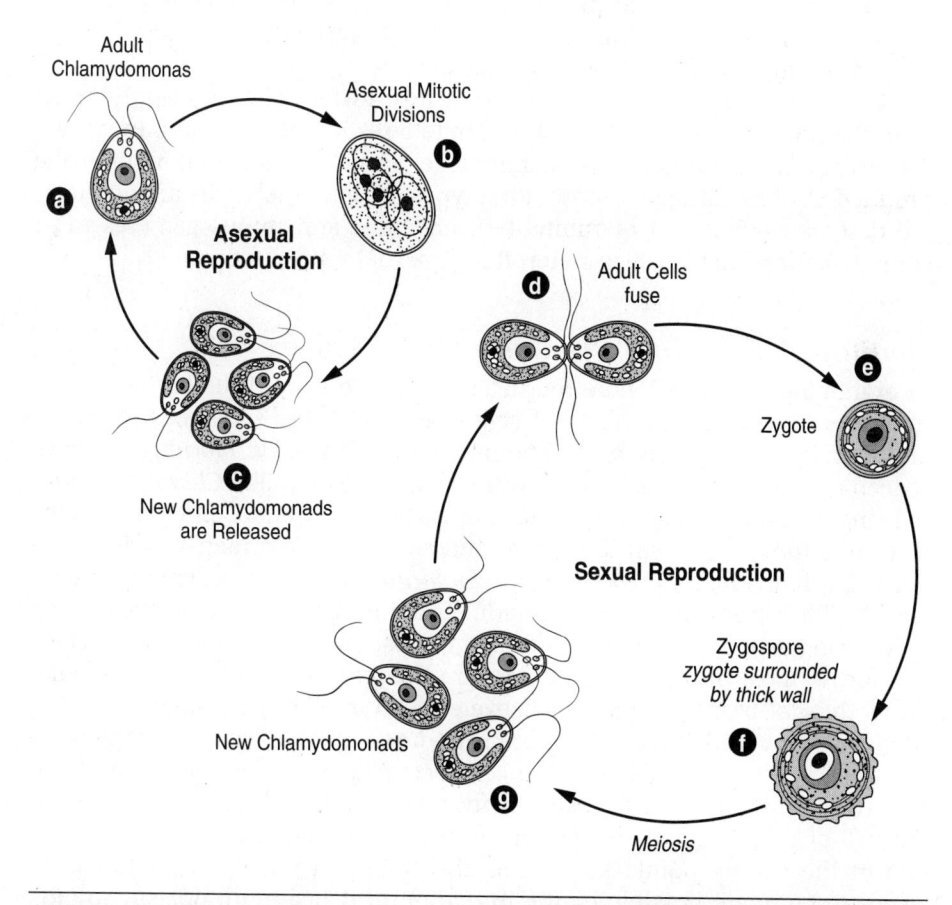

**Figure 13-2** (a) *Chlamydomonas*. The name derives from the chloroplast forming a cloak around the nucleus. (b) Asexual reproduction: mitotic divisions within the cell wall. (c) New Chlamydomonads are released. (d) Sexual reproduction: cells fuse in pairs to form zygotes, (e) and (f), which can then undergo meiosis to produce new *Chlamydomonas* cells, (g).

### ❈ Notes ❈

Reproduction in *Chlamydomonas* is both asexual and sexual. In asexual reproduction, the cell undergoes mitotic divisions within the cell wall. There may be as few as two divisions or as many as four. The daughter cells are at first contained within the parent cell wall; after a short time, however, the wall breaks down, and the miniature Chlamydomonads escape and grow to full size.

A change of environmental conditions sets the stage for the sexual process. Again, the protoplast divides a number of times by mitosis, producing a number of cells within the cell wall. When the cell wall breaks down, these small cells escape, come together, and fuse in pairs. Because the gametic cells that fuse are alike in size and shape, this is called **isogamous** fusion. The cells come in contact at their flagellar ends and fuse to create a diploid **zygote**, which at first has four flagella. The flagella then fall off, and a hardened outer covering is produced. This results in the formation of a **zygospore** able to survive adverse conditions. When living conditions again become favorable, the zygospore undergoes meiosis to produce four haploid **zoospores**. These mature into the new Chlamydomonads. Note that the only diploid cell in this life cycle is the zygospore.

In some situations, the **gametes** that fuse to form the zygospore come from the same cell. This is called **homothallic** sexual reproduction. In certain species of *Chlamydomonas*, this cannot take place. Rather, fusion requires the production of isogametes from mating types of cells. Such cells and gametes are designated plus (+) or minus (−), and plus and minus gametes must come together. This is called **heterothallic** sexual reproduction.

### *Gonium, Pandorina, Pleodorina, and Volvox*

In examining figure 13-1 note that leading upward toward the left is a series of algae of increasing numbers of cells. The first in the series has four cells and is called *Gonium pectorale*. The next form, *Pandorina morum*, can have sixteen or thirty-two cells. In both of these the cells are like *Chlamydomonas* in being flagellated and in other characteristics. The essential difference seems to be that the cells are stuck together and embedded in a matrix, rather than occurring individually. *Pleodorina* (or *Eudorina*) is another colonial form of alga. It differs from the others primarily in the number of cells in the colony, there being as many as 128. *Volvox* is the most complex of the several colonial forms; most of the cells, however, are *Chlamydomonas*-like. The colonial forms thus far mentioned are collectively called the Volvacine line.

In *Gonium* and *Pandorina*, there is no interdependence among the cells. Each cell is an individual, and the cells occur together only because of the matrix. A coordination of flagella activity, however, does develop. In *Eudorina*, the ball of cells becomes hollow in the center, and a polarity develops; one end of the colony points forward as the colony swims. This coordination seems to be made possible by the formation of delicate protoplasm strands that interconnect the cells.

Both asexual and sexual forms of reproduction occur in the Volvacine line. In *Gonium*, each cell can divide to form a new colony. Because the cells have walls, the divisions that form the new colonies take place within the original cell wall. Only after the new colony has formed does the wall break down and allow release. After the breakdown of the parent cell wall, no further divisions occur. The new colony that forms within the parent wall is called a **coenobium**. A coenobium can be defined as a colony of unicellular organisms surrounded by a common wall.

Asexual reproduction in *Pandorina* is somewhat different. The cells divide internally to produce zoospores, which then break out of the parent "box" all at the same time (thus the term *Pandora's box*). The zoospores then divide to form new colonies. Sexual reproduction in *Pandorina* is very similar to that described for *Chlamydomonas* except that the gametes that come together differ in size. This is referred to as **anisogamous** sexual reproduction (see figure 13-3). Two types of gametes are produced: larger and smaller. When fusion occurs, only larger and smaller gametes come together. The resulting zygote, then, is diploid, and is the only cell in the life cycle that has the doubled chromosome number. Cell divisions of the zygote are meiotic and result in the formation of zoospores.

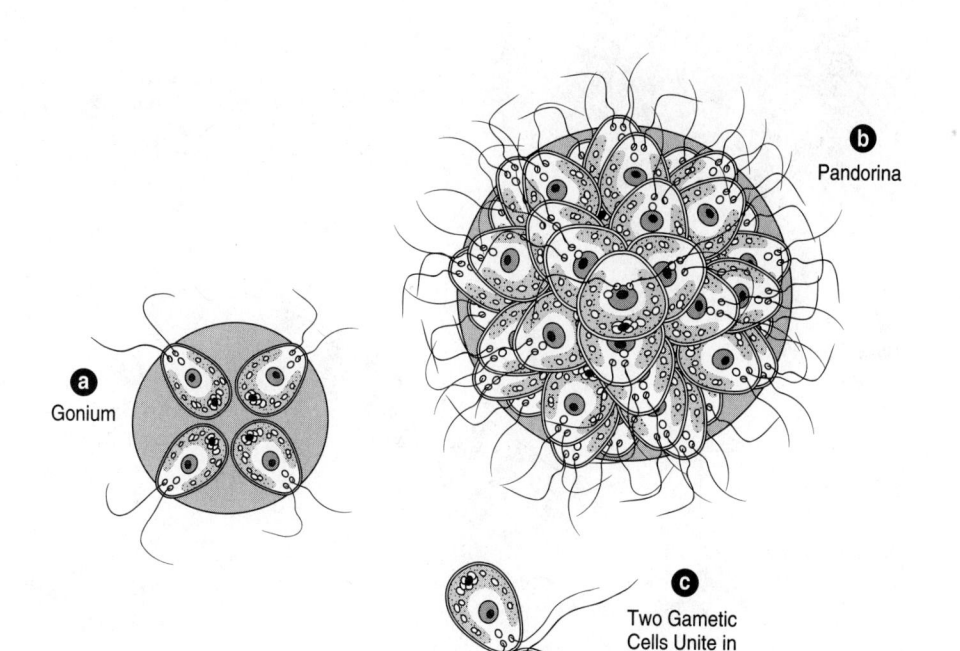

※ **Notes** ※

**Figure 13-3** The Volvacine line: (a) *Gonium* and (b) *Pandorina*. Anisogamous sexual reproduction may occur. (c) Two gametic cells unite. The zygote is the only diploid cell. When the zygote germinates, meiosis occurs, producing zoospores that grow into new colonies.

## ✻ Notes ✻

*Volvox* marks the high point in the evolution of the Volvacine line. The number of cells in the colony can range from 20,000 to as many as 50,000. Each cell in the colony is *Chlamydomonas*-like. A cell in the colony may enlarge, break away internally from the surface of the sphere, undergo mitotic divisions, and form a new colony. Several of these may be seen in the hollow center of the mother colony, where an amazing event occurs. When first formed, the minute colonies all have their flagella directed inward toward the center of the hollow ball, which contains a hole. Then, all at the same time, the entire colony turns itself inside out so that the flagella are on the outside of the sphere.

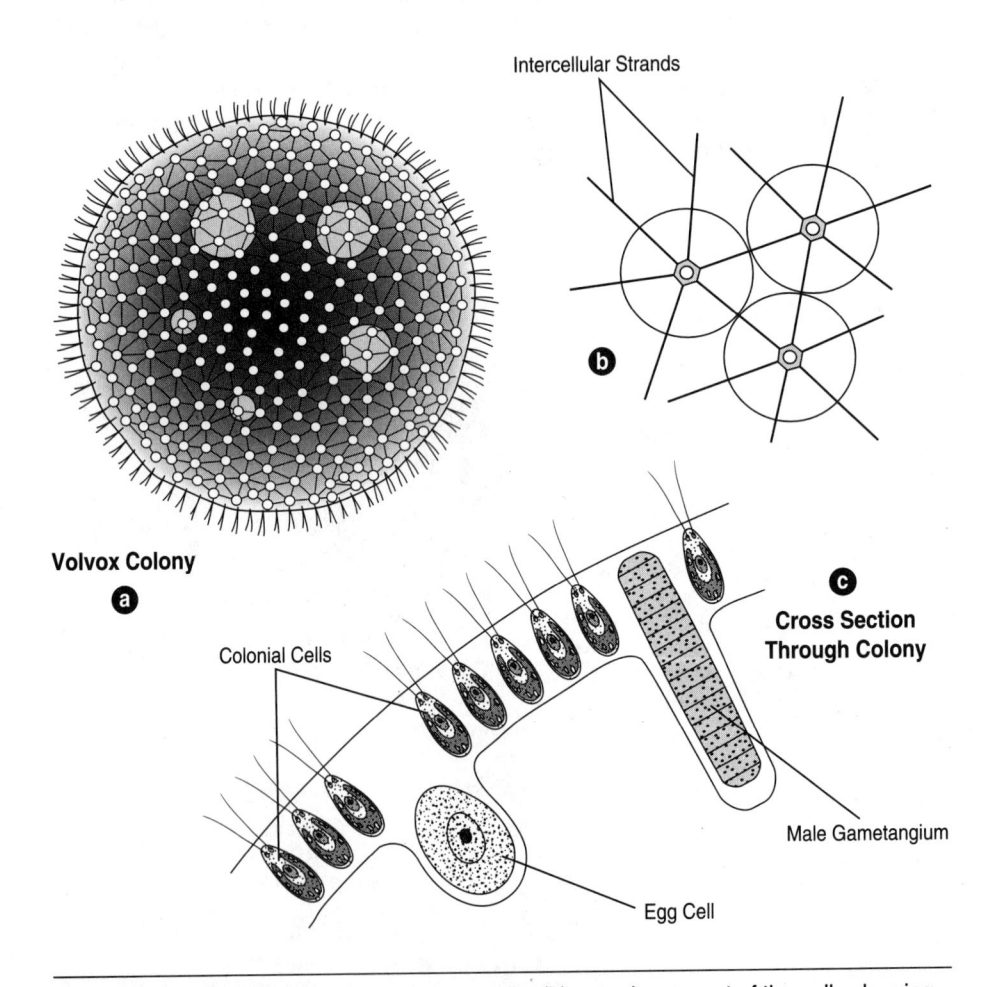

**Figure 13-4** *Volvox*: (a) a colony of many cells; (b) an enlargement of the cells showing intercellular strands and surrounding matrix; (c) colony portion showing an individual colonial cell, an egg cell, and male gametangium.

Sexual reproduction in *Volvox* involves clearly differentiated gametes: **sperm** and **egg**. This condition is called **oogamy**. In some species of *Volvox*, both sperm and eggs are found in the same colony and are thus referred to as **monoecious** (literally, "living in one house"). In other species, the colonies are either male or female; sperm cells are formed in one colony, egg cells in another. This condition is referred to as **dioecious** (literally, "living in two houses"). It is tempting to infer that such differentiation into separate sexes accounts for this condition in subsequent evolutionary history. The fact that *Volvox* appears to be a dead end in evolution contradicts this inference, however. The differentiation of sexes as we know it in the modern world apparently took another route.

## The Tetrasporine Line

In the Tetrasporine line, the cells are uninucleated and primarily nonmotile. If motile cells do occur, they do so only during reproductive phases. Cells in the Tetrasporine line of green alga grow in filaments or sheets. *Pleurococcus, Ulothrix, Oedogonium, Spirogyra,* and *Ulva* are several examples of this line.

### *Pleurococcus*

*Pleurococcus* (*Protococcus*) is perhaps the most abundant algal cell in the world. Boy Scouts use the knowledge that this alga grows on the north sides of tree trunks to find their way out of the woods! *Pleurococcus* appears to divide only by mitosis, forming small clusters of cells. The cells are quite like those of *Chlamydomonas* but without flagella.

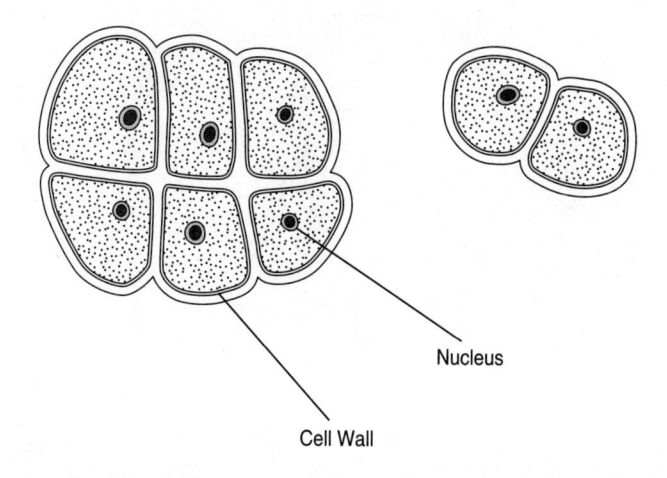

**Figure 13-5** *Pleurococcus.*

### 🐾 Notes 🐾

### *Ulothrix*

*Ulothrix* is a filamentous green alga, the cells of which each have a nucleus surrounded by a girdle-shaped chloroplast. The filaments may be found attached by a basal cell to a rock or other substrate in cold, running water. The cells reproduce mitotically to increase the length of the chain. *Ulothrix* cells produce four to eight zoospores, and each zoospore bears four flagella. The zoospores escape through a small hole in the cell wall. They swim around for a time before settling onto a substrate and undergoing divisions to form a new filament. The cells that give rise to the zoospores are called **sporangia**,

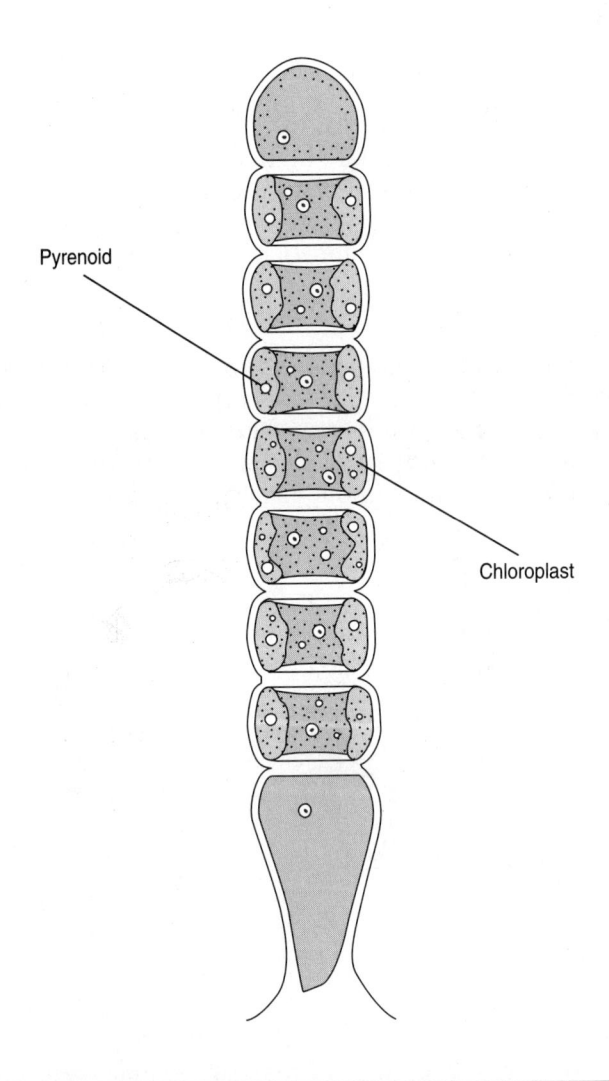

**Figure 13-6** *Ulothrix.*

or zoosporangia. Isogametes are also produced in cells called **gametangia**. Each such gamete possesses two flagella rather than four. Gametes from different filaments come together to form zygotes (*Ulothrix* being heterothallic). The zygotes are thus temporarily four flagellated. When the zygotes germinate, they become either motile or nonmotile spores. The spores, in turn, grow into new filaments. The zygotes have thick, resistant walls that enable the cells to withstand adverse conditions such as winter weather.

*Ulothrix* exhibits a form of cell differentiation in that the **basal cell**, or **holdfast**, functions in a way different from that of the other cells. Rootlike extrusions that extend from the basal cell attach to rock or some other substrate. In contrast with the colonial arrangement, this suggests the beginning of a multicellular individual.

## *Oedogonium*

*Oedogonium* is another filamentous, uninucleated green alga of the Tetrasporine line. The filament of *Oedogonium* has a specialized basal cell called a holdfast. The chloroplasts of the vegetative cells are netlike, taking the shape of tubes and surrounding the protoplasts. There are several pyrenoids. This form is found in lakes, slow-moving streams, and stagnant water. More than 250 species have been described, approximately one-half of which occur in the United States.

When cells divide to increase the length of the filament, a specialized process takes place at the distal end of the cell (the end opposite the attached end of the filament). This process results in a band, or a thickening of the wall, called a *cap*.

Both asexual and sexual methods of reproduction occur in *Oedogonium*. The most conspicuous cell in a filament is the one that forms the zoospore. The entire contents of the cell become round and develop a circle of cilia. When the cell wall breaks open, the zoospore swims around before settling down and growing into a new filament of *Oedogonium*. The cell containing the zoospore will regularly break open at the place where the cap occurs (that is, at the distal end of the cell).

Sexual reproduction in *Oedogonium* is heterogamous; both eggs and sperms are produced. Certain cells become **oogonia**, or bearers of eggs. These cells enlarge, their contents becoming round, and each produces a single, nonmotile egg cell. The egg is thus contained within the cell wall and within the oogonium. A small hole in the oogonial wall allows the entrance of a sperm cell.

The sperm cell is produced in a special cell called the **antheridium**. Antheridial cells are small, and two sperm cells are produced in each cell. The sperm cells are ciliated and, when released, swim toward the oogonium. A chemical attractant secreted by the oogonial wall induces the sperm to swim in the proper direction. One of the sperms may enter the oogonium through the opening and fertilize the egg. When fertilization occurs, the

## ❧ Notes ❧

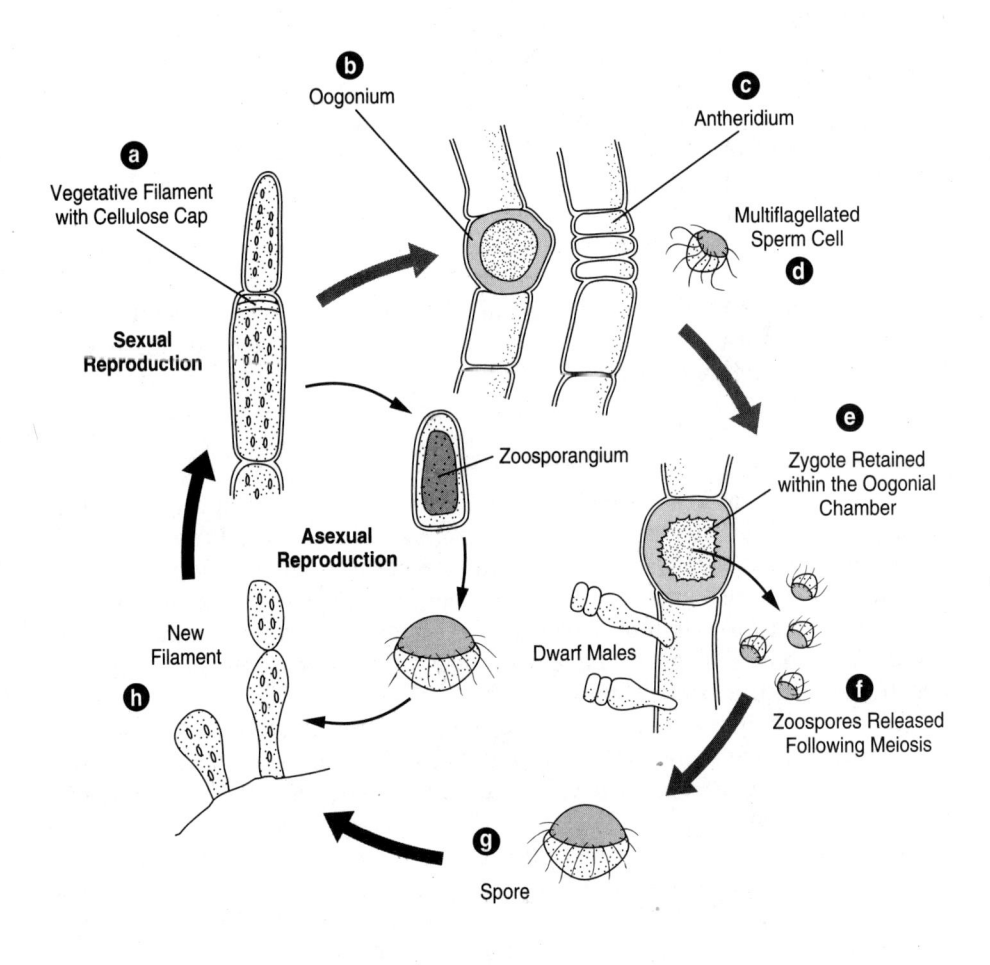

**Figure 13-7** *Oedogonium*: (a) vegetative filament with cellulose cap; (b) filament with oogonium bearing an egg; (c) filament with antheridia; (d) multiflagellated sperm cell; (e) a zygote retained within the oogonial chamber. When the zygote germinates, meiosis occurs, producing zoospores, (f). Dwarf males grow close to an oogonium. Asexual reproduction is shown in the smaller circle.

opening in the oogonial wall closes, and the zygote develops a resistant covering. The zygote thus can resist harsh conditions for a period of months. Note that the zygote is the only diploid cell that occurs in *Oedogonium*. When germination occurs, meiosis takes place, and four haploid zoospores are formed. These grow into new *Oedogonia*.

Before proceeding, consider that reproduction was initially accomplished solely by fission; that, in time, asexual reproduction was accomplished by spores; that there then came isogamous followed by heterogamous sexual reproduction; and that then the sexes became inherent in different individuals.

An important development among *Oedogonium* is the evolution of minute male plants that can attach to the cells adjacent to oogonia. These **dwarf males** occupy positions of opportunity; the sperm cells they produce need travel a minimal distance in order to accomplish their mission (see figure 13-7).

## *Spirogyra*

The handsome, spiral-shaped chloroplasts of *Spirogyra* always seem to bring a stir of elation to botanists. *Spirogyra* occurs in long filaments of cells, the chloroplasts of which hug the inside surface of the cell membrane. The centrally placed nucleus is held in position by strands of cytoplasm. The chloroplasts have numerous pyrenoids. Depending on one's taste for names, *Spirogyra* has been labeled both "Mermaids Tresses" and "Frog Spit." The filaments have a gelatinous sheath, which lends them a slimy feel. There are approximately sixty-five species of *Spirogyra* in the United States. Various forms grow in shallow ponds, slow-moving water, or in stagnant pools.

Asexual reproduction in *Spirogyra* is achieved by fragmentation of the chain of cells, which may result from changes in temperature or acidity or from mechanical means. Although there are male and female protoplasts, sexual reproduction is isogamous. When physiologically different filaments come to lie side by side and parallel to one another, small protuberances grow on the surfaces of the cells on the sides facing the neighboring filament (see figure 13-8b and c). The protuberances of the adjacent filaments extend outward and come in contact. The walls at the point of contact then break down, and a bridge between neighboring cells (called a *conjugation tube*) forms. Because the appearance of the two filaments at this moment is like a ladder, the event is called **scalariform** (ladderlike) **conjugation**. All the protoplasts up and down the line then become round, and those of one filament move across the bridges virtually in unison to fuse with the protoplasts of adjacent cells, thus forming zygotes. The movement is said to be **amoeboid**. In that they move, the protoplasts of one filament thus behave as sperm cells; and in that they remain in place, the protoplasts of the other filament take on the role of eggs. Because the bulk of the receiving cell doubles, this process must be accompanied by the expulsion of water. This is achieved by **contractile vacuoles**.

As in *Oedogonium*, a diploid zygote forms. The zygote secretes a thick wall and may remain dormant for some time. Upon germination, meiotic division results in the formation of four haploid spores. But at this point, something happens: three of the four spores always degenerate and disappear, leaving only one that can grow into a new spirogyra plant. This is of particular significance in that this event occurs in higher plants. Unlike other genera thus far discussed, the *Spirogyra* spore is nonmotile.

## ❈ Notes ❈

136 ♦ Chapter 13

❊ **Notes** ❊

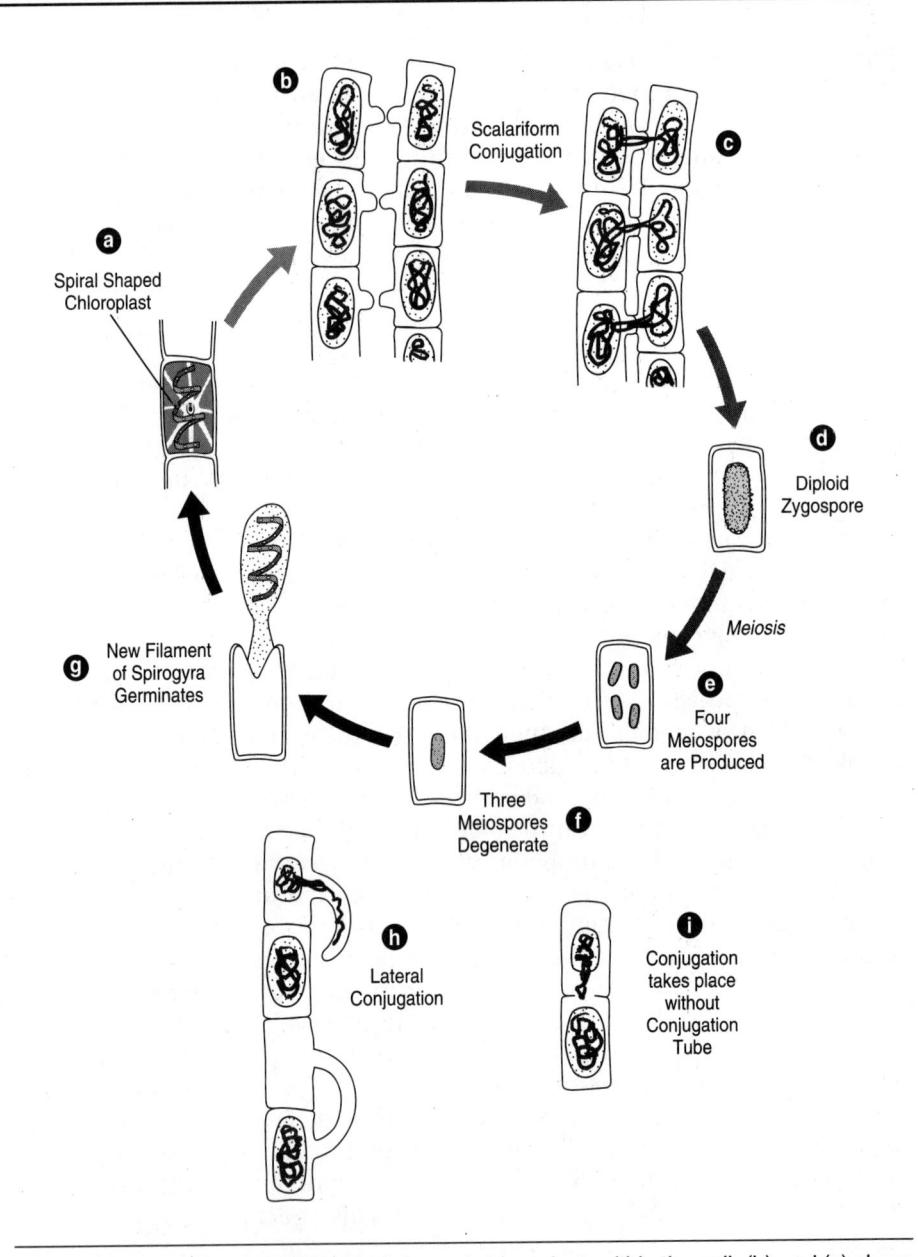

**Figure 13-8** *Spirogyra*: (a) the spiral-shaped chloroplast, within the cell. (b) and (c) show scalariform conjugation. Two filaments of cells come to lie parallel with end walls precisely aligned. Outgrowths of each cell extend medially, make contact, and produce bridges over which the entire protoplast of a cell can migrate to the neighbor. In this manner, a diploid zygospore, (d), is produced. (e) Meiosis within the spore wall follows. Three of the four meiospores degenerate and disappear leaving one, (f), to germinate and produce a new filament of *Spirogyra*, (g). (h) Another form of conjugation known as lateral conjugation. A protoplast moves downward to a neighboring cell below, leaving every alternate cell empty. (i) Conjugation takes place without the formation of a conjugation tube and by the protoplast moving through the cell end wall.

*Spirogyra* also exhibits another, less frequently seen kind of conjugation. This form of conjugation occurs between adjacent cells of the same filament (see figure 13-8h and i). A cell produces a conjugation tube, but, having no neighboring filament to which to reach out, the tube instead grows downward to the next cell below. The protoplast then moves down to form a zygote. This is called **lateral conjugation**. It is readily apparent that in scalariform conjugation, one filament becomes entirely empty of cell contents, whereas in lateral conjugation, every alternate cell becomes empty.

Sometimes, the cells of a spirogyra begin the process of gamete formation (getting ready for sexual union), but the process is halted, and no fusion occurs. We may think of this as an egg being denied fusion with a sperm. The egg cell, however, is still capable of growing into a new spirogyra. The production of a new individual by an egg cell without that egg cell being fertilized by a sperm cell is called **parthenogenesis**.

## *Ulva*

Thus far in our discussion of the Tetrasporine line, we have examined algae that grow only in single filaments. *Ulva*, or sea lettuce, grows in broad sheets. Consider how this can occur. If the orientation of the mitotic spindle were always the same as a cell goes through divisions, the result would be a single line of cells—a uniseriate filament. If, however, the spindle were to shift at right angles to the previously described position, a thin sheet of cells would result—a sheet one cell thick. This scenario occurs in the green alga *Monostroma*. If the cells in such a sheet were to change the polarity of the mitotic spindle, the result could be a sheet two cells in thickness. Such is the case with *Ulva*. When *Ulva* begins its growth from a single cell, the beginning divisions produce a filament. Soon, however, the orientation of the spindle shifts to a second and then a third plane, thus producing a broad **thallus** of two layers of cells.

In *Ulva*, asexual reproduction by spores and sexual reproduction by gametes alternates with the passing generations. The plant that produces gametes is the **gametophyte**, and the plant that forms spores is the **sporophyte**. Thus, in the alternation of generations, the gametophyte generation is followed by a sporophyte generation, and the sporophyte generation then produces the next gametophyte generation (see figure 13-9). The gametophyte plants and sporophyte plants look alike but produce different things.

In the gametophyte generation, the sexes are separate: a male thallus and a female thallus (or, if the species is isogamous, plus and minus). Many cells function as gametangia, producing biflagellated gametes. These gametes fuse in pairs. The members of each pair come from different thalli, and their fusion produces a zygote. The zygote then grows into a new thallus that looks the same as the gametophyte plant. Because the divisions of the zygote are mitotic, however, the new thallus is diploid (2N). This plant is a sporophyte. It produces quadriflagellated zoospores in sporangia. Meiosis then occurs, and the zoospores are haploid (1N). The zoospores must be of two sorts because some of them produce male gametophytes and some produce female gametophytes.

❋ **Notes** ❋

**❧ Notes ❧**

Whereas the gametophyte generation is haploid, then, the sporophyte generation is diploid. And whereas gametes are produced by mitosis, spores are produced by meiosis. Because the two generations look alike, this can be referred to as alternation of **isomorphic** generations. Alternation of generations, though not of isomorphic generations, is the universal condition in higher plants.

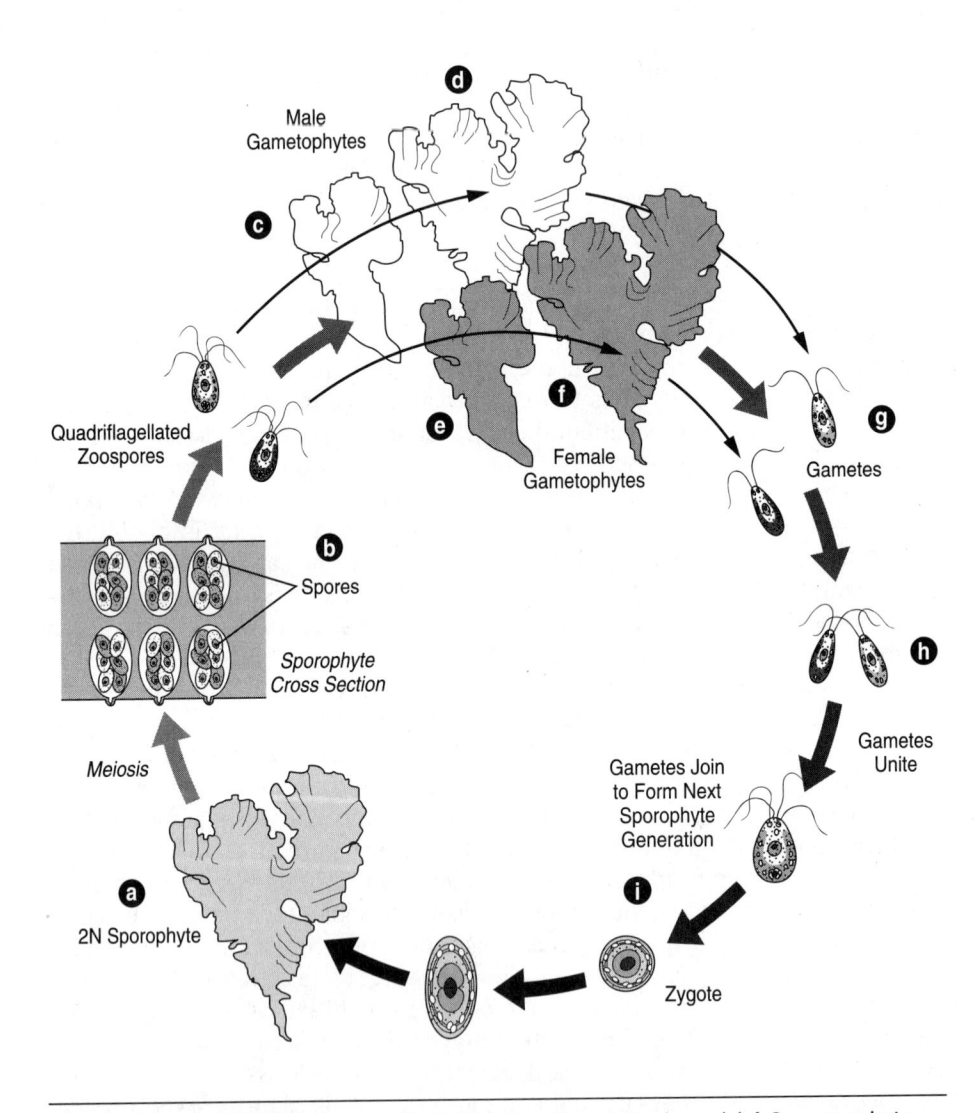

**Figure 13-9** *Ulva*, showing alternation of isomorphic generations: (a) A 2N sporophyte that produces quadriflagellated zoospores by meiosis; (b) sporophytes in cross section. Sporophytes are of two sorts, producing male gametophytes, (c) and (d), or female gametophytes, (e) and (f). The gametes, (g), unite, (h), and produce the next sporophyte generation, (i).

## The Siphonous Line

The last of the postulated lines of evolution among the green algae is the Siphonous line. Members of this line may have unbranched or branched filaments. Their most striking feature, however, is that they are **multinucleated**. These forms seem to have lost the capacity to make cross walls following the divisions of the nucleus, which results in the multinucleated, or **coenocytic**, condition. The coenocytic condition occurs both in algae and in fungi but is not found in higher green plants.

*Codium* is an example of this line. The thallus bears gametangia of two sorts: those producing larger gametes and those producing smaller gametes. The form is anisogamous. When the gametangia break open, biflagellated gametes emerge and come together in pairs, each of a larger gamete and a smaller gamete. Each pair of gametes forms a four-flagellated zygote that readily begins growth to form a new plant. The divisions are mitotic. The plant produced is diploid, and the only haploid cells in the life cycle are the gametes (meiosis occurring in gametogenesis). There is clearly no alternation of generations.

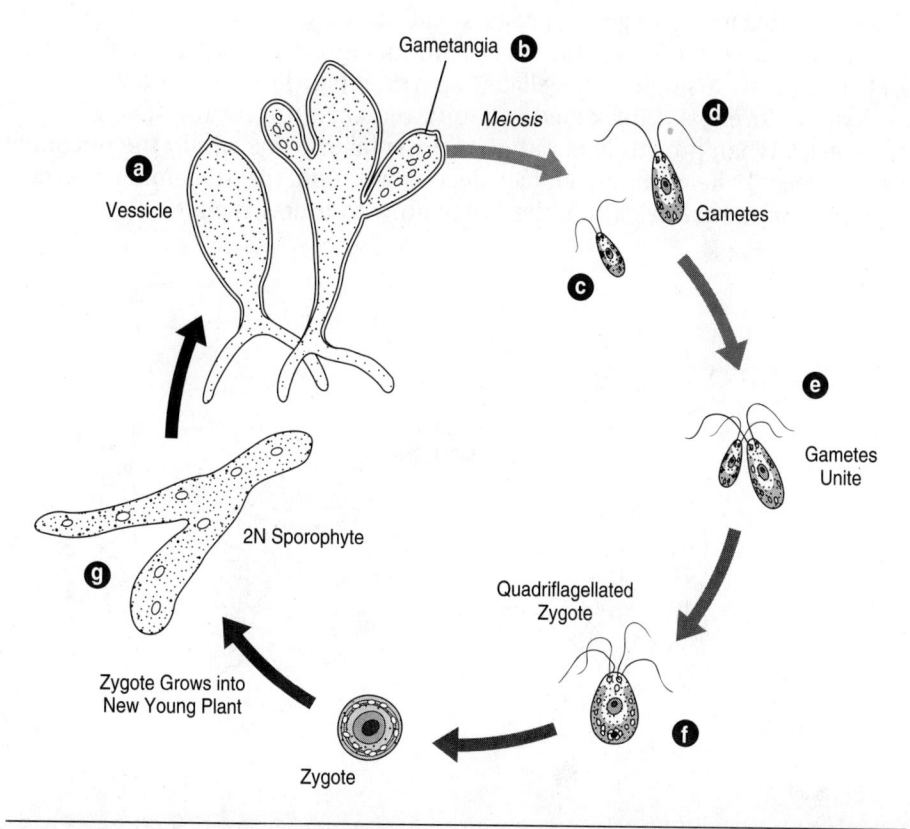

**Figure 13-10** *Codium*: (a) a vesicle bearing gametangia, (b); (c) and (d) gametes that differ in size. (e) The union of gametes produces a quadriflagellated zygote, (f), that grows into a new young plant, (g). Meiosis occurs in the production of gametes.

### ❋ Notes ❋

## Green Alga or Bryophyta?

There remains a presumed green alga different enough from all others that some algologists place it in a separate class. While it is generally believed to belong with green algae, a few would classify it as Bryophyta. It is an ancient form, tracing its beginnings to Silurian time.

The representative form is *Chara*. The cell walls contain cellulose and are also much impregnated with lime. *Chara* is a submerged, bottom-dweller of clear water. It possesses whorls of branches and both vertical and horizontal stems, and is anchored by rhizoids. Single, large cells extend from one set of branches to the next and may have several nuclei. Growth occurs from apical cells, as it does in higher plants. The apical cells divide by mitosis, and several kinds of cells including oogonia and antheridia are differentiated further down the stem. A singular feature of the Charophyceae is that the gametangia are surrounded by jackets of sterile cells.

Figure 13-11 shows gametangia borne on branches. The oogonium is surrounded by a jacket of elongated, curved cells. It contains a single egg. In maturity, the sheath cells separate near the tip to allow the entry of sperm. The antheridia are arranged in clusters and show a complex structure. Within the sheath of sterile cells, the sperm-producing cells occur in chains, and each cell forms a single, biflagellated sperm. When fertilization takes place, the zygote develops a thickened wall and remains in a resting state for several weeks. When growth is initiated, meiosis takes place within the oogonial wall. Three of the resulting nuclei degenerate, and the remaining nucleus produces a protonema, which then may grow to a new thallus.

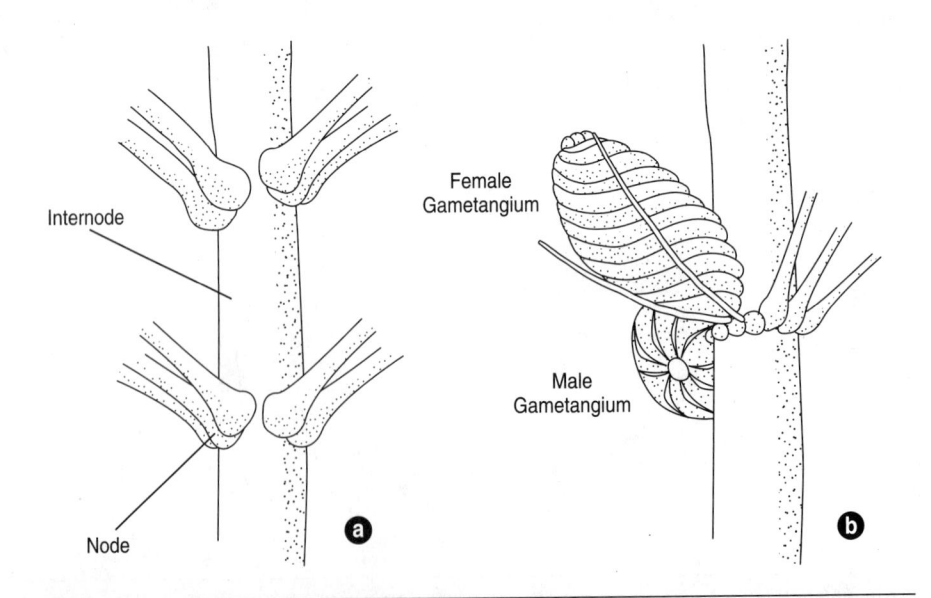

**Figure 13-11** *Chara*, a stonewort: (a) a portion of stem showing nodes and internodes; (b) male and female gametangia.

## Questions for Review

1. At what stage in the life cycle does the chromosome number change from N to 2N? From 2N to N?
2. Green algae may be classified according to what three presumed lines of evolution?
3. Which one of the lines in the answer to question 2 do you think might be the ancestor of vascular plants?
4. Define the terms *isogamy* and *heterogamy*.
5. Define the terms *homothallic* and *heterothallic*.
6. Define the terms *antheridium* and *oogonium*.
7. Describe two types of conjugation in *Spirogyra*.

## Suggestions for Further Reading

Blume, J.L. 1956. The ecology of river algae. *Botanical Review* 22:291–341.

Chapman, V.J. 1957. Marine algal ecology. *Botanical Review* 23:320–350.

Chase, F.M. 1941. Useful algae. *Smithsonian Institution Annual Report* 401–452.

Dawson, E.Y. 1961. The rim of the reef. *Natural History* 70:8–16.

Lewin, J.E., and R.L. Guillard. 1963. Diatoms. *Annual Review Microbiology* 17:377–414.

Papenfuss, G.F. 1951. Phaeophyta. In Smith, G.M. (ed.) *Manual of Phycology*. Waltham, MA: Chronica Botanica.

Prescott, G.W. 1954. *How to Know the Freshwater Algae*. Dubuque, IA: William C. Brown and Company.

Round, F.E. 1965. *The Biology of the Algae*. London: Edwin Arnold, Ltd.

Smith, G.M. 1950. *Freshwater Algae of the United States*. New York: McGraw-Hill Book Co.

Tseng, C.K. 1944. Agar: A valuable seaweed product. *Scientific Monthly* 58:24–32.

※ Notes ※

# 14

# Phaeophyta: The Brown Algae

There is great variation among Phaeophyta, the brown algae. A few are microscopically sized, most are conspicuous, and some can reach lengths of more than 150 feet. The large brown algae are called **kelp**. Pigments in the brown algae are chlorophyll a, chlorophyll c, beta-carotene, xanthophylls, and fucoxanthin, a pigment that gets its name from the brown alga *Fucus* and is not found elsewhere. The pigments are contained in plastids. The cell wall has an inner cellulose layer and an outer gelatinous coat composed mainly of **algin**, which has a number of commercial uses (noted in a following section). The product of photosynthesis in many brown algae is a soluble polysaccharide called **laminarin**. Alternation of generations occurs in the life cycles of brown algae except those of the order Fucales. The motile reproductive cells are distinctive, being laterally biflagellated. A longer flagellum extends anteriorly, and a shorter one extends posteriorly. While the anterior flagellum is tinsel type, the posterior one is not. In the Fucales order, this arrangement is reversed.

✼ Notes ✼

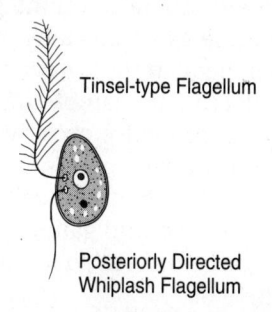

**Figure 14-1** A biflagellated gamete showing an anterior tinsel-type flagellum and a posteriorly directed whiplash flagellum.

※ Notes ※

## Products from Brown Algae

Algin, which occurs in the cell walls of brown algae, is widely used in industry as an emulsifying agent and a stabilizer. It is used to create smooth body and texture in ice cream; to prevent the coarsening of ice cream in storage; in the preparation of jellies, icings, and whipping cream; in the manufacture of chocolate milk; as a creaming agent in the manufacture of rubber; as a dental impression material; in the sizing of cloth; in paints, cosmetics, hand lotions, lubricating jellies, printing ink, fire retardant compounds, insecticides, and varnishes; and to finish leather, among other uses!

During the eighteenth and nineteenth centuries, kelp were used as sources of soda and potash. Because they have the capacity to concentrate iodine from the sea water, they were for a long period the prime sources of this element.

## Reproduction in Brown Algae

Asexual reproduction is accomplished by either fragmentation or spores. When cells divide, centrioles are apparent in the mitotic figure. This is significant because it is not ordinarily seen in plant cells. Sexual reproduction is of three types: isogamous, anisogamous, and oogamous. Three examples of life cycles are considered following.

### *Ectocarpus*

We will use *Ectocarpus* to represent the isogamous form of sexual reproduction in brown algae. The center circle of figure 14-2 shows asexual reproduction while sexual reproduction is shown in the larger circle. Although the sporophyte is 2N and the gametophyte is N, the two generations look alike. Two types of sporangia are produced on the sporophyte: unilocular sporangia (having one compartment) and plurilocular sporangia (having many compartments). The unilocular sporangia each at first contain a single cell. This cell goes through repeated divisions, the first of which are meiotic and, thus, reduce the chromosome number. These first divisions are followed by a number of mitotic divisions, which generate a number of biflagellated haploid zoospores. The zoospores then grow into gametophytes. The gametophytes, of course, are haploid. Plurilocular gametangia are produced on the gametophytes. Biflagellated isogametes are produced in these gametangia by mitosis. The gametes fuse in pairs, thus producing zygotes that can then grow into new sporophyte plants.

While the preceding is clearly an example of alternation of generations, *Ectocarpus* can go through a reproductive cycle without alternation of generations. Look again at the sporophyte plants in figure 14-2 and note the plurilocular sporangia. These contain many cells that mature into diploid zoospores, which, instead of growing into gametophytes, become new sporophytes. Because there is no gametophyte generation in this sequence, alternation of generations is bypassed.

Some species of *Ectocarpus* are monoecious, having both plus and minus gametangia on the same plant. Other species are dioecious, requiring a fusion of gametes from different plants to form zygotes.

*Phaeophyta: The Brown Algae* ♦ 145

❋ **Notes** ❋

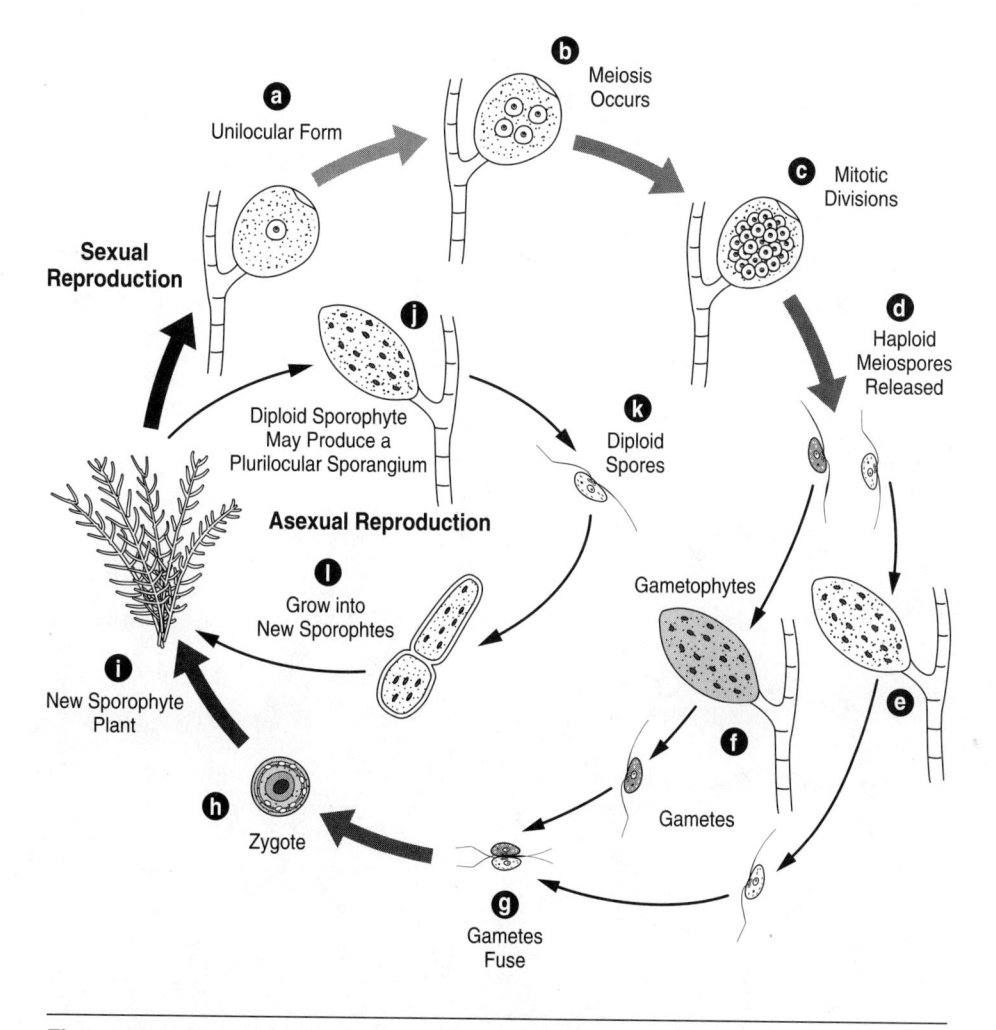

**Figure 14-2** *Ectocarpus* life cycle. At right are sporophyte generations exhibiting two types of sporangia: (c) unilocular and (j) plurilocular. In the unilocular form, (a), meiosis occurs, (b), followed by mitotic divisions, (c). Haploid meiospores (d) are released, producing gametophytes, (e) and (f). Gametes fuse, (g), producing a zygote, (h), which grows into a new sporophyte plant, (i). The sporophyte at (i) is diploid and may produce a plurilocular sporangium (j). This sporangium produces diploid spores, (k), which grow to new sporophytes, (l), thus circumventing alternation of generations.

## *Laminaria*

The sporophyte generation of *Laminaria* tends to reach considerable size and is differentiated into a rootlike holdfast, a stipe, and a flat blade. These plants attach with such firmness to rocks that when torn away, portions of the rock often come away also. Because the blade portion is much damaged by buffeting wave action, it is replaced by new growth in the spring. Each year, a zone of dividing (meristematic) cells where the stipe and the blade meet renew the

## Notes

blade. While the blade is annual, the stipe is perennial. When the plant is mature enough to produce sporangia, dark patches called **sori** appear on the surface of the blade. Sori are collections of sporangia. Meiosis takes place in the production of zoospores. These zoospores are minute, biflagellated cells of two types: those that produce male gametophytes and those that produce female gametophytes. One-half of the zoospores are of the one type, one-half are of the other.

The microscopically sized gametophytes actually evaded detection until 1916. Each gametophyte is composed of only a few cells. The antheridia that

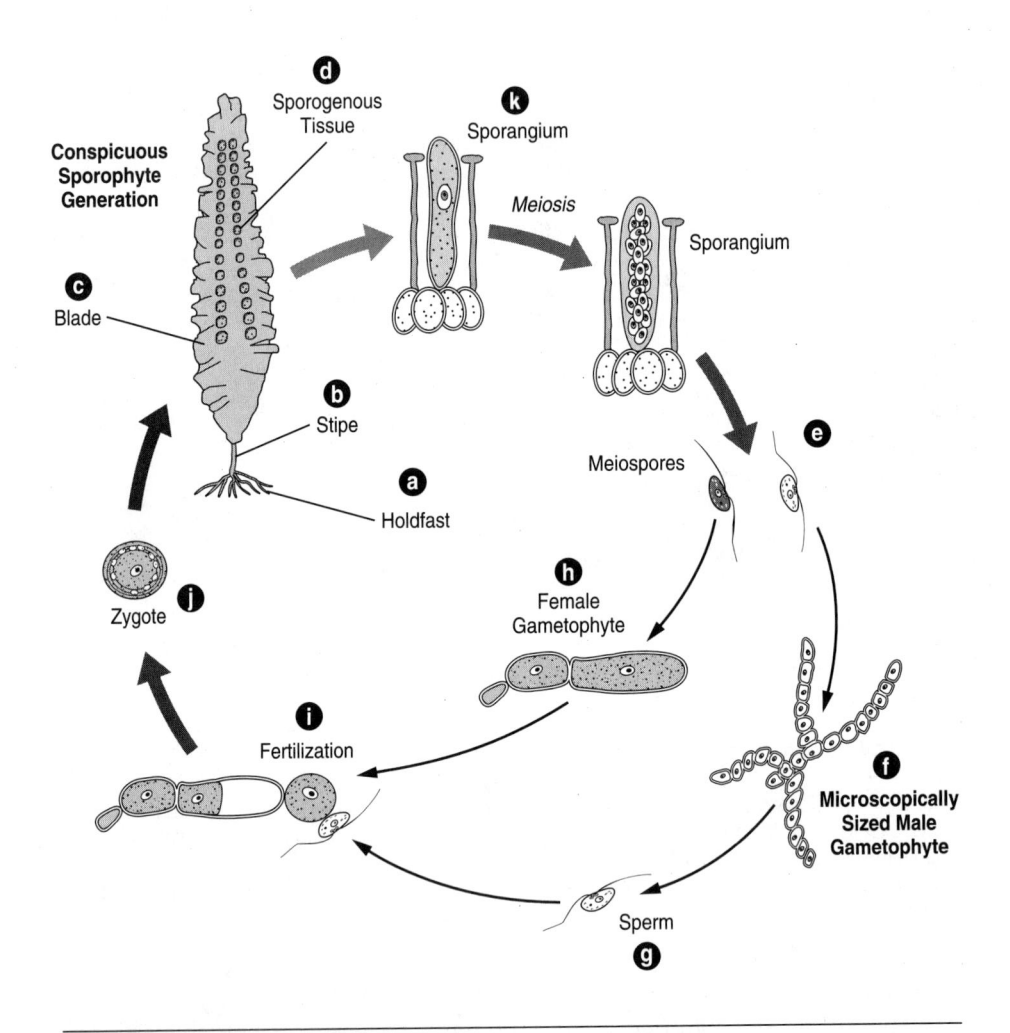

**Figure 14-3** *Laminaria*: at left, the conspicuous sporophyte generation consisting of (a) holdfast, (b) stipe, and (c) blade with (d) sporogenous tissue. (k) An enlargement of a sporangium intermingled with paraphyses. The spores at (e) are meiospores, which grow into microscopically sized gametophytes. The male gametophyte, (f), and the female gametophyte, (h). One egg cell, (i), is fertilized by the sperm, (g), to produce the zygote, (j), which grows into a new sporophyte.

are borne on the male gametophytes each produce a single sperm cell. Likewise, the oogonia of the female plants each house a single egg. Sperm cells swim to the egg cell at the opening of the oogonium. The zygote thus formed grows into the familiar sporophyte plant.

## *Fucus*

*Fucus* exists as a flat, dichotomously branching thallus bearing a short stipe and growing approximately one foot in length. The ends of certain branches become swollen; these swellings are called **receptacles**. The receptacles contain spherical

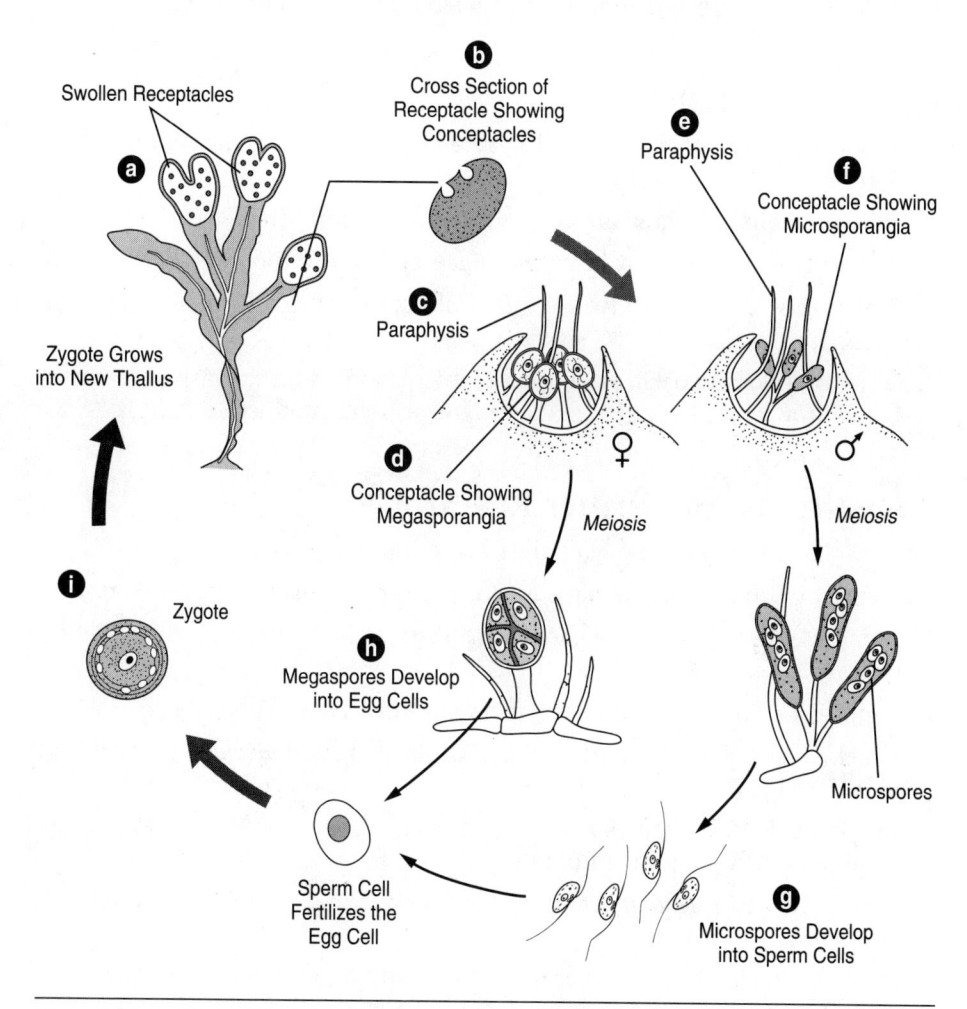

**Figure 14-4** *Fucus.* The plant is dichotomously branching with swollen receptacles, (a), having conceptacles, (b). The conceptacles are of two types: microsporangiate with microsporangia, (f), and paraphyses, (e); and megasporangiate with megasporangia, (d), and paraphyses, (c). The microspores develop directly into sperm cells (g), and the megaspores develop directly into egg cells (h). The zygote (i) grows into a new thallus. Because the spores develop directly into gametes, there is no alternation of generations.

### Notes

chambers called **conceptacles**, within which sporangia grow. The sporangia are of two types: microsporangia and megasporangia. In *Fucus spiralis*, both types of sporangia occur in the same conceptacle. In *Fucus vesiculosis*, the microsporangia and megasporangia are found on different plants. Microspores divide meiotically to form sperm cells, and megaspores produce egg cells by reduction divisions. Although both eggs and sperm are produced, no gametophytes are generated. The spores divide to produce gametes. The gametes are then released into the water, where fertilization occurs. The gametes are the only cells that have the haploid chromosome number. Although both spores and gametes are formed, it is not appropriate to label this alternation of generations.

## Questions for Review

1. Why is the term *algae* not valid in reference to a natural system of classification?
2. What are kelp, and how are they economically important?
3. What pigment is present in brown algae and not found elsewhere?
4. *Ectocarpus* produces two kinds of spores: _____ and _____.
5. *Fucus* produces two kinds of spores: microspores and megaspores. Microspores grow into _____, and megaspores produce _____.

## Suggestions for Further Reading

Blume, J.L. 1956. The ecology of river algae. *Botanical Review* 22:291–341.

Chapman, V.J. 1957. Marine algal ecology. *Botanical Review* 23:320–350.

Chase, F.M. 1941. Useful algae. *Smithsonian Institution Annual Report* 401–452.

Dawson, E.Y. 1961. The rim of the reef. *Natural History* 70:8–16.

Lewin, J.E., and R.L. Guillard. 1963. Diatoms. *Annual Review Microbiology* 17:377–414.

Papenfuss, G.F. 1951. Phaeophyta. In Smith, G.M. (ed.) *Manual of Phycology*. Waltham, MA: Chronica Botanica.

Prescott, G.W. 1954. *How to Know the Freshwater Algae*. Dubuque, IA: William C. Brown and Company.

Round, F.E. 1965. *The Biology of the Algae*. London: Edwin Arnold, Ltd.

Smith, G.M. 1950. *Freshwater Algae of the United States*. New York: McGraw-Hill Book Co.

Tseng, C.K. 1944. Agar: A valuable seaweed product. *Scientific Monthly* 58:24–32.

# 15

# Rhodophyceae: The Red Algae

Rhodophyceae is a very large class, encompassing approximately 4,000 species. Some of these are found in northerly latitudes, but most are found further south. Color varies among the red algae and is certainly not always red. A number of red algae are used as food. *Porphyra* is used for food in Japan. *Chondrus* is used in beer manufacture and making desserts. *Rhodymenia* is used in making candy. While most species of red algae are small (not more than one centimeter across), a few can grow to be two feet or more in width.

No motile cells are found in the red algae, including among the gametes and spores. The pigments are chlorophyll a, possibly chlorophyll d, carotenes, and phycocyanin and **phycoerythrin** (two pigments also present in the blue-green algae). The product of photosynthesis is a kind of starch found only in this group: **floridian starch**. Both uninucleated and multinucleated forms are known. *Griffithsia* has thousands of nuclei per cell.

All red algae compose the single class Rhodophyceae, which is divided into two subclasses: the Bangiophycidae and the Floridiophycidae.

## Bangiophycidae

The Bangiophycidae subclass is the smaller of the two subclasses and probably the more primitive. *Porphyra* is an example of this subclass. In some

❈ Notes ❈

## ❋ Notes ❋

species of *Porphyra*, the thallus is made of a single layer of cells; in others, there are two cell layers. Both asexual and sexual means of reproduction occur. In asexual reproduction, monospores are produced. These monospores may then grow into new plants. Sexual reproduction is more involved. Figure 15-1 shows a monostromatic (having only one cell layer) form of *Porphyra*. Any cell of the thallus can play the role of an antheridium and, by repeated divisions, produce up to 128 spermatia. Recall that there are no motile cells in red algae. Whereas motile male gametes are called sperm cells, nonmotile

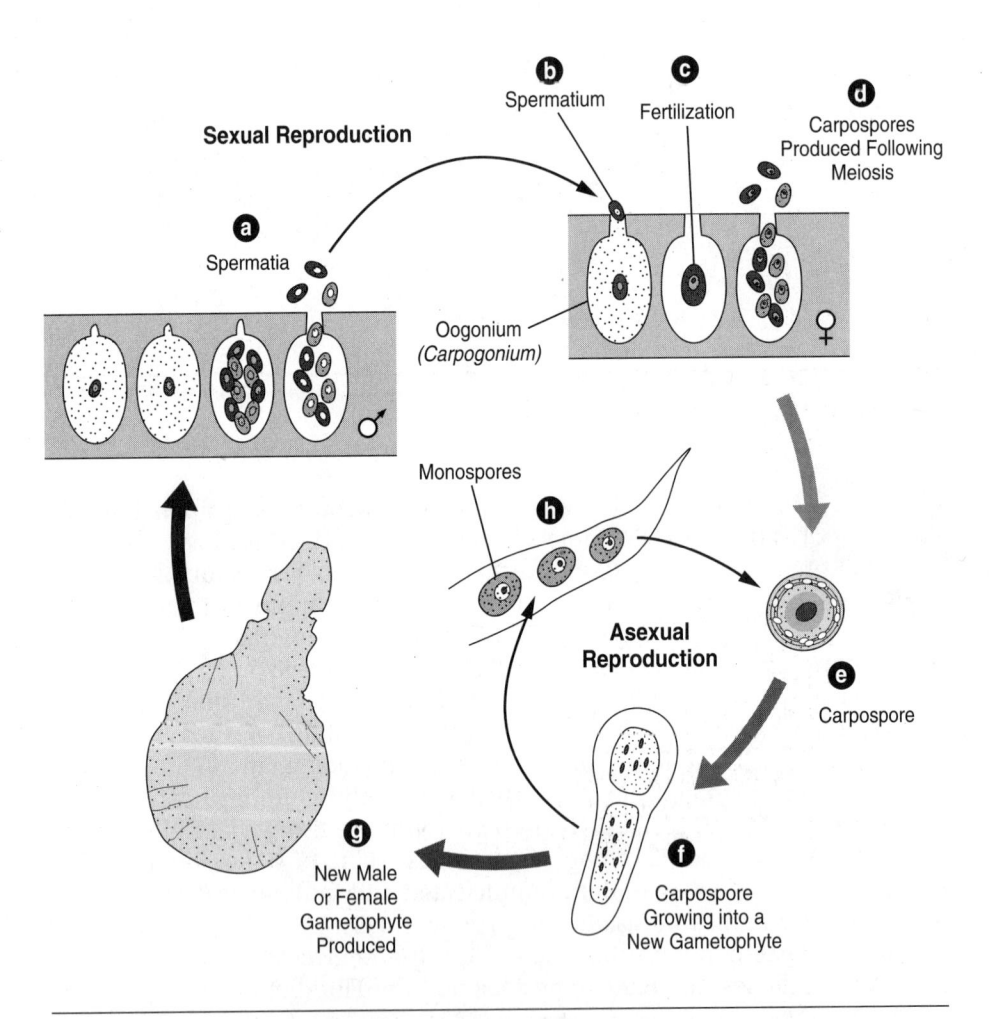

**Figure 15-1** A monostromatic form of *Porphyra*: (a) spermatia; (b) an oogonium, or carpogonium, with spermatium at the end of the trichogyne; (c) fertilization. (d) Meiosis occurs and carpospores are produced. The carpospores grow into new gametophytes, (e), (f), and (g). At (h) an asexual pathway generates monospores, which can grow into new plants.

male gametes are called **spermatia**. Other cells on the same plant, or in other species on different plants, become oogonia, each bearing a single egg cell. A tubular emergence called a **trichogyne** develops from the oogonium to the exterior of the thallus. (The trichogyne also occurs among the Ascomycetes, a class of fungi.) For reasons that will soon become apparent, the oogonium can also be called a **carpogonium**.

When the spermatia are released from the antheridium, they float around in the water. One or a few of the spermatia will come in contact with the trichogyne on the oogonium. A male nucleus may then make its way, by being drawn in through this receptive emergence, to the egg nucleus. Fusion then takes place followed promptly by meiosis and several mitotic divisions. The resulting nuclei are surrounded by cytoplasm and are then known as **carpospores**. Although carpospores in other examples may be diploid, these are haploid. Under favorable conditions, the carpospores may grow directly into new foliose plants. There is no alternation of generations. In some species, the carpospores are of two types: that half that produces male gametophytes and that half that produces female gametophytes.

When living conditions are less favorable, a different course of events may unfold. Instead of producing foliose plants, the growing carpospores produce minute plants that liberate monospores, which, in turn, may then form foliose plants.

## Floridiophyceae

Floridiophyceae is by far the larger of the two Rhodophyceae subclasses, encompassing approximately 3,900 species. A prominent characteristic among this subclass is the presence of cytoplasmic strands that connect adjacent cells. Such strands are called **primary pit connections**. As shown in figure 15-2a and b, the nucleus of the cell divides, and the nuclei migrate to opposite ends of the cell. A constriction then develops in the cell wall. Thus, division is not completed; while the cells are separate cells, they are not separated cells.

Figure 15-2 shows yet another rather amazing event that may occur, resulting in **secondary pit connections**. The nucleus of the cell divides, and one of the daughter nuclei migrates to the lower edge of the cell (figure 15-2d). Here, in response to the presence of the nucleus, a protuberance grows out from the cell wall (figure 15-2e). The nucleus then moves into this outgrowth. As this outgrowth continues to expand, it comes in contact with the neighboring cell wall, which is thus dissolved. The result is an open channel to the cell below through which the nucleus continues its passage into the lower cell (figure 15-2f). Why doesn't the nucleus either go through the opening that is already there or stay in its original place? Although the answers to such questions are unknown, the events are

## ❧ Notes ❧

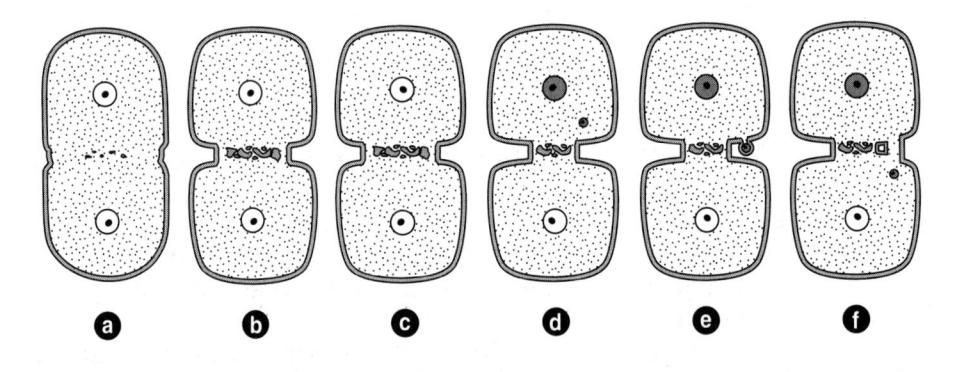

**Figure 15-2** Pit connections. At (a), (b), and (c), the primary pit connection is formed by incomplete division of the cell. (d) The upper nucleus has divided, and one nucleus migrates to the lower corner, where an outgrowth of the cell membrane, (e), develops. (f) A new bridge has formed to the lower cell and the nucleus has migrated across the bridge to the lower cell.

enchanting to observe. Figure 15-2 shows secondary pit connections only between sister cells, but such connections also occur between cells that are not sisters.

*Polysiphonia* is an example of the subclass Floridiophyceae, and its life cycle follows. Alternation of generations combined with some other features add to the complexity of the life cycle. The male and female gametophytes are separate. Numerous spermatogonia are produced on the male gametophyte where they produce large numbers of spermatia. The female gametophyte bears oogonia (carpogonia), each with a single egg cell and a trichogyne. When the spermatia are released in the surrounding water, some will come in contact with the tip of the trichogyne, where the wall breaks down to allow in the spermatium. Fertilization results in the formation of carpospores, which extend out from the carpogonium. At the same time, an outgrowth of protective envelope, called the **cystocarp**, occurs. The carpospores resulting from fertilization are diploid. When released, these carpospores give rise to **tetrasporophytes**. Cells of the tetrasporophytes go through meiotic divisions to produce haploid tetraspores, which, in turn, can grow into male and female gametophytes.

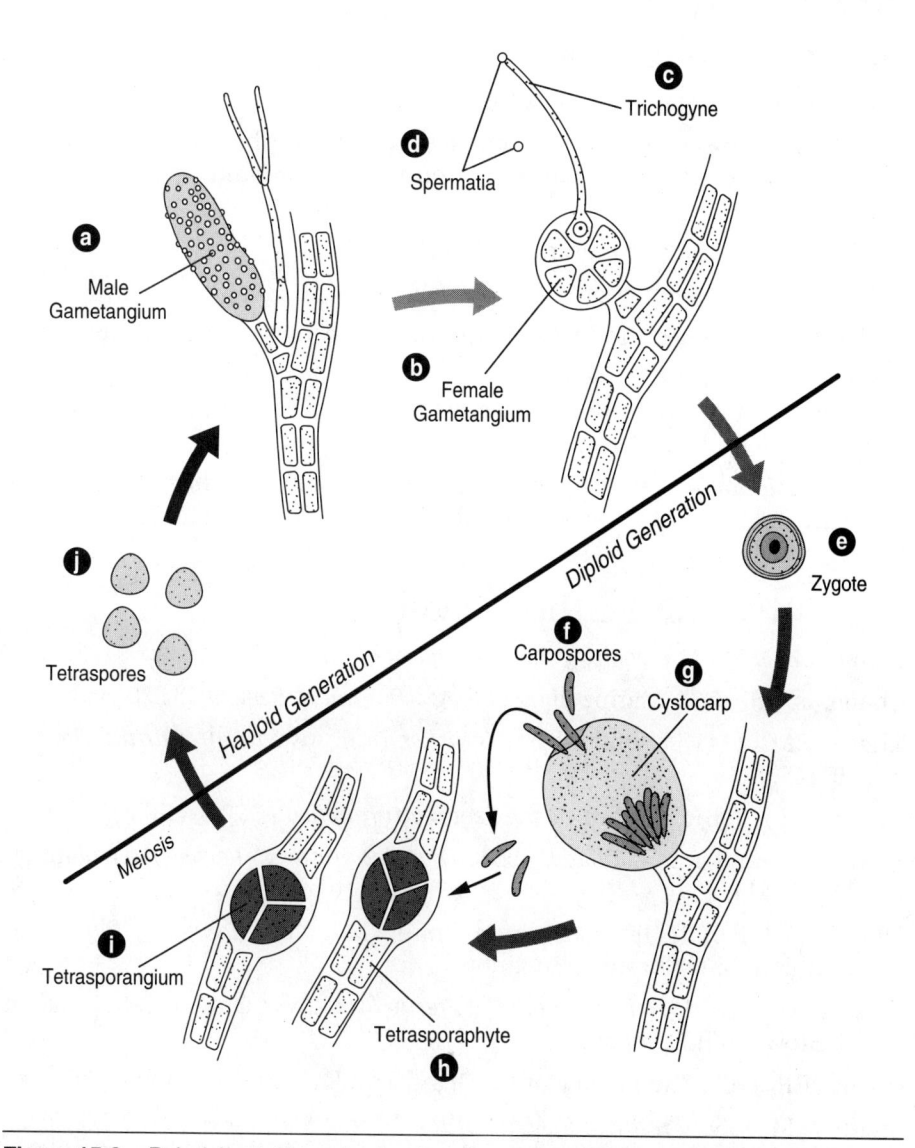

**Figure 15-3** *Polysiphonia* life cycle showing alternation of diploid and haploid generations. A broken line separates the diploid forms from the haploid forms. (a) Male gametangium with spermatia; (b) female gametangium with trichogyne, (c); (d) spermatia, one in contact with the tip of the trichogyne; (e) zygote; (f) carpospores; (g) cystocarp; (h) tetrasporophyte; and (i) and (j) tetraspores.

❋ Notes ❋

## Questions for Review

1. Where do red algae occur most abundantly?
2. Red algae possess a structure that seems to relate them to certain fungi. What is the structure, and to what group of fungi are red algae considered to be related?
3. What red algae may be useful as food?
4. What kind of starch is produced by red algae?
5. The male gametes produced in red algae are nonmotile and are called _____.
6. To reach an egg cell, the male gamete migrates through a tubular emergence called the _____.
7. In *Polysiphonia*, fertilization of the egg results in the formation of _____, which are diploid; these in turn give rise to _____.

## Suggestions for Further Reading

Blume, J.L. 1956. The ecology of river algae. *Botanical Review* 22:291–341.

Chapman, V.J. 1957. Marine algal ecology. *Botanical Review* 23:320–350.

Chase, F.M. 1941. Useful algae. *Smithsonian Institution Annual Report* 401–452.

Dawson, E.Y. 1961. The rim of the reef. *Natural History* 70:8–16.

Lewin, J.E., and R.L. Guillard. 1963. Diatoms. *Annual Review Microbiology* 17:377–414.

Papenfuss, G.F. 1951. Phaeophyta. In Smith, G.M. (ed.) *Manual of Phycology.* Waltham, MA: Chronica Botanica.

Prescott, G.W. 1954. *How to Know the Freshwater Algae.* Dubuque, IA: William C. Brown and Company.

Round, F.E. 1965. *The Biology of the Algae.* London: Edwin Arnold, Ltd.

Smith, G.M. 1950. *Freshwater Algae of the United States.* New York: McGraw-Hill Book Co.

Tseng, C.K. 1944. Agar: A valuable seaweed product. *Scientific Monthly* 58:24–32.

# 16

# Other Algae

I f using a system of classification that was acceptable a number of years ago, we might say that we have already considered all the algal groups: the Cyanophyta, the Chlorophyta, the Phaeophyta, and the Rhodophyta. But the classification system employed in this and other recent texts includes additional algal phyla: Xanthophyta, Euglenophyta, Chrysophyta, and Pyrrophyta. These are sometimes not considered in a beginning botany course, but if there is enough time, there is still enchantment to be found in these several groups.

❧ Notes ❧

## Xanthophyta: The Yellow-green Algae

The pigments in Xanthophyta are chlorophyll a, possibly chlorophyll e (although there is some uncertainty related to a suspicion that its presence may be connected to limitations in extraction methods), and an abundance of carotenoid pigments. Motile cells have two unequal flagella: a tinsel-type flagellum that extends anteriorly and a whiplash flagellum that trails posteriorly. There are approximately 400 species.

The life cycle of *Vaucheria* is detailed as an example of this phyla. *Vaucheria* is a coenocytic form. In asexual reproduction, a large multinucleated zoospore is produced in a terminal sporangium. In forming the sporangium, a cross wall develops at the base of the sporangium, thereby separating the sporangium from the rest of the cytoplasm. When the zoospore is ready to be released, a hole develops at the apical end of the sporangium. The zoospore then emerges and can grow into a new filament.

In sexual reproduction, both antheridia and oogonia are formed. In some forms, this happens on different filaments, in other forms, on the same

## Notes

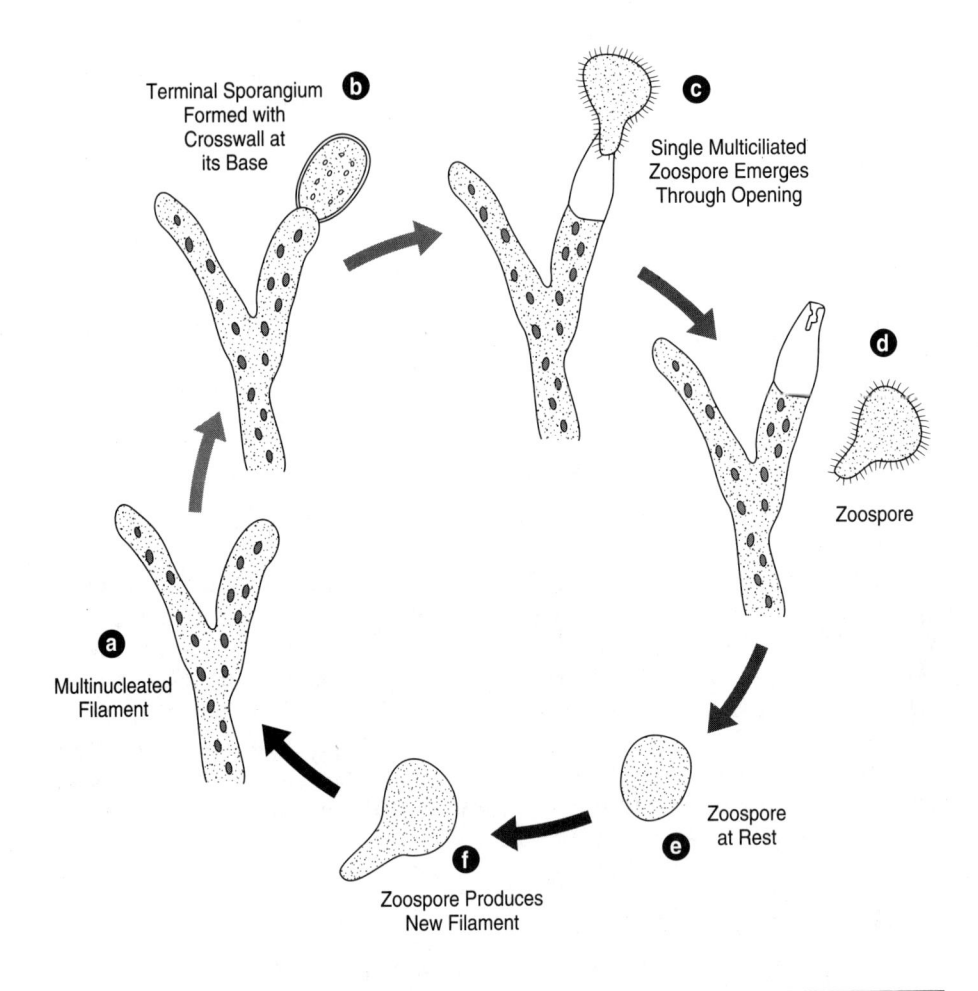

**Figure 16-1** Asexual reproduction in *Vaucheria*. (a) The multinucleated filament. (b) A terminal sporangium forms and a cross wall develops at the sporangium's base. (c) A single, multiciliated zoospore emerges through an opening. (d) Zoospore at rest, (e), and producing a new filament, (f).

filament. *Vaucheria* is distinctly oogamous. Figure 16-2 shows an oogonium arising from the filament. The oogonium is at first multinucleated, but in maturity (after a wall forms to separate it from the remainder of the cytoplasm), it contains a single, uninucleated egg. Nearby is an antheridium, in which a large number of biflagellated sperm cells form. When the sperm cells emerge, one of them may enter a pore in the oogonium and fuse with the egg cell. It is interesting to note that in this coenocytic filament where many nuclei reside, some of the nuclei travel up into oogonia and, in time, develop

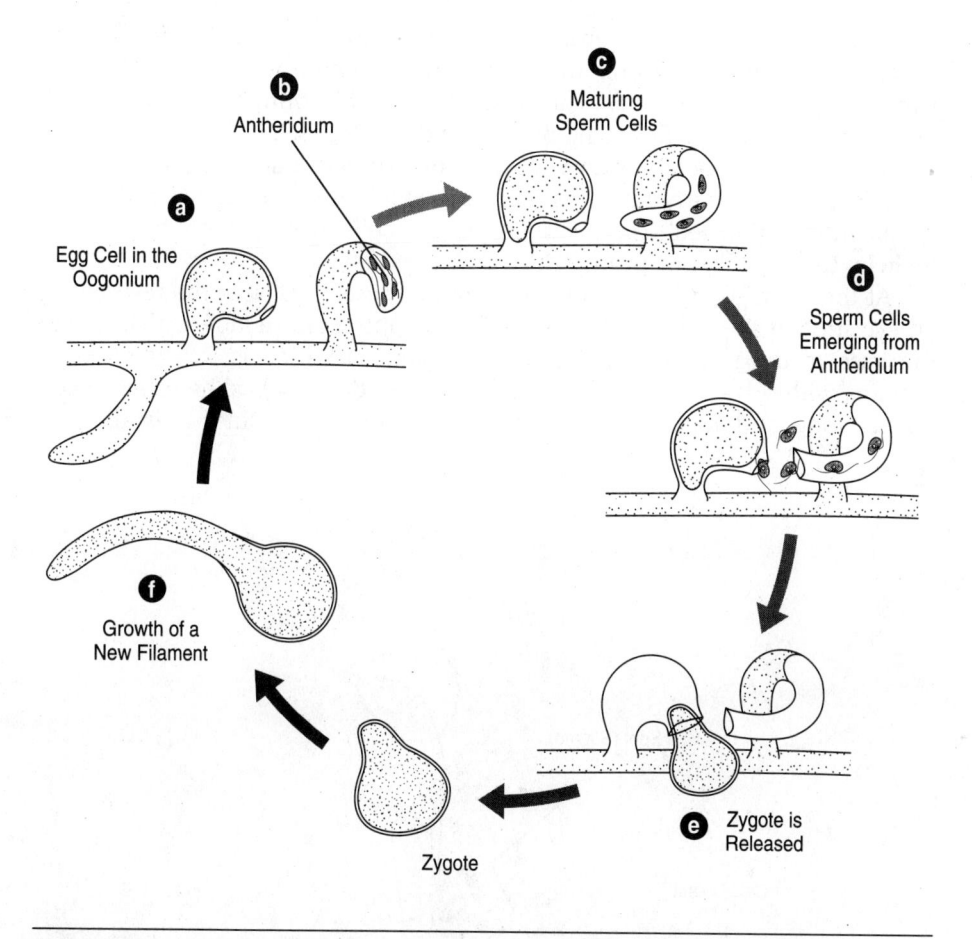

**Figure 16-2** Sexual reproduction in *Vaucheria*. (a) An egg cell in the oogonium; (b) antheridium; (c) maturing sperm cells; (d) sperm cells emerging from the antheridium; (e) and (f) the zygote and growth of a new filament.

into female gametes, while some others from very nearby migrate into the antheridium, where they adopt the role of sperm cells.

## Euglenophyta

Students of botany as well as those of zoology often are required to study the organisms of Euglenophyta because these organisms possess both animal and plant characteristics. These organisms are currently (if temporarily) considered part of the Protista kingdom. Yet, they are still called algae. The number of Euglenophyta species is estimated to range from 450 to 800.

**Notes**

A classic example of the group is *Euglena*, a single-celled, flagellated, generally green (although several forms are colorless, and one form is red) organism lacking a rigid cell wall. The pigments are chlorophyll a, chlorophyll b (as in green algae and flowering plants), and carotenes. They store not starch but a starchlike product, **paramylum**, and oil droplets. Rather than a cell wall, there is a plasma membrane, just beneath which are fine, parallel strips that spiral around the cell. These structures are collectively referred to as the **pellicle**. Cellulose is not produced.

At the anterior end is a gullet leading to a reservoir. Near the reservoir is an eye spot; at the base of the reservoir is a contractile vacuole. Arising from the base of the reservoir are two flagella, one of which is so short that it remains within the reservoir (and long escaped detection). The other flagellum is long and has numerous tiny hairs along one side. It is used in

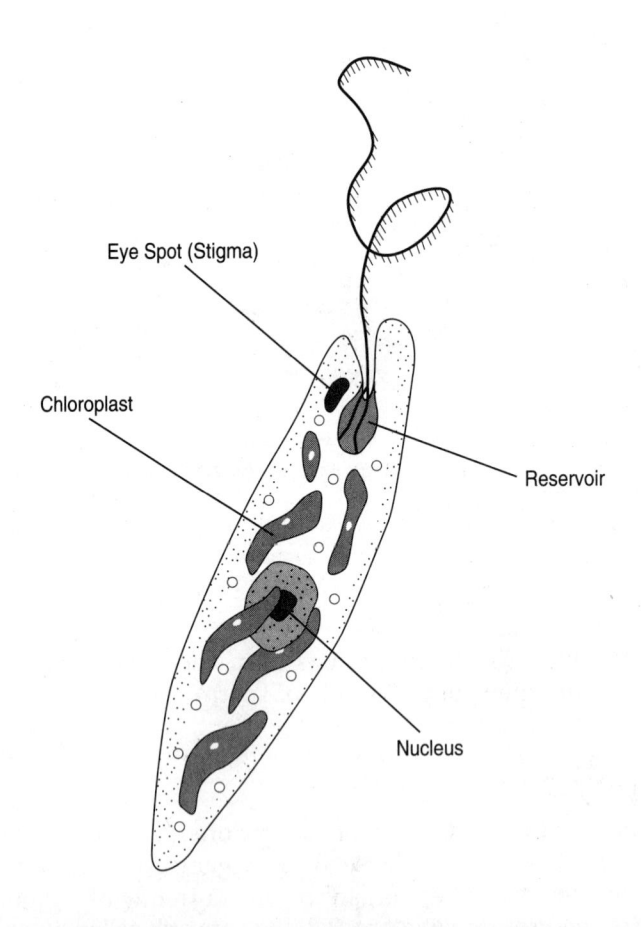

**Figure 16-3** *Euglena*: Eye spot (stigma), reservoir, nucleus, and chloroplast.

propulsion. When *Euglena* divides, the process is by longitudinal fission; no sexual process occurs.

 *Euglena* possesses several chloroplasts and a centrally placed nucleus. These organisms are plantlike in that they carry on photosynthesis. They are also, however, able to thrive in the dark, if suitable nutrients are provided in the water. A good recipe for a medium in which to grow *Euglena* in the laboratory calls for boiling an extract of horse manure. In certain conditions, *Euglena* is observed to divide more rapidly than do its chloroplasts and, thus, it becomes colorless. Such colorless forms are able to survive and reproduce if provided with suitable nutrients. Dr. Knut Norstog[1] notes that photosynthetic forms of

## ❋ Notes ❋

---

[1] Norstog, Knut, and Robert Long. 1976. *Plant Biology*. Philadelphia: W.B. Saunders Co.

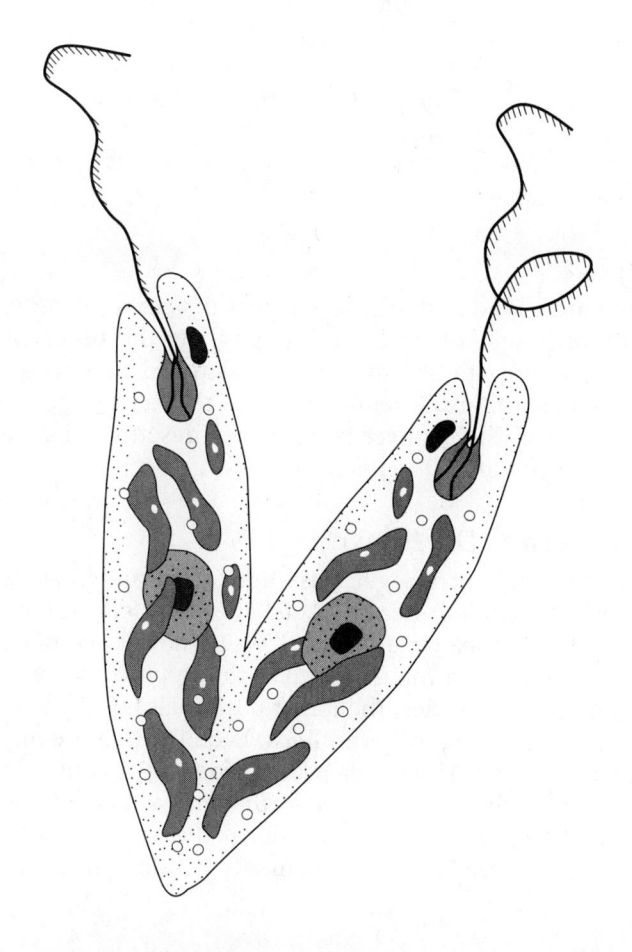

**Figure 16-4**   *Euglena* divides by longitudinal fission.

### Notes

*Euglena* can be converted to colorless forms by treatment with streptomycin, ultraviolet light, or heat. Dr. Arthur Cronquist[2] asserts that *Euglena* manufactures the essential amino acid lysine using the same pathway as is noted in fungi.

## Chrysophyta

The Chrysophyta are characterized as unicellular and have the pigments chlorophyll a, sometimes chlorophyll c, and perhaps chlorophyll e, carotenes, diadinoxanthin, and fucoxanthin. The products of photosynthesis are oils and the polysaccharide **leucosin**. Cell walls, if present, may have silica, calcium, and some organic material, but little or no cellulose.

A student of algae soon comes to appreciate the uncertainties researchers face in attempting to classify these organisms. Some algae are relegated to the kingdom Monera, some to the kingdom Protista, and most to the kingdom Plantae. When it comes to Chrysophyta, the different points of view regarding classification are particularly striking. One point of view purports there to be three classes of Chrysophyta, another, two. In this text, two classes are considered: Chrysophyceae (the yellow-brown or golden algae) and Bacillariophyceae (diatoms).

### *Chrysophyceae*

There are both amoeboid and flagellated types of Chrysophyceae. The flagella are two in number and unequal; one is pinnate, the other whiplash. The amoeboid forms are quite like the amoebas studied in zoology except that they have chloroplasts. Sexual reproduction is rare and isogamous. Approximately 300 species of Chrysophyceae have been described. Figure 16-5 shows two forms: *Dinobryon* and *Distephanus*.

### *Bacillariophyceae (Diatoms)*

Bacillariophyceae is a much larger class than Chrysophyceae, with 5,500 to (if fossil forms are included) nearly 10,000 species. Modern forms are found in both fresh water and sea water. The cell wall is composed of two parts; one part overlays the other in a manner similar to the parts of a pill box or petri dish. These are called **frustules**, the larger outer half being the **epitheca**, and the smaller inner half being the **hypotheca**. Each frustule is composed of two parts: a valve and a band. The bands overlap, forming a girdle. The walls have a pectic substance richly impregnated with silica, which creates handsome patterns of minute perforations and striations. The silica account for as much as 90 percent of the cell wall. The pigments in the cell are chlorophyll a,

---

[2]Cronquist, Arthur. 1961. *Introductory Botany*. New York: Harper and Bros.

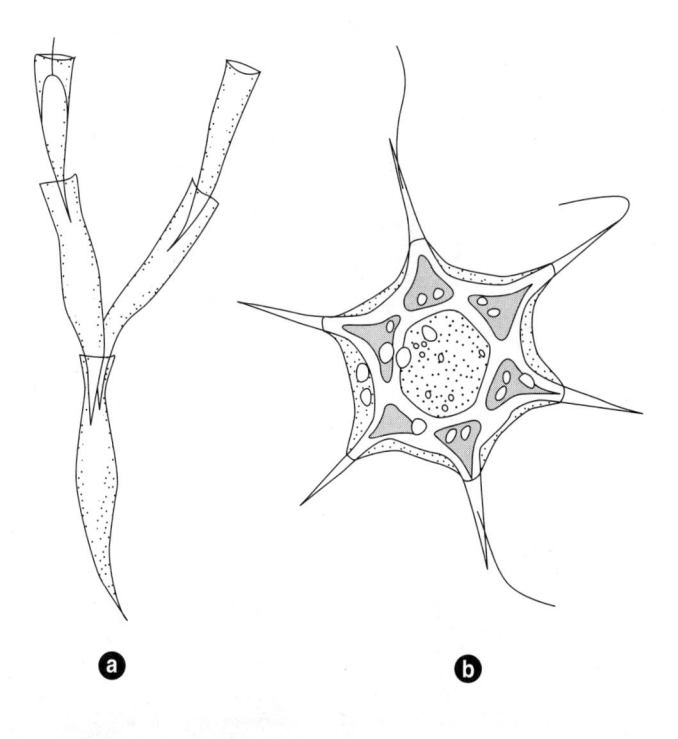

**Figure 16-5** (a) *Dinobryon* and (b) *Distephanus*

chlorophyll c, carotenoids, and xanthophylls. Fucoxanthin is the most abundant pigment.

The class is divided into two orders: the Centrales, which are radially symmetrical, and the Pennales, which are bilaterally symmetrical. The Pennales have no flagella, but they can glide along forwards and backwards by a halting kind of movement (which has yet to be explained). A fine groove called the **raphe** runs along the length of the frustule. A spherical nodule is in the center and a polar nodule is at either end of the raphe. The raphe and nodules may somehow contribute to the ability to move. Each cell has a single nucleus at the periphery of the cytoplasm and numbers of band-shaped or star-shaped chloroplasts. The products of photosynthesis are oil and a complex carbohydrate called **chrysolaminarin**.

When the cells divide, the frustules separate. Each frustule of the parent cell then functions as the epitheca of the new cell, a new hypotheca being formed within it. One can readily surmise that if a hypotheca functions as an epitheca in each succeeding generation, a gradual diminishing of cell size will result. This does, in fact, happen until a certain minimal size occurs; then, a sexual process of reproduction takes place, thereby restoring the original cell size.

162 ◆ Chapter 16

❋ Notes ❋

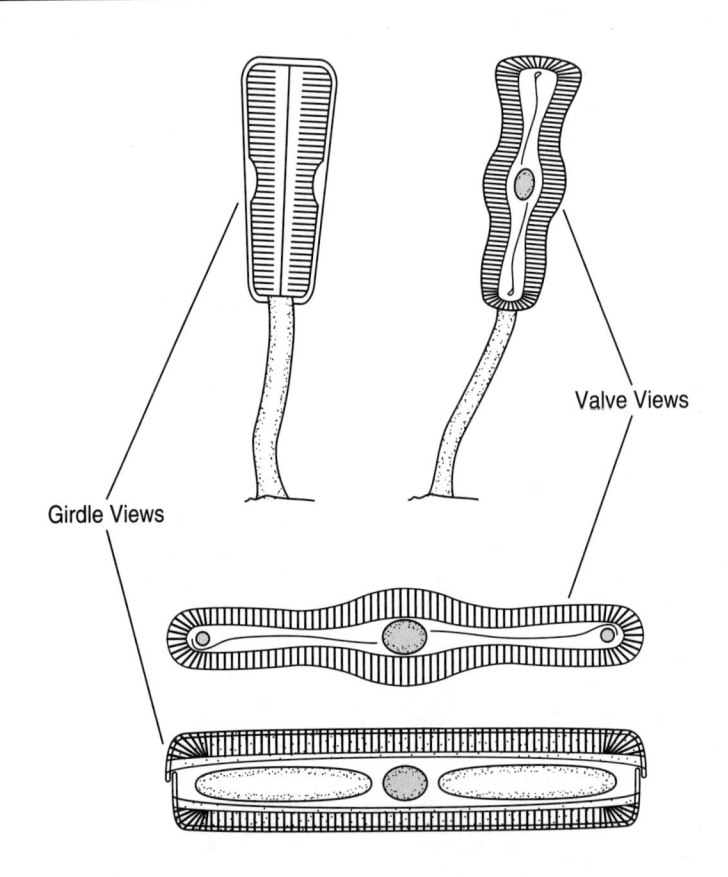

**Figure 16-6** *Pinnularia*, a diatom

In sexual reproduction, pennate diatoms produce amoeboid male gametes, which move through a mucilaginous matrix. Both male and female gametes move toward each other. Diatoms are diploid (the first algae known to be so), and the gametes are the only haploid cells. This is similar to animal reproduction and uncommon in algae.

In the Centrales order of diatoms, the male gametes are flagellated (having a single pinnate flagellum) and are thus able to swim to the egg cells.

In ancient times, diatom skeletons settled to the bottom of the sea in enormous numbers, creating deposits many meters thick. In later geological times, these deposits lifted out of the ocean and formed mountainous deposits of diatomaceous earth. The city of Richmond, Virginia is built on a layer of diatoms fifty feet deep. Perhaps the most famous deposit of diatoms is found in Lompoc, California, where they lay 300 meters deep. Given that

*Other Algae* ♦ 163

❊ **Notes** ❊

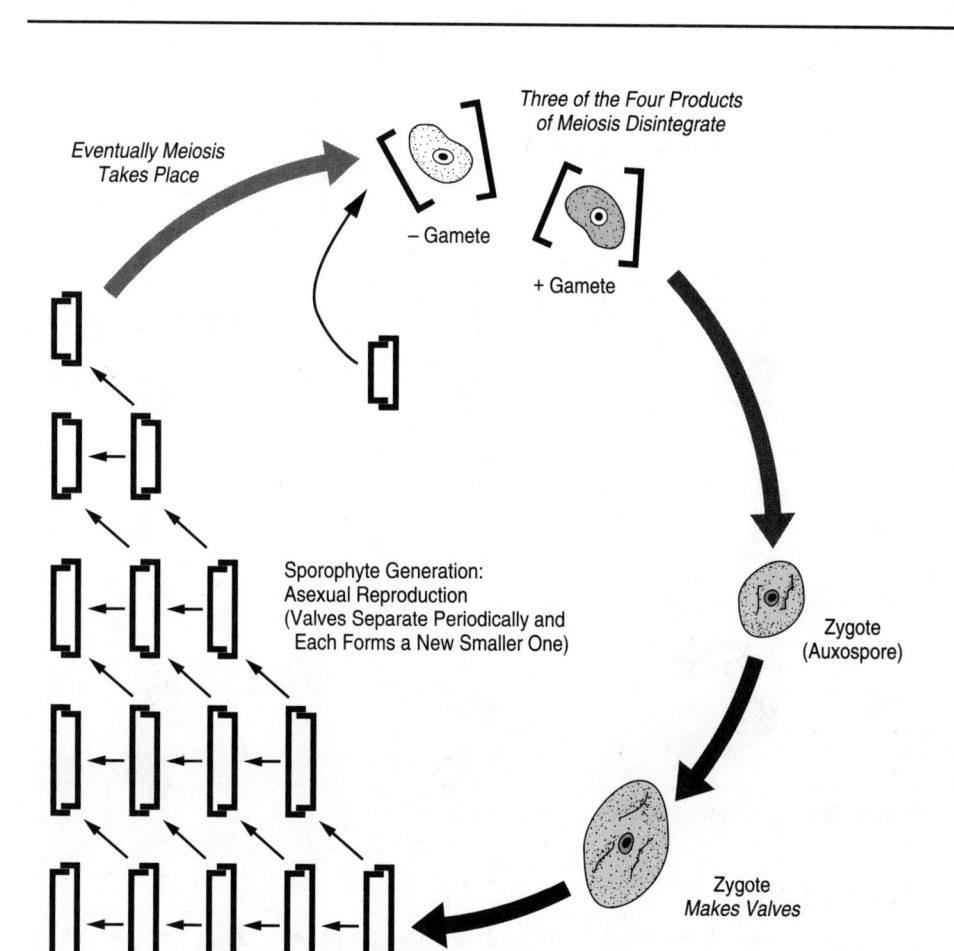

**Figure 16-7** Diatom reproduction. Each daughter cell receives one of the valves of the parent cell, the older valve always forming the epitheca (top lid) of the new individual. When the epitheca forms the top lid, the new form remains the same size. When the hypotheca forms the top lid, size diminishes. A continued reduction in size brings about sexual reproduction, which returns the cells to original size.

there are 40 million diatoms per cubic inch, this area certainly boasts an impressive number of diatoms!

Diatomaceous earth is used commercially in scouring powder, toothpaste, silver polish, sugar refining, filtering systems, and the manufacture of reflective paint. Because certain pollutants greatly increase their numbers while others have the reverse effect, populations of living diatoms are sometimes analyzed to obtain clues regarding ecological conditions.

### ❧ Notes ❧

**Figure 16-8** Each species of diatom has a characteristic pattern.

## Pyrrophyta

Two classes are recognized in the phylum Pyrrophyta: Dinophyceae (dinoflagellates) and Cryptophyceae (cryptomonads). Because all are unicellular, they are classified in the kingdom Protista. There are approximately one thousand mostly marine, mostly motile species. A few are colonial or filamentous. Chlorophylls a and c are present, as are carotenoids and xanthophylls. Peridinin, a reddish-brown pigment, is largely responsible for the color of these organisms. The product of photosynthesis is starch. The method of cell division is unusual. The chromosomes inside the nuclear membrane remain visible at all times, not seeming to lose their organization, as happens in normal mitosis. Further, the nuclear membrane remains intact during cell division. Some organisms have cellulose walls; those that don't instead have pellicles. Motile forms have two flagella. A longer flagellum lies in a groove called the **sulcus**, and a shorter flagellum lies perpendicular to the first in a groove called the **girdle**. The positioning of the flagella causes the cell to swim in a rotating motion.

### *Dinophyceae: The Dinoflagellates*

The prefix *dino-* here refers to the whirling motion of the cells as they swim along. The rate at which they can move is thousands of times faster than is noted in diatoms. The chief method of reproduction is longitudinal cell division, whereby one cell gets one of the flagella and a portion of the body covering, and then, in growing, supplies the missing parts. These species are similar to *Euglena* in the sense that they can absorb nutrients from the water and, while photosynthetic, can thrive in darkness. It has been suggested that they can also ingest small, solid, food particles. The dinoflagellates may be studied in zoology as well as in botany.

The **dinoflagellates** have two distinctive traits: they are luminescent, and they are poisonous. Light production is related to the pigment **luciferin** being acted on by the enzyme **luciferase**, a conversion that produces light without heat. This light reaction seems to be stimulated by disturbance such as when a boat goes through the water; at night, light is seen at the bow. *Noctiluca* and *Gonyaulax* are examples of bioluminescent dinoflagellates. Dinoflagellates produce severe toxins that can kill fish and sometimes people. These toxins act on the nervous system, causing paralysis. Interestingly, mussels that consume such dinoflagellates seem not to suffer any damaging effects. Given that the toxins of such dinoflagellates concentrate in the bodies of mussels, however, great harm would befall any person who consumed them.

❋ **Notes** ❋

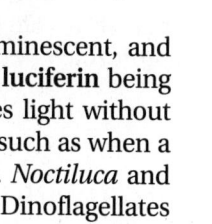

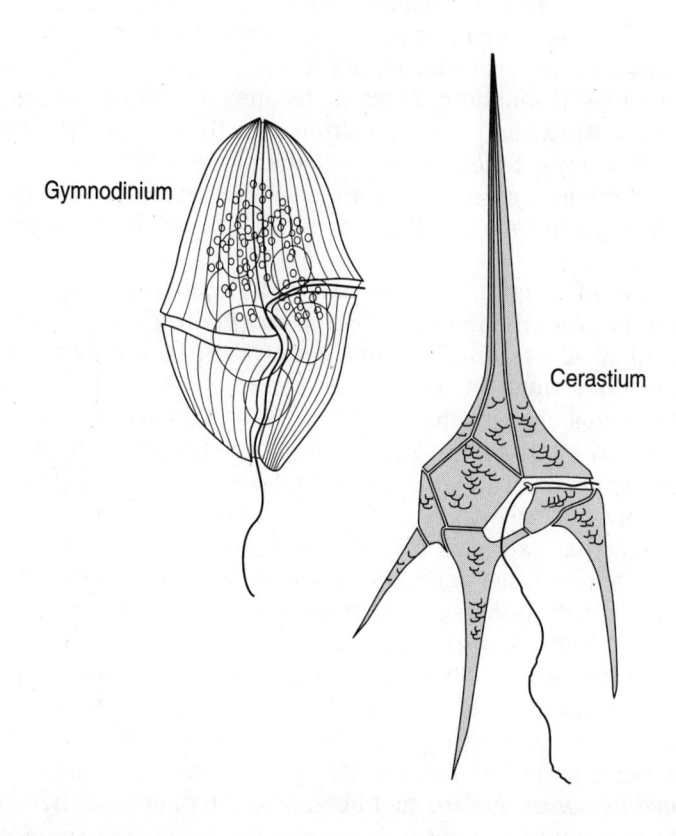

**Figure 16-9** Dinoflagellates.

✺ Notes ✺

### Cryptophyceae: The Cryptomonads

There are approximately a hundred species of Cryptophyceae. In addition to the already noted characteristics of the phylum, members of this class have bilaterally compressed cell walls. While photosynthetic, some seem also to be heterotrophic (able to take in nutrients through the cell membrane). They seem to require an external source of vitamins (again suggesting a resemblance to *Euglena*). They are considered by some to be closely related to protozoa.

## *Acetabularia*: A Green Alga

While working with the green marine alga *Acetabularia* in the early 1930s, Joachim Hammerling observed that the cell continued to produce a particular protein for up to two months after the nucleus had been removed. This seemed to indicate that the mechanism of control of protein synthesis resided in the cytoplasm. Hammerling wanted to determine the extent to which the nucleus controlled cellular functions.

*Acetabularia*, romantically called "the mermaid's wineglass," can reach a height of two to ten centimeters; yet it is a single-celled form possessing only one nucleus throughout most of its life. As shown in figure 16-10, it has a foot, a stalk, and a cap. During growth, the nucleus enlarges to approximately twenty times its original diameter, making it a convenient candidate for nuclear studies. Different species exhibit different kinds of caps. *Acetabularia mediterranea* has a round cap, while *A. crenulata* has a jagged cap.

Using microdissection techniques in a first series of experiments, Hammerling removed the nucleus of an *A. mediterranea* and replaced it with the nucleus of an *A. crenulata*. He then cut off the cap and waited for the cell to regenerate a new one. The regenerated cap took the same form as the original cap. When this cap was then also removed and a second-generation cap regrew, it took an intermediate form between the original cap and the cap of *A. crenulata*. When this cap was in turn removed and a third-generation cap regrew, the new cap was entirely of the *crenulata* form—in accord with the kind of nucleus that had been placed in the cell.

Although it took three generations of regenerating the caps of *Acetabularia*, we now clearly understand that the nucleus governs what goes on in the cytoplasm. In the first-generation regrowth, the cytoplasm already possessed programming from the original *mediterranea* nucleus. In the second-generation regrowth, residual chemical information from the original nucleus mixed with new programming chemicals from the replacement (*crenulata*) nucleus. In the third-generation regrowth, the influence of the original (*mediterranea*) nucleus had been exhausted, and the new nucleus of *A. crenulata* was entirely in control.

Other Algae ♦ 167

❋ **Notes** ❋

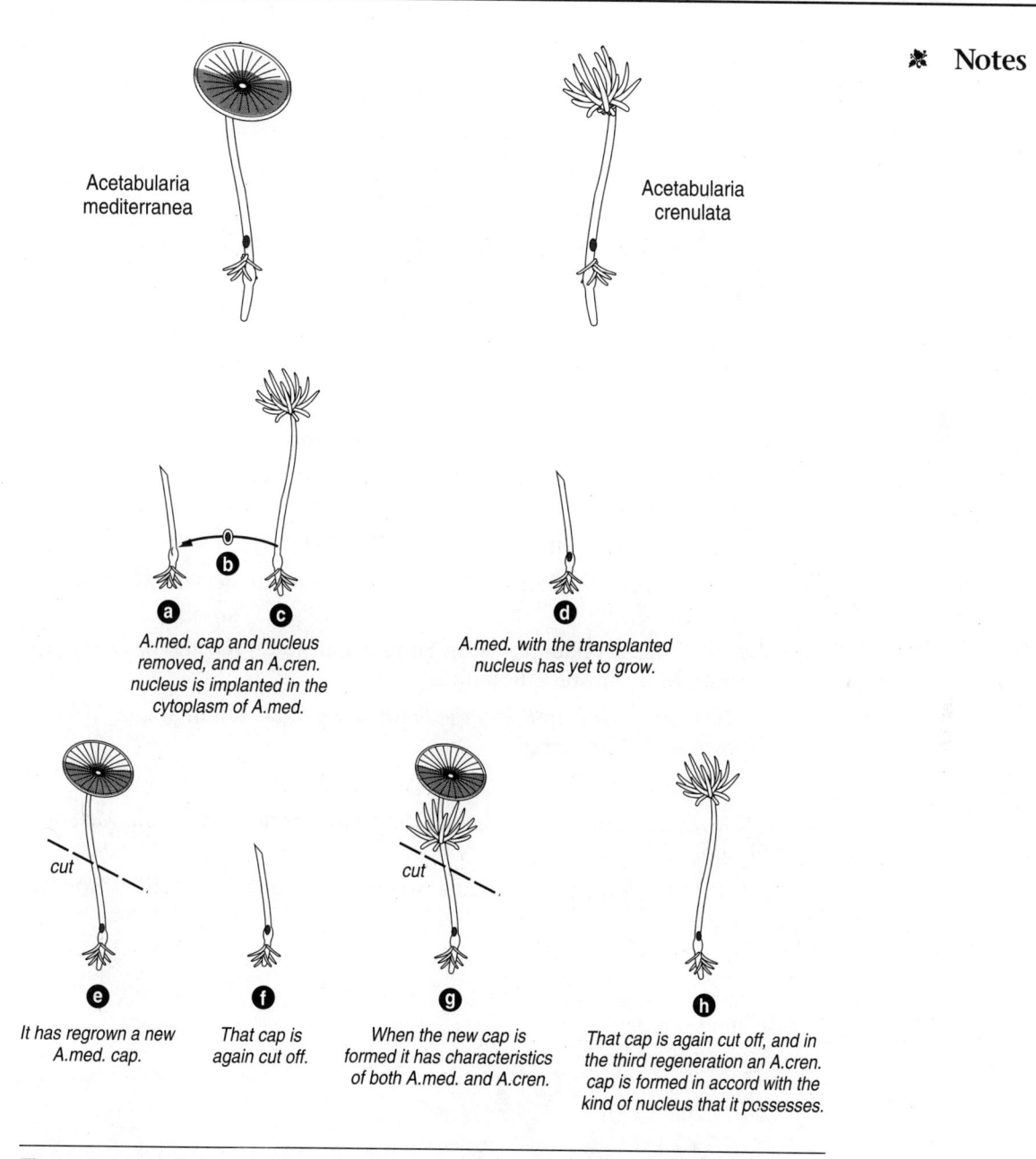

**Figure 16-10** At top left, *Acetabularia mediterranea*; at the top right, *A. crenulata*. (a) The cap of *A. mediterranea* is removed, as is its nucleus. (b) The nucleus of *A. crenulata* is implanted in the cytoplasm of *A. mediterranea*. (c) *A. crenulata*, the source of the implanted nucleus. (d) *A. mediterranea* with the transplanted nucleus has yet to grow. (e) *A. mediterranea* regrows an *A. mediterranea* cap, which is again cut off, (f). (g) When the new cap forms, it has characteristics of both *A. mediterranea* and *A. crenulata*. (h) When this cap is removed, the third regeneration is an *A. crenulata* cap, in accord with the transplanted nucleus.

### ❋ Notes ❋

## Questions for Review

1. What organisms are useful in making scouring powder?
2. How are the outer walls of diatoms constructed?
3. Define the terms *coenocytic, zoospore, mitospore, stigma,* and *epitheca.*
4. Some algae are able to produce light. This is called _____.
5. Describe an experiment performed with *Acetabularia* and seeming to confirm that the nucleus is a center of control.

## Suggestions for Further Reading

Blume, J.L. 1956. The ecology of river algae. *Botanical Review* 22:291–341.

Chapman, V.J. 1957. Marine algal ecology. *Botanical Review* 23:320–350.

Chase, F.M. 1941. Useful algae. *Smithsonian Institution Annual Report* 401–452.

Dawson, E.Y. 1961. The rim of the reef. *Natural History* 70:8–16.

Lewin, J.E., and R.L. Guillard. 1963. Diatoms. *Annual Review Microbiology* 17:377–414.

Papenfuss, G.F. 1951. Phaeophyta. In Smith, G.M. (ed.) *Manual of Phycology.* Waltham, MA: Chronica Botanica.

Prescott, G.W. 1954. *How to Know the Freshwater Algae.* Dubuque, IA: William C. Brown and Company.

Round, F.E. 1965. *The Biology of the Algae.* London: Edwin Arnold, Ltd.

Smith, G.M. 1950. *Freshwater Algae of the United States.* New York: McGraw-Hill Book Co.

Tseng, C.K. 1944. Agar: A valuable seaweed product. *Scientific Monthly* 58:24–32.

# 17

# Fungi

Fungi are found nearly everywhere: in water; on decaying wood; on fish; in the mouths of animals; as parasites in plants and animals; and as saprophytes on all sorts of substances. There are great numbers of species, many of which have complex life cycles.

## Fungi Classification

How should the fungi (**fungus**) be characterized? While it is a concern in classification to put together those organisms that are related to one another by descent from a common ancestor, this is certainly not the case with fungi. While these organisms do share certain traits, this does not imply kinship.

All fungi lack chlorophyll. They can be **unicellular**, multicellular, or acellular. The cell walls often have chitin similar to the chitin found in insects and crustacea. Cellulose may also be present.

In the five kingdom system, the fungi are awarded the status of kingdom. It is not a perfect arrangement and it does not suggest that the things called *fungi* are related to each other by common origin; rather, it is a way to keep track of these organisms. Some researchers subdivide Fungi into five classes, some six, and some seven: Chytridiomycetes, Oomycetes, Zygomycetes, Ascomycetes, Basidiomycetes, and for want of a better place to classify them, the Fungi Imperfecti and perhaps the Lichens (which, though certainly not fungi, do keep the company of fungi). Sometimes, the first three of these—Chytridiomycetes, Oomycetes, and Zygomycetes—are considered one class: the Phycomycetes. This scheme, which designates these three as subclasses, is the one adopted in this text.

Fungi are classified according to the kind of fruiting body produced in the sexual phase of reproduction. Phycomycetes produce zygospores; Ascomycetes produce **asci**, and Basidiomycetes produce **basidia**. Both asexual and sexual methods of reproduction take place, although different terminology is used to

❋ Notes ❋

describe these functions in fungi. The asexual method is called the **imperfect** stage, and the sexual method is called the **perfect** stage. Zygospores, asci, and basidia are produced during the perfect stages of the reproductive cycle.

For purposes of illustration in this discussion, two groups of fungi will be considered: Myxomycetes (the slime molds) and Eumycetes (the true fungi). (This ignores the cellular **slime molds**, which include Acrasinomycetes and Labyrinthulomycetes.)

## Myxomycetes

The organisms of Myxomycetes present a challenge in classification because during one part of the life cycle they exhibit animal-like characteristics, and during another part, they exhibit plantlike traits. A large, naked, noncellular mass of protoplasm possibly several inches in diameter and called a **plasmodium** can be found growing on damp forest debris. It may be variously colored yellowish, brown, green, and orange. It is capable of amoeboid movement and can glide slowly to get away from light. While it will shun light at one stage in its life cycle, it will move into the light at another stage. When a plasmodium reaches a certain maturity, sporangia grow upward from it, and it thus takes on a plantlike role.

*Stemonitis* is an example. The vegetative phase of this slime mold can be found on rotting logs in the woods. It can be recognized by the creamy drops of liquid that it exudes to its surface. A large mass of protoplasm is found just under the drops. It is coenocytic and capable of amoeboid movement. In this way, it can avoid light by "flowing" to the underside of the log. This is the plasmodium of *Stemonitis*. The nuclei are diploid. Upon maturity, a number of fruiting bodies develop. Nuclei from the plasmodium migrate up into the fruiting bodies, which will become sporangia. Here, the nuclei undergo meiosis; the resulting spores are thus haploid. A wall called the **peridium** forms on the surface of the sporangium. Within the body of the sporangium, where the spores reside, is a mass of filamentous threads called the **capillitium** (see figure 17-1). The peridium disintegrates upon drying and releases the spores, which transform into amoeboid **swarm spores**. The swarm spores soon undergo another change, developing cilia. In this form, they are able to go through a number of divisions. Then, after a short time, they lose their cilia and become swarm spores again. The swarm spores come together in twos, producing diploid zygotes, which, in turn, coalesce to produce a new coenocytic plasmodium. The coalescence of the swarm spores is due to the influence of the hormone **acrasin**. (Interestingly, an acrasinlike substance that accomplishes the same thing can be extracted from the urine of pregnant women.)

Microscopic examination of the plasmodium reveals a rhythmic flow of the cytoplasm; the cytoplasm flows for a moment, comes to a brief rest, and flows again in the opposite direction. While the cytoplasm flows forwards and backwards, the entire plasmodium can move continuously in one direction at a rate of nearly one inch per hour. It feeds on bits of organic material and bacteria along the way.

*Fungi* ♦ 171

❊ **Notes** ❊

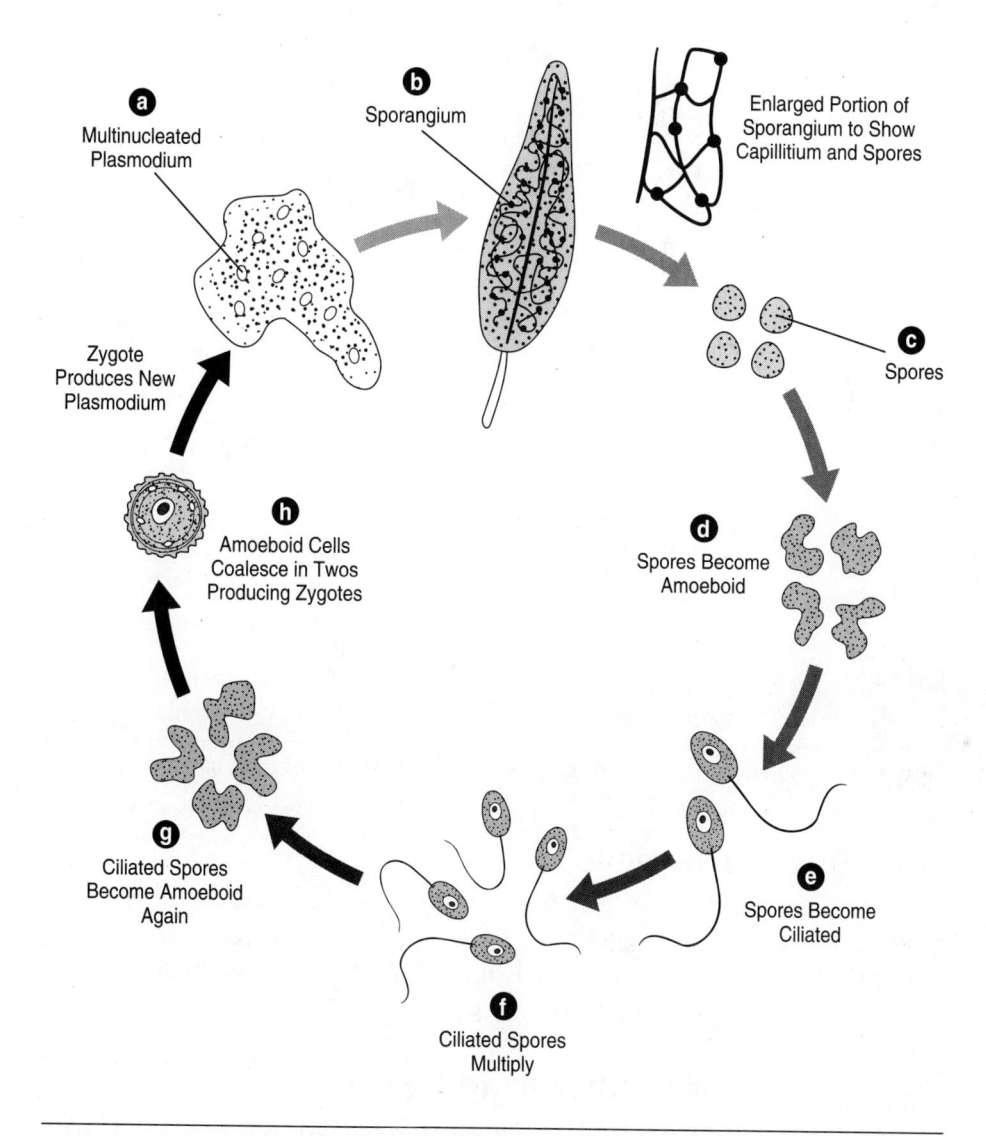

**Figure 17-1** The life cycle of *Stemonitis*. (a) The multinucleated plasmodium and (b) the sporangium. At the right of the sporangium, a portion of sporangium much enlarged to show capillitium and some spores. (c) Spores, which become amoeboid, (d), and then ciliated, (e). The spores then multiply, (f), and again become amoeboid, (g). The amoeboid cells coalesce in twos, producing zygotes, (h), which, in turn, produce a new plasmodium.

*Plasmodiophora brassicae* is a slime mold that infects cabbage plants. Common names for the disease are **finger and toe disease** and **club root**. It is reported that one cubic centimeter of soil can contain one hundred million spores of this slime mold. The disease is controlled by liming the soil.

❋ **Notes** ❋

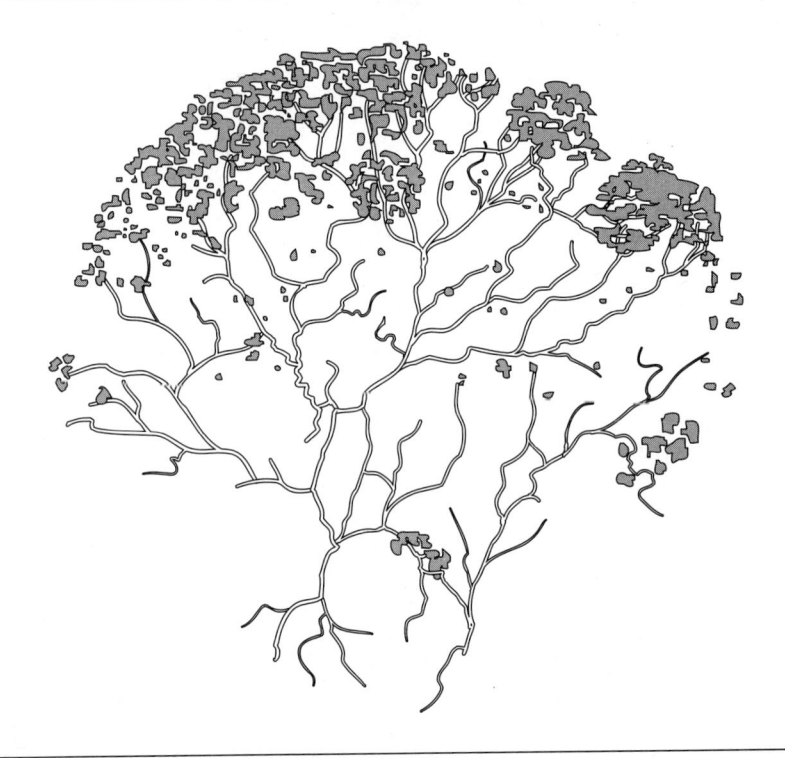

Figure 17-2  An enlargement of the plasmodium showing protoplasmic streaming.

## Questions for Review

1. At what stage in the life cycle is a myxomycete coenocytic?
2. What influences the coalescence of swarm spores in *Stemonitis*? From what source other than the fungus may this substance be obtained?
3. Name a myxomycete that causes disease in plants.

## Suggestions for Further Reading

Alexopoulos, Constantine. 1962. *Introductory Mycology*. New York: John Wiley and Sons, Inc.

Christensen, Clyde M. 1965. *The Molds and Man.* Minneapolis: University of Minnesota Press.

Cooke, R. 1977. *The Biology of Symbiotic Fungi.* New York: John Wiley and Sons, Inc.

Hawker, L.E. 1967. *Fungi.* New York: Hillary House Publishers.

Rose, H. 1960. Yeasts. *Scientific American* 202(2):136–146.

Smith, A.H. 1963. *The Mushroom Hunter's Field Guide.* Ann Arbor: University of Michigan Press.

# 18

# Phycomycetes

As indicated in chapter 17, the Phycomycetes class of fungi has three subclasses: Chytridiomycetes, Zygomycetes, and Oomycetes. All are coenocytic and nonseptate, and many are tubular. The only cross walls are those formed at the bases of sporangia and gametangia. In the perfect stage, a zygospore or an oospore is produced. The name *Phycomycetes* means "alga-like fungi," although this does not mean that these fungi derive from algal ancestors. Rather, many mycologists consider these fungi to have evolved from protozoa.

❈ Notes ❈

## Chytridiomycetes

If dead leaves or onion bulb scales are immersed in a little water[1] containing a bit of soil, an abundance of chytrids will likely appear in a short time. Their spores seem to be everywhere. They are the simplest of the fungi, reproducing primarily by zoospores. Some species exhibit alternation of generations. The simple, globe-shaped *Chytridium sphaerocarpum* (figure 18-1), which extends rhizoids into the substrate and produces unicellular sporangia and zoospores, and *Allomyces arbuscula* (figure 18-2), which shows alternation of generations, are two examples of chytrids. Chytrids consist of only a spherical cell and a few rhizoids that penetrate the host tissue.

---

[1] Chlorinated tap water excluded.

## ✱ Notes ✱

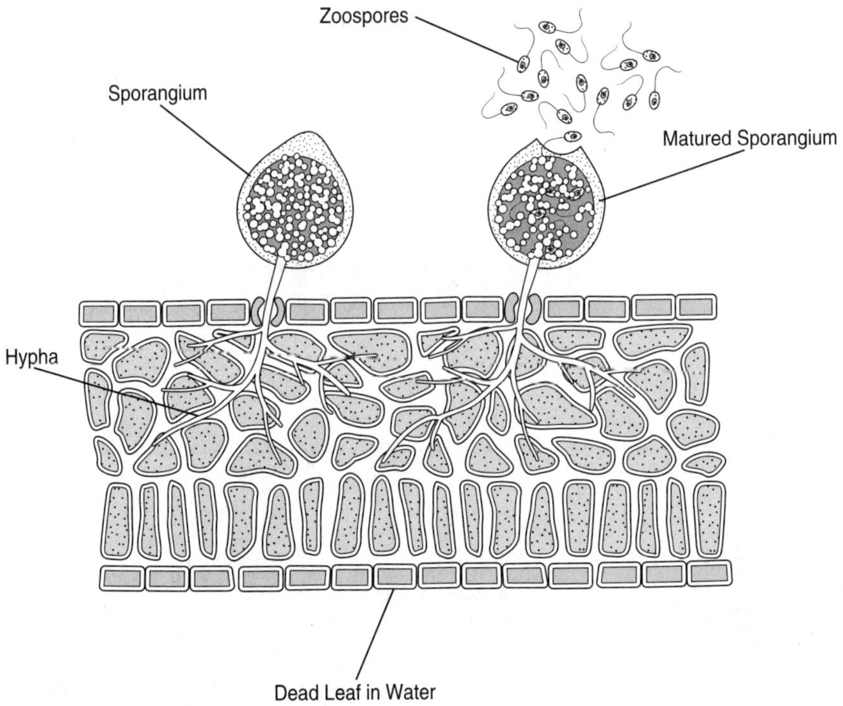


**Figure 18-1** *Chytridium sphaerocarpum* growing on a dead leaf in water.

Figure 18-2 shows two types of sporangia on the *Allomyces* sporophyte plant: thick-walled sporangia and thin-walled sporangia. In the thick-walled sporangia, meiosis occurs and the resulting spores, called **meiospores**, are haploid. The meiospores grow into gametophytes. In the thin-walled sporangia, mitosis occurs, and the resulting spores, called **mitospores**, are diploid. Mitospores do not produce gametophytes but, rather, grow into sporophytes. If a culture of *Allomyces* were to be dried out, the thin-walled sporangia would not survive and the thick-walled sporangia would. Such treatment would therefore ensure alternation of generations. The rendering of the gametophyte in figure 18-2 shows two kinds of gametangia: the antheridia, situated below, and the oogonia, located above. Whereas the antheridia (male gametangia) are orange in color, the female gametangia are colorless. When the gametes unite, a biflagellated zygote forms, which, in turn, grows to a sporophyte thallus. Sperm cells are attracted to the egg by a hormonal substance having the romantic name **sirenin**.

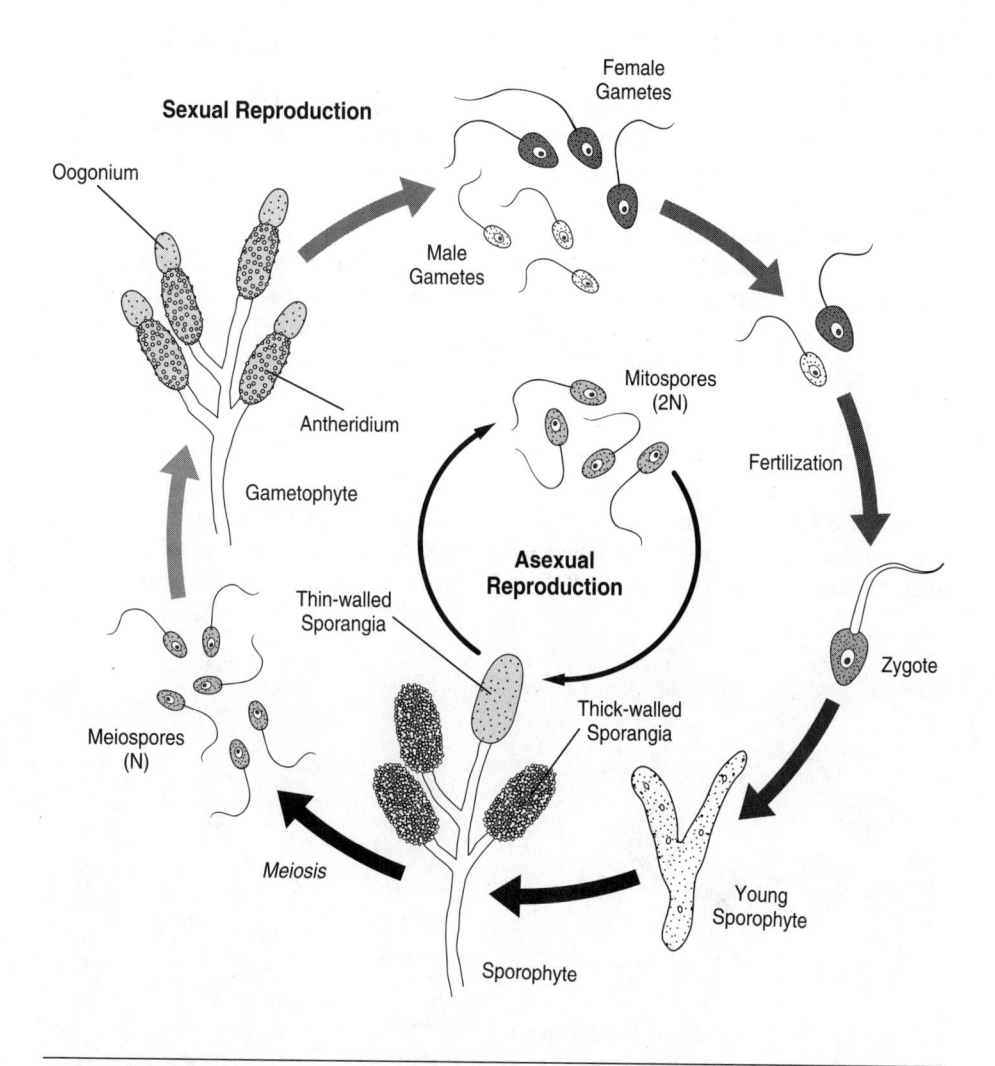

**Figure 18-2** *Allomyces arbuscula*. Two kinds of sporangia are produced: thin-walled sporangia, which produce mitospores, and thick-walled sporangia, which produce meiospores. The mitospores produce sporophytes, and the meiospores produce gametophytes. The antheridia are orange in color. A fertilized egg grows into a sporophyte.

# Zygomycetes

There are no flagellated cells in Zygomycetes. Rather, there is an extensive coenocytic mycelium within the substrate as well as an aerial mycelium that produces sporangia. The walls are chitinous. Black bread mold (*Rhizopus nigricans*) is the most common form of Zygomycetes and demonstrates both

### Notes

the vegetative production of mitospores and a sexual process (called conjugation) whereby zygospores are produced.

If a piece of bread is placed in a moist container, a cottony, white mycelium will soon grow throughout the substance. Rootlike **haustoria** penetrate the substrate while the upright **hyphae** send **sporangiophores** upward. When sporangia are produced, the entire mass becomes black. As the sporangia enlarge, cross walls form, thus separating the sporangium from the rest of the mycelium. Such a cross wall is called a **columella**. The contents of the sporangium form a large number of uninucleate mitospores. The rupture of the sporangium wall releases the spores. The columella then expands, forming a new sporangium wall. Additional nuclei then move into the new sporangium, and a new columella is made. In this manner, a series of sporangia form and rupture at the end of a single sporangiophore.

Sexual reproduction requires that mating types of *Rhizopus* grow near each other. The fungus is heterothallic. When a hypha of a plus (+) strain grows close to a hypha of a minus (−) strain, projections grow out from the hyphae and come in contact with each other. The nuclei of the plus and minus strains fuse to produce a diploid zygospore, which develops a

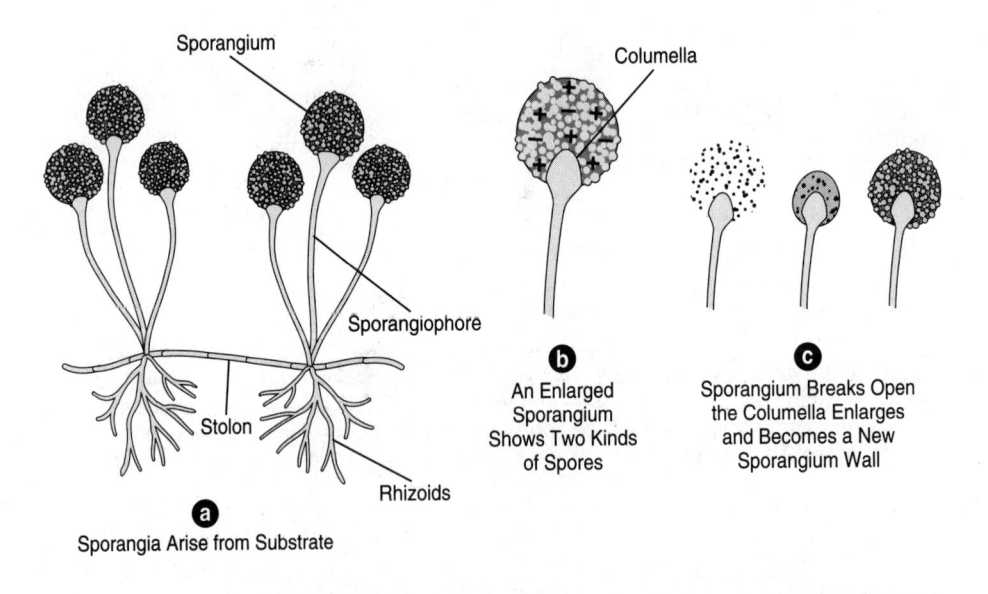

**Figure 18-3** Asexual reproduction in *Rhizopus nigricans*. (a) Sporangia arise from the substrate. (b) An enlarged sporangium showing two kinds of spores: plus and minus. (c) When a sporangium breaks open, the columella enlarges and becomes a new sporangium wall. A new columella is then produced.

**a** + and − Spores are Planted on Nutrient Media

**b** Mycelium Expands, + and − Filaments Become Closely Associated

**c** Fusion Produces a Zygospore

**d** Zygospore Germinates to Form a New Hypha

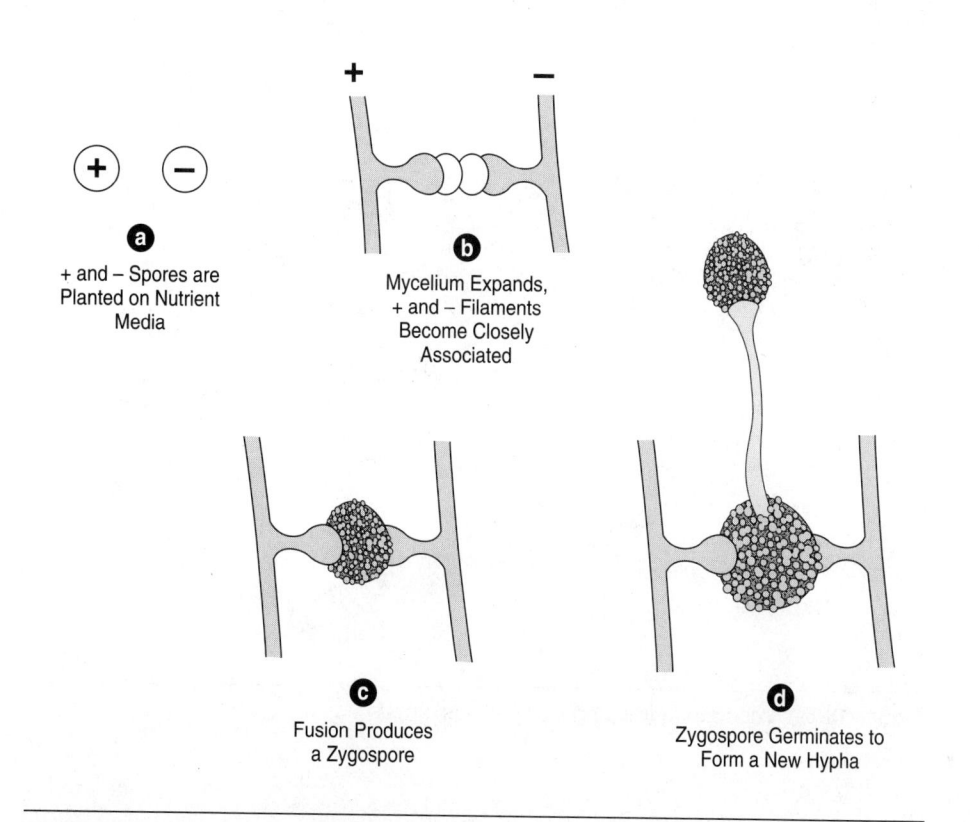

**Figure 18-4** Sexual reproduction in *Rhizopus nigricans*. (a) Plus and minus spores are planted on a nutrient medium. (b) The mycelium expands, and plus and minus filaments become closely associated. (c) Fusion produces a zygospore, which can germinate to produce a new hypha, (d). The first divisions of the zygote nucleus are meiotic.

hardened outer covering and is able to remain dormant for some time. Upon germination, meiosis takes place, and the haploid condition is restored.

Some members of Zygomycetes grow as parasites on protozoa and other small creatures. The fungus develops hyphae in the body of the host, where it can absorb nutrients. There is also a soil-dwelling form that captures microscopically sized roundworms in hyphal loops. If a nematode were to swim through a loop, the loop would suddenly clamp down on the worm, hold it, and produce haustoria (which would, in turn, grow through the worm's body). When such fungi are grown in the laboratory, they do not produce loops unless nematodes are placed in the growth medium.

### ❋ Notes ❋

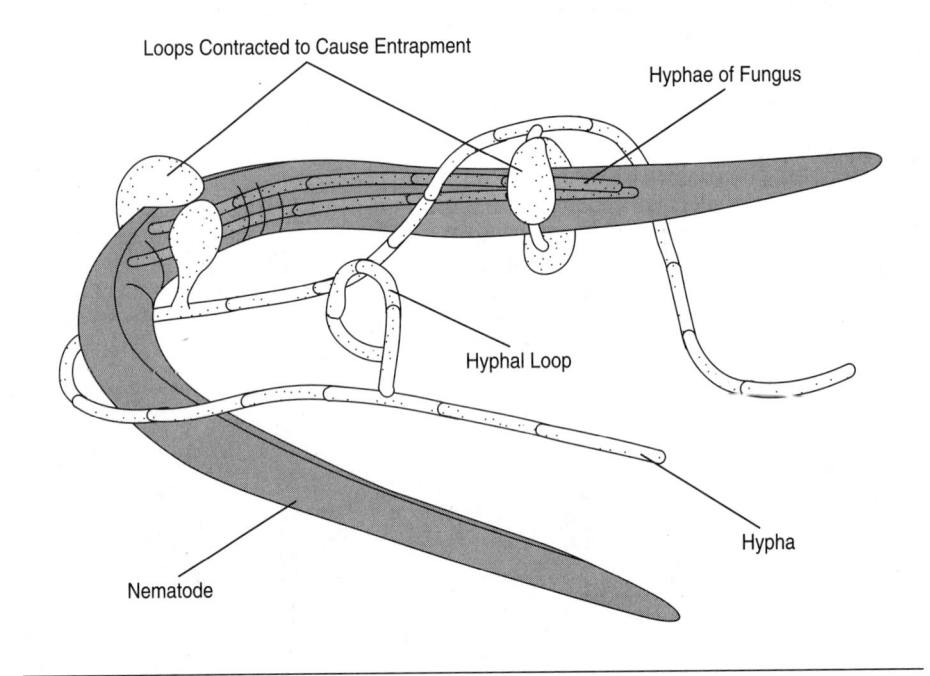

**Figure 18-5** A nematode-trapping, carnivorous fungus.

## Oomycetes

Whereas the Zygomycetes are mostly terrestrial, the Oomycetes are mostly aquatic and are thus often called "water molds." The term is somewhat misleading, however, because not only do numerous other fungi also grow in water, but Oomycetes can frequently be found growing in soil. Sexual reproduction is heterogamous. The walls are mainly of cellulose. Flagellated mitospores and meiospores are produced.

*Saprolegnia* is representative of this subclass. It can be found growing on the bodies of fish, which suggests that it is a parasitic form. While this may be the case, it is also possible that the fungus grows only on dead tissues or mucus. If this is so, *Saprolegnia* is not truly parasitic. If fresh pond water is brought into the laboratory, strained, put in a finger bowl, and baited with a dead fly, *Saprolegnia* will grow on the dead fly. (Other fungi can also grow on dead flies.) *Saprolegnia* produces a branched, multinucleated mycelium. Elongated mitosporangia may be produced at the ends of branches. A **septum** separates the sporangium from the rest of the mycelium, and biflagellated zoospores are produced. These zoospores are described as pear shaped (**pyriform**), with two anteriorly placed flagella. The zoospores escape through a pore at the end of the sporangium, become encysted, and, after a time, germinate to produce zoospores again. These secondary zoospores are

### ※ Notes ※

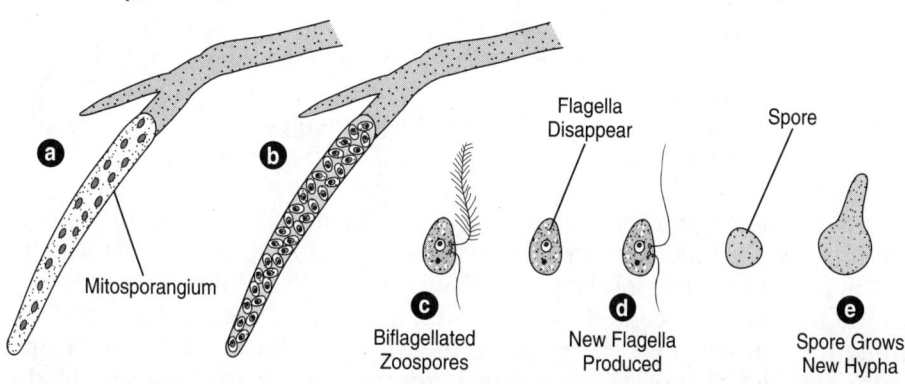

also biflagellated, but with the flagella laterally placed, and are described as kidney shaped (**reniform**). They can each grow a new mycelium after making contact with a suitable substrate, such as the body of a fish.

In sexual reproduction in *Saprolegnia*, a globose oogonium containing several eggs forms at the end of a short branch. Antheridia are produced on

**Asexual Reproduction**

**Sexual Reproduction**

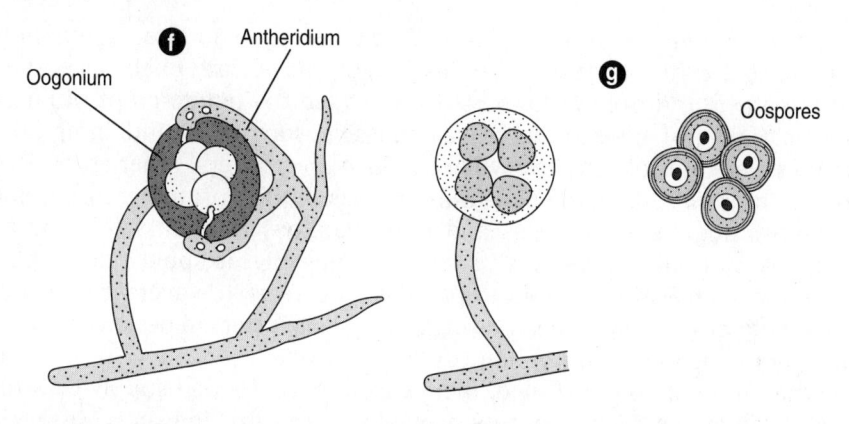

**Figure 18-6** *Saprolegnia*. The top row shows asexual reproduction. (a) A mitosporangium, which breaks up into a number of biflagellated zoospores, (b), each having a tinsel-type flagellum and a whiplash flagellum, (c). (d) The flagella disappear and new flagella are produced. (e)These flagella are lost when the spore grows new hyphae. The bottom row shows sexual reproduction. (f) An oogonium, consisting of a single cell containing several eggs. An antheridial hypha grows upwardly nearby, producing an antheridium that comes in contact with the oogonium. (g) The wall is broken down, and fertilization produces oospores.

## ❧ Notes ❧

nearby hyphae. The antheridia grow to make contact with and penetrate the oogonial wall. Male nuclei are discharged into the oogonium and fuse with the eggs to produce zygotes. The zygotes develop resistant walls and remain dormant for some time. When they germinate, they produce biflagellated zoospores, which can then produce new mycelia. This form of sexual reproduction bears certain resemblances to that of *Vaucheria*, leading some investigators to suggest that they are related forms.

*Albugo candida* is another Oomycete. This fungus attacks members of the mustard family of plants and is known as a "white rust." Both asexual and sexual reproduction occur. In asexual reproduction, chains of multinucleated sporangia are cut off from the end of a hypha and are dispersed in the air. When the sporangia break open, each nucleus emerges as a biciliated spore. Note that rather than dispersing spores directly, *Albugo candida* disperses sporangia. The spores are ciliated when they emerge from the sporangium, but soon lose their cilia, come to rest, and develop into new mycelia.

In the sexual phase of reproduction, oogonia and antheridia are produced, developing on separate hyphae in the intercellular spaces of the host tissue. The infestation occurs immediately under the epidermis of the stem of the host tissue. The invasion is probably accomplished through a **stomate**. In the oogonium, a central nucleus develops into an egg, and one nucleus from the antheridium unites with the egg cell. Both the oogonia and antheridia are coenocytic. A zygote is formed, and a resting stage of the zygote, called a **chlamydospore**, follows. When the chlamydospore germinates, it gives rise to numerous zoospores, each of which can develop a new mycelium.

*Phytophthora infestans* is an Oomycete that played an integral role in the tragic history of the Irish people. This fungus causes **late blight** in potatoes and tomatoes, and from 1845 to 1847 devastated the potato crops of Europe and Ireland. The consequence was massive starvation and emigration. Potatoes were the sole culture. While the Irish grew several other crops, these crops were exported. Home consumption was limited primarily to potatoes, sometimes augmented by eggs and milk. Potatoes were easy to grow and nutritious. Although there was usually a family pig, it could not be eaten because it often had to be sold to pay the rent. Thus, the average consumption of potatoes in good times was eight pounds per person per day, and children went to school with potatoes in their pockets.

When the potato crop failed, there was little other food supply on which to rely. The leaves of all the potato plants turned to slime in a single day. The potatoes, too, became slime. Nobody knew what caused the blight. While the source was known to be America, the cause was unknown. When rent was due and could not be paid, the Irish people were forced out of their mud huts and into the cold and rain. They took to eating leaves and mud and cattail sprouts. At night, numbers of people went into the cemeteries to harvest nettles, which could be eaten in their tender young stages. They ate turnip tops, sand eels, and seaweed. Sports, dancing, and music came to a halt. They bartered away their household items, their fishing equipment, their handiwork, the family pig, and even the manure pile. Finally, they lay down at the side of the road to die. More than one million people died of starvation and other diseases related to malnutrition during the Great Potato Famine.

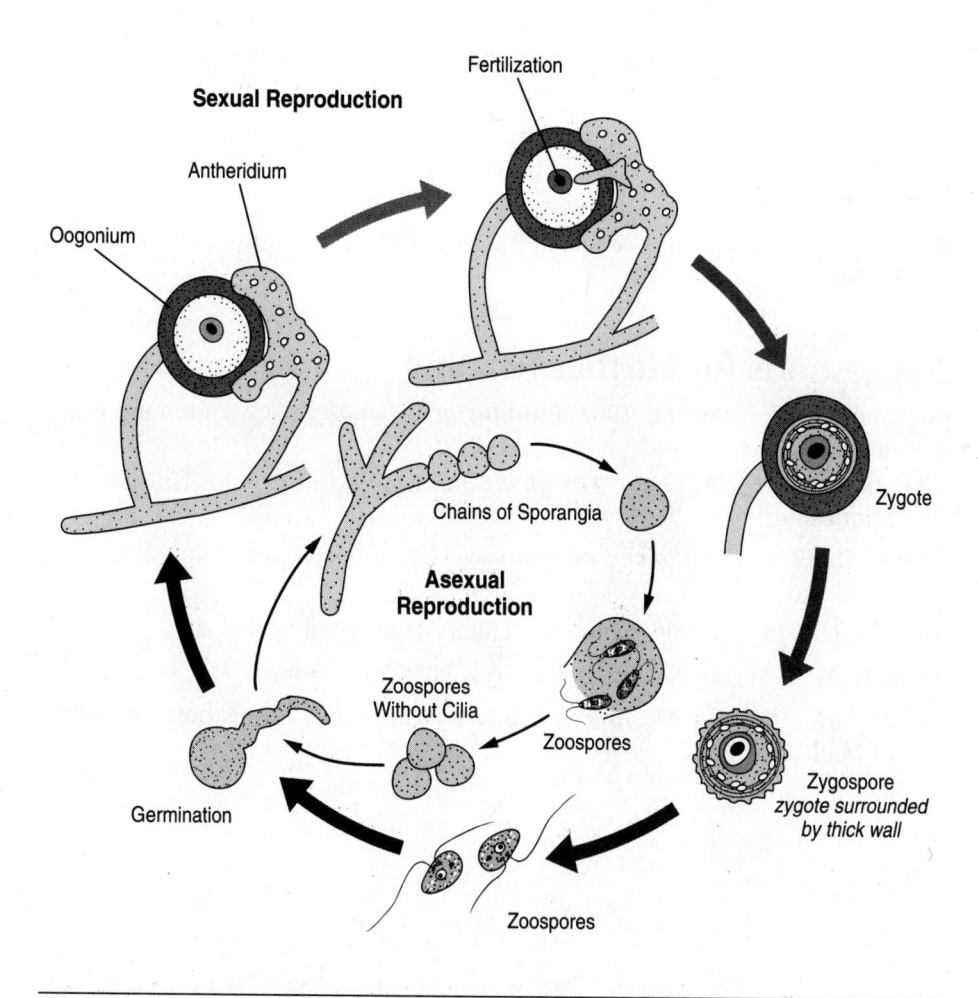

**Figure 18-7** *Albugo candida.* In asexual reproduction, chains of sporangia break away from the hyphae. When mature, the sporangium ruptures, releasing zoospores, which swim about for a time, lose their cilia and grow into a new mycelium. In sexual reproduction, fertilization occurs when the male gamete of the antheridium penetrates the oogonium to fertilize the egg cell. The zygospore thus formed produces biciliated zoospores which grow a new mycelium.

※ Notes ※

## Questions for Review

1. Define the terms *zygospore, sporangiophore,* and *columella.*

2. Define the terms *hypha, mycelium,* and *saprophyte.*

3. The Phycomycetes do not form cross walls and, therefore, are _____.

4. One subclass of the Phycomycetes is the Zygomycetes. Black bread mold is an example. Give its scientific name and describe its methods of reproduction.

5. The water mold *Saprolegnia* is representative of the subclass Oomycetes. Describe its asexual and sexual methods of reproduction.

6. What is the name of a fungus that infects mustard-family plants? What does it disperse in its asexual reproduction, and what do these break open to release?

7. Name the fungus that caused the Great Potato Famine in Ireland in the 1840s.

## Suggestions for Further Reading

Alexopoulos, Constantine. 1962. *Introductory Mycology.* New York: John Wiley and Sons, Inc.

Christensen, Clyde M. 1965. *The Molds and Man.* Minneapolis: University of Minnesota Press.

Cooke, R. 1977. *The Biology of Symbiotic Fungi.* New York: John Wiley and Sons, Inc.

Hawker, L.E. 1967. *Fungi.* New York: Hillary House Publishers.

Rose, H. 1960. Yeasts. *Scientific American* 202(2):136–146.

Smith, A.H. 1963. *The Mushroom Hunter's Field Guide.* Ann Arbor: University of Michigan Press.

# 19

# The Ascomycetes

**Notes**

The Ascomycetes subclass of fungi is very large, encompassing perhaps as many as 37,000 species and including the blue and green molds of cheeses, *Penicillium*, truffles, and **yeasts**. The most distinguishing characteristic of Ascomycetes is the formation of an elongated, spore-bearing structure called the ascus. An ascus typically contains eight **ascospores**. Ascus formation is associated with the perfect phase in reproduction. Ascomycetes fungi are septate, having cross walls in the hyphae; and the cross walls are incomplete, having perforations that allow a continuous cytoplasm from cell to cell. These perforations also allow the passage of nuclei from cell to neighboring cell, a process that has been observed to take place with great rapidity. The presence of cross walls helps

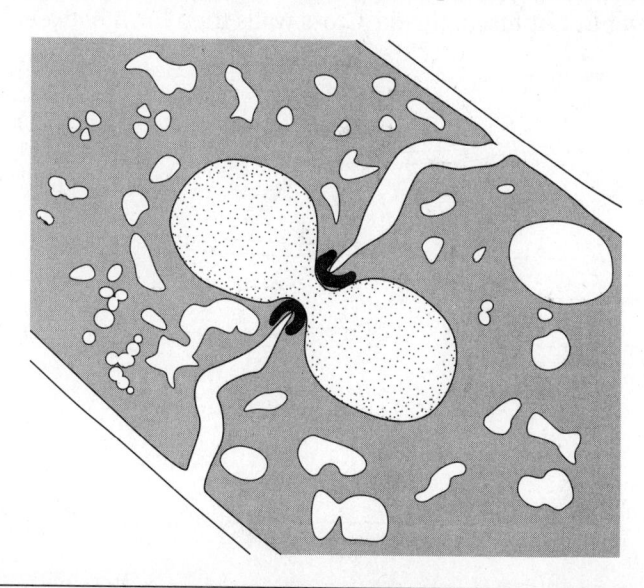

**Figure 19-1** The end wall of a cell, showing a central perforation through which a nucleus is passing. Drawn from an electron micrograph.

✽ **Notes** ✽

distinguish Ascomycetes from Phycomycetes, where cross walls are absent. Another striking feature of Ascomycetes is the growth of a long, tubular, receptive emergence on the oogonium (**ascogonium**). This tubular emergence is called the trichogyne. The fact that a trichogyne is also observed in red algae (but not elsewhere) suggests a possible kinship between these two groups.

## Reproduction in Ascomycetes

Asexual reproduction occurs by the formation of **conidiospores**. Conidiospores are produced either at the ends of hyphae by **abstriction**, or all at once by the fragmentation of the hypha.

The perfect stage in the reproductive cycle involves the formation of asci, and while the details vary in different species, *Pyronema confluens* is fairly typical of the subclass (figure 19-3). *Pyronema confluens* appears to grow only in burned-over areas. When male and female mycelia intermingle, a saucer-shaped fruiting body called an **apothecium** develops. A mat of hyphae called the **hymenium** grows in the floor of the apothecium (see figure 19-4c).

Antheridia and oogonia arise from separate hyphae in the hymenial layer. The multinucleated oogonium, here more properly called the ascogonium, produces a tubular emergence called the trichogyne. The antheridium arises nearby. The trichogyne makes contact with the antheridium. At the point of contact, the walls break down and allow the passage of the male nuclei through the tube to the ascogonium, where they pair with the female nuclei. Fusion of the nuclei, however, is delayed. At this point, hyphae, which are destined to become asci, grow upwardly from the ascogonium. A pair of nuclei migrates into the ascogenous hypha. The hypha develops a bend called a **crozier**. The nuclei, paired but not fused, divide. Cross walls then form between the daugh-

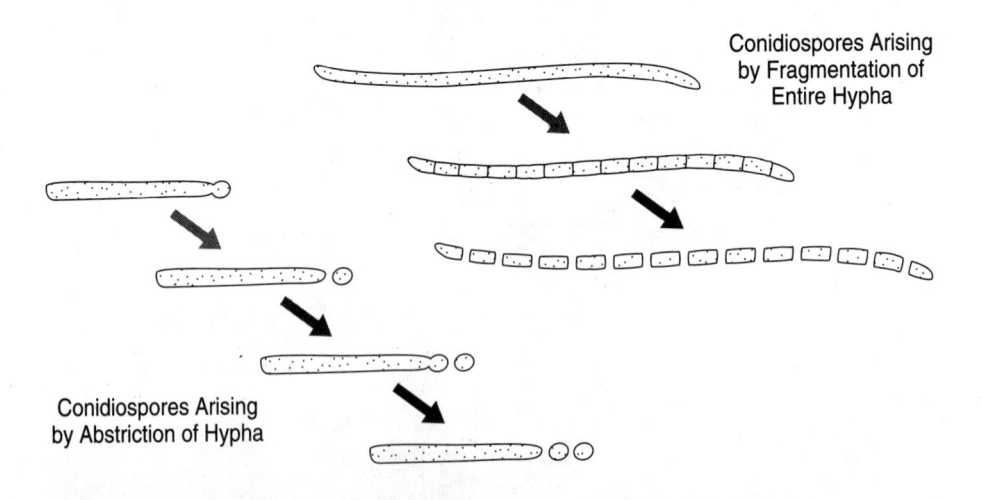

**Figure 19-2** Conidiospores may arise from the end of a hypha by either abstriction or the fragmentation of the entire hypha.

ter nuclei. The terminal cell is thus still binucleated (**dikaryotic**). These nuclei then fuse, at which time the cell is regarded as a zygote. Following nuclear fusion, the diploid zygote undergoes, first, meiosis and, then, one mitotic division, resulting in an ascus containing eight ascospores. During the same time, other hyphae in the floor of the apothecium send up sterile branches that can be seen mingled with the asci. These sterile filaments are the **paraphyses**. A small circular opening develops at the upper end of the ascus, allowing the escape of the ascospores. Were all of the ascospores to **germinate**, four would yield male mycelia and four would yield female mycelia.

❀ **Notes** ❀

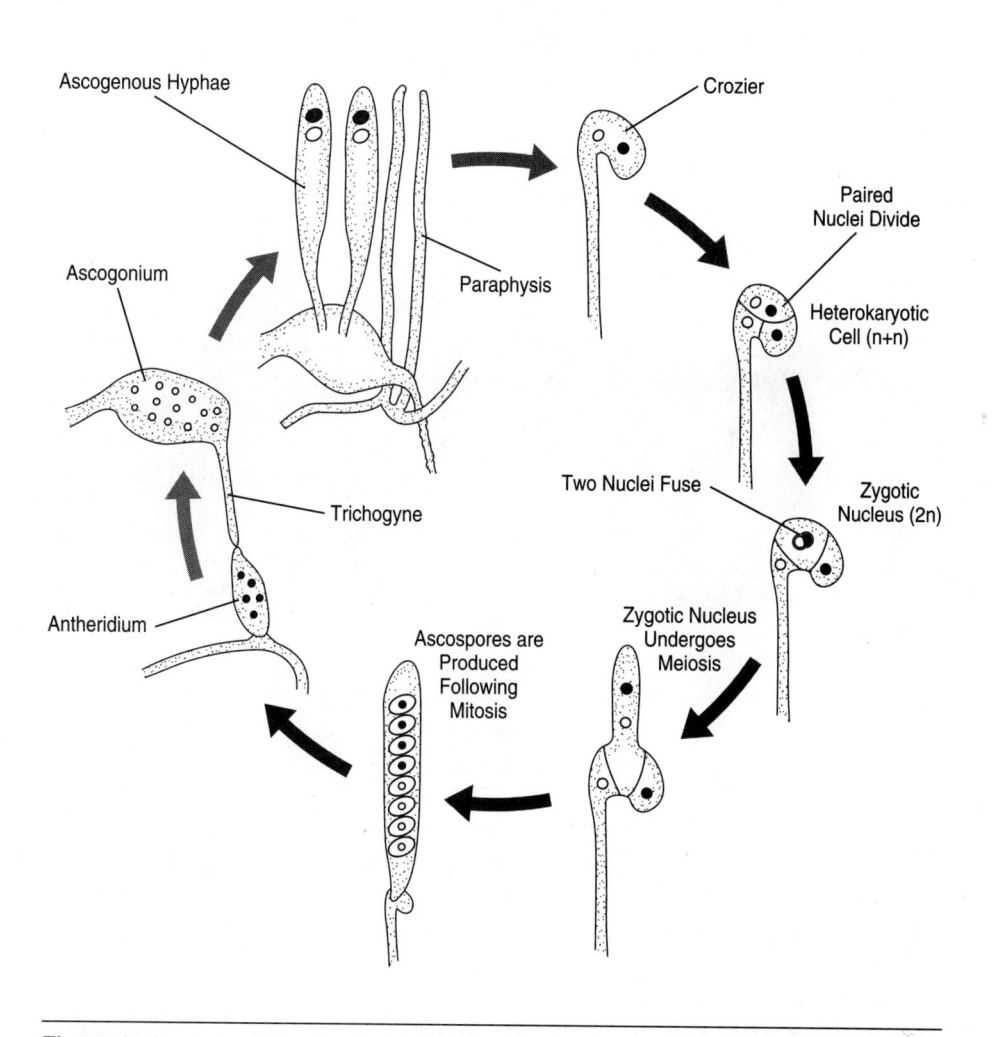

**Figure 19-3** Ascus formation in detail. At left, the trichogyne from the ascogonium makes contact with the antheridium. Ascogenous hyphae arise from the ascogonium and develop a crozier. The paired nuclei divide and cross walls form between daughter nuclei. Two nuclei in the apical cell fuse followed by meiosis and a single mitotic division to produce eight ascospores.

## Notes

## Fruiting Bodies

The fruiting body in *Pyronema* is the apothecium. Other fruiting bodies occur in other species. In some species, the fruiting body containing the asci and the paraphyses may be entirely closed. This is called a **cleistothecium** and is shown in figure 19-4a. In one genus, *Exoascus*, there is no evident fruiting

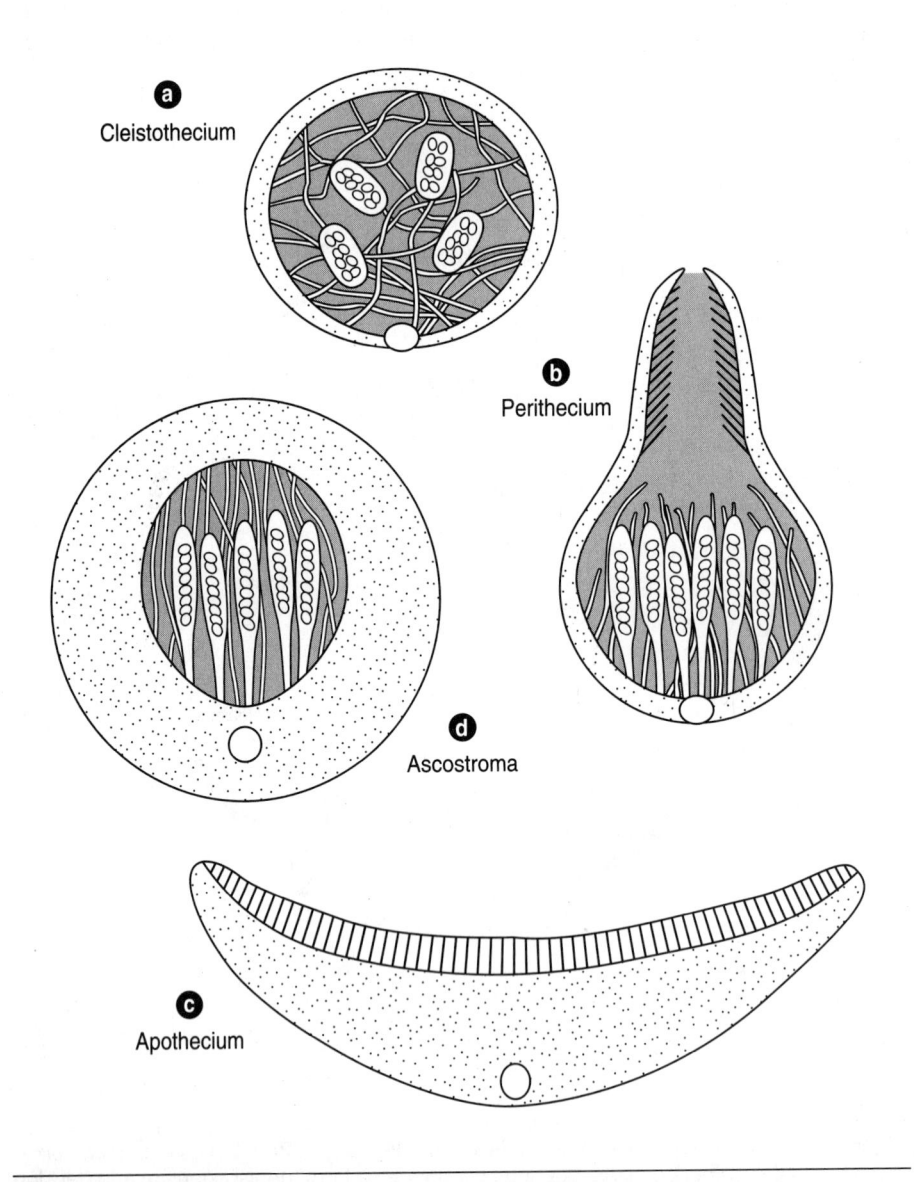

**Figure 19-4** Several ascocarps: (a) a cleistothecium; (b) a perithecium; (c) an apothecium; and (d) an ascostroma.

body. Rather, the asci and paraphyses develop on an open surface. A fruiting body that is nearly but not quite enclosed is called a **perithecium** (figure 19-4b).

## Yeasts

The classification of yeasts as Ascomycetes is surprising because the yeasts are known for reproducing by budding. The yeast cell typically develops a bud on the side, the nucleus divides, and a daughter nucleus migrates into the bud. The cells may either divide to form two separate cells or remain attached to each other. This asexual process of budding can occur when the nuclei are haploid or when they are diploid.

Under certain circumstances, ascus formation is noted in yeasts, however. Either two yeast cells come together and fuse to form a diploid nucleus (figure 19-6a), or one yeast cell buds, the nucleus divides, and the two daughter nuclei come together again to fuse (figure 19-6b). Both cases result in a diploid condition. These events can be followed by meiotic nuclear divisions within the cell, resulting in a tetrad of four haploid meiospores or ascospores (although the typical number of spores produced in ascomycetes is eight). In yeasts, ascospores are produced in a rounded ascus rather than an elongated, saclike structure. One of the traits of yeasts is that they do not produce a mycelium. This is uncommon among ascomycetes.

*Saccharomyces cerevisiae* is a yeast of great economic importance, mainly due to the kinds of enzymes it produces. Enzymes are proteins that enable and improve chemical reactions. They are highly specific; it takes a certain kind of an enzyme to act on a certain kind of substance to yield a certain kind of chemical reaction. **Zymase**, which occurs in yeast cells, converts glucose into ethyl alcohol and carbon dioxide. Invertase, another yeast enzyme, converts sucrose into fructose and glucose.

Yeasts are used in the alcoholic fermentation of sugars, breaking down the sugars into carbon dioxide and alcohol or, if oxygen is present, into

❋ **Notes** ❋

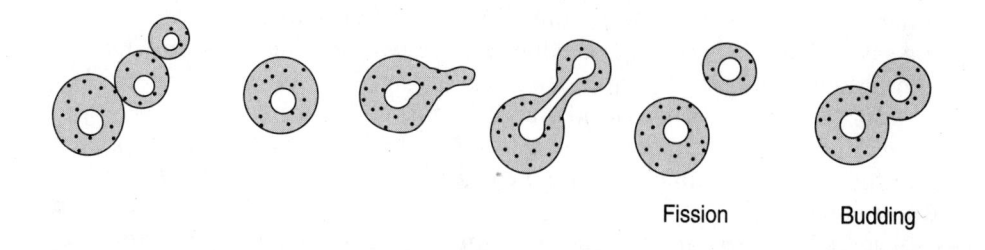

**Figure 19-5** Yeasts in the process of budding and fission.

## Notes

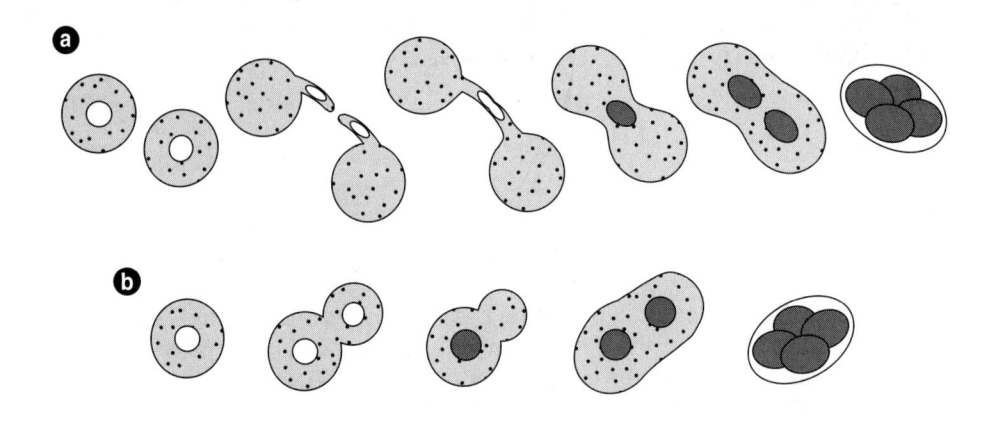

**Figure 19-6** Two methods of sexual union of nuclei in yeasts. (a) Two cells come together to form a diploid nucleus, which then produces ascospores by meiosis. (b) A nucleus divides and the two daughter nuclei then fuse again to produce a diploid condition. Ascus formation follows.

carbon dioxide and water. In going through fermentation, baker's yeast is inhibited in its activity when the alcohol content reaches 4 or 5 percent. In a sense, this is a disadvantage to the yeast because alcohol formation slows the growth of the organisms. It is easy to see that by their forming alcohol they have built the cause of their own demise. Brewer's yeast, on the other hand, is a strain of *Saccharomyces cerevisiae* in which growth is able to continue until the alcohol content reaches between 14 and 17 percent. Different strains of yeast react differently.

When a mole of sugar is decomposed into alcohol and carbon dioxide by yeast and in an anaerobic environment, 28,000 calories of energy are liberated. If the environment is instead aerobic, the end products are carbon dioxide and water and the amount of energy liberated is 674,000 calories.

Aside from *Saccharomyces cerevisciae,* the yeast of the brewing industry, other species of *Saccharomyces* are used in making other beverages including sake, ginger beer, rum, and wine. *Saccharomyces* is also rich in vitamin B and protein (composed of approximately 50 percent), making it a nourishing feed for livestock. During World War II, thousands of tons of yeast were produced in Germany as a source of protein. Suitable methods of yeast production for protein manufacture could make yeasts valuable resources for third world countries.

Not all species of yeast deserve the high accord given *Saccharomyces cerevisiae*. Some species are pathogenic, causing skin and lung diseases;

others cause **blastomycosis**, an invasion of yeastlike cells in the skin, mucous membranes, lungs, and, sometimes, internal organs.

## Pathogenic Ascomycetes

*Microsphaera alni* is an ascomycete that infects the leaves of the common lilac. It produces conidiospores at the ends of hyphae by abstriction. Ascus production occurs in a completely enclosed fruiting body called a **cleistocarp**. The formation of asci is not associated with a fusion of gametic nuclei. When an infected lilac leaf is examined, the conspicuous element is the whitish mycelium with its powdery profusion of conidiospores. It is a powdery mildew. Careful examination with a lens discloses small, black dots; these are the cleistocarps. In figure 19-7 a cleistocarp having anchoring filaments has ruptured, and asci-bearing ascospores are being extruded.

*Endothecia parasitica* is another ascomycetous plant pathogen. It has caused the near extinction of chestnut trees in the United States. The fungus was inadvertently brought to this country on stock brought from the Orient near the year 1900. From the initial point of infection in Long Island, New York, it spread throughout the country and destroyed nearly all the chestnut trees. Although a few resistant trees remain and a comeback is possible, it is not yet much evident.

*Ceratostomella ulmi* causes Dutch Elm disease, so-called because it was first described in the Netherlands. This fungus infects the American Elm tree. The infection began in this country in 1930; again, the source was imported stock. It has spread widely and caused vast destruction among American Elm trees. The destiny of the American Elm appears to be the same as that of the American Chestnut.

🌸 **Notes** 🌸

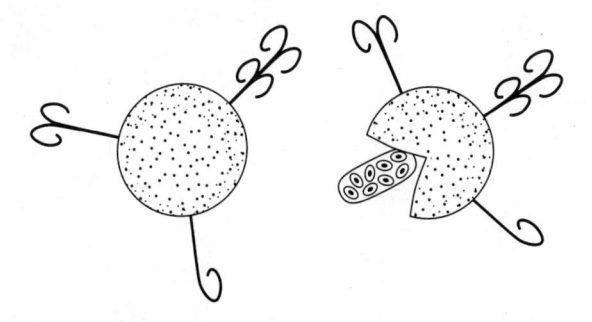

**Figure 19-7** *Microsphaera alni*: at right, the cleistocarp has ruptured, releasing an ascus having eight ascospores.

### Notes

## *Penicillium* and *Aspergillus*

*Penicillium* and *Aspergillus* are widespread blue and green molds that grow on decaying fruits, vegetables, jellies, and preserves. Because ascus formation has been observed in very few species, many mycologists classify these forms in the taxonomic catchall of Fungi Imperfecti. Some mycologists, however, relegate them to Ascomycetes. Only conidiophore formation is discussed following.

In *Aspergillus,* the apex of a sporangiophore develops a bulbous swelling, from the surface of which large numbers of short branches extend. Conidiospores are pinched off at the ends of these branches. The microscopic appearance of the conidiophores, with their chains of spores, reminded an observer of an **aspergillum** (a device resembling a perforated globe on a short handle and used in sprinkling holy water); thus the name.

One species of *Aspergillus* is used to convert sugar to citric acid, a substance used in flavoring. *Aspergillus* is also used to produce gallic acid (used in the manufacture of dyes and indelible black ink) and to make soy sauce.

In *Penicillium,* the conidiophore divides to form two or more branches, each of which terminates in a chain of conidia. The Latin word *penicillium,*

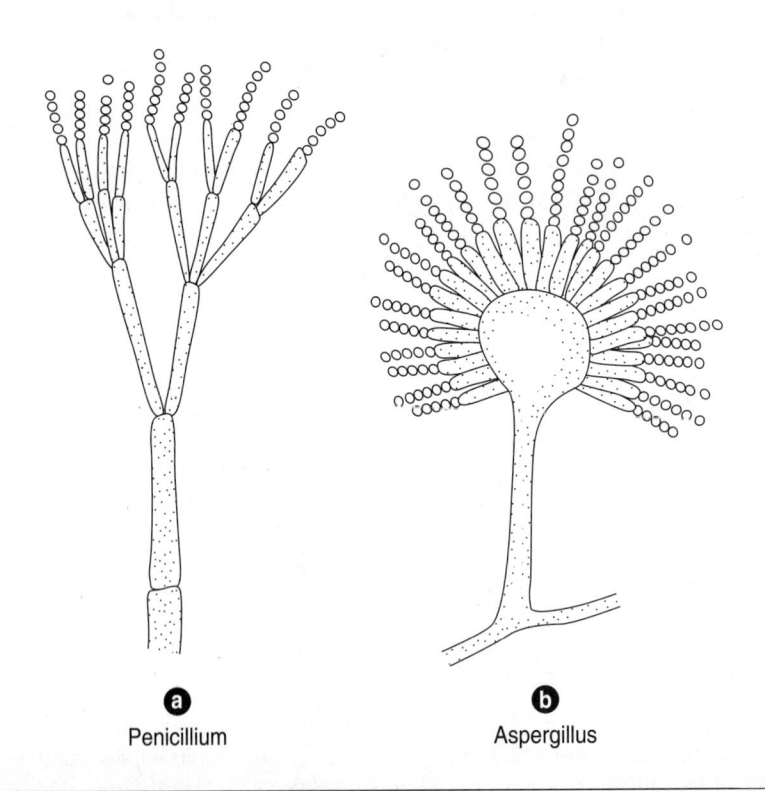

**Figure 19-8** (a) *Penicillium* and (b) *Aspergillus*

meaning "little brush," is the basis of the genus name. *Penicillium camemberti* and *Penicillium ruquefortii* are well known for their effect on cheese. The great fame of *Penicillium*, however, relates to the discovery of the antibiotic **penicillin**, which came originally from *P. crysogenum*. British microbiologist and physician Alexander Fleming first discovered penicillin in 1928, when he noted the inability of bacteria to grow in a culture medium close to a contaminant of the *Penicillium* mold. It was not until 1940 that the drug was isolated in pure form and its promise to medicine was demonstrated. By that time, the exigencies of war made large-scale production of the new "wonder drug" of supreme importance. A team of British and United States workers began by culturing the original mold; the amount of penicillin that could be extracted was extremely small and the cost was very great, however. The team thus appealed to the public to send material that had become moldy with greenish or bluish mold. Garbage flowed to them from all over the country. In a short time, the team had a mold, found on a spoiled cantaloupe, that would produce twenty-five times more penicillin than could be obtained from the original culture. They were next successful in finding a strain that yielded eighty times the original amount. Eventually, by the use of x-radiation techniques, molds were cultured that yielded 225 times the original culture. By D-Day, when allied forces invaded France on the coast of Normandy, enough penicillin had been obtained to treat several million patients. It was an enormous undertaking.

## Morels and Truffles

Although most **mushrooms** used as food belong to Basidiomycetes, several belong to Ascomycetes. Such is the case with **morels** and **truffles**. *Morchella esculenta* is highly regarded for its delicate flavor. These mushrooms have been esteemed since ancient times. Plutarch (c. 46–after 119) thought they were produced by thunder. Eating them can present certain hazards, however. Specifically, some people have been adversely affected when consuming morels along with alcoholic beverage. The fruiting bodies of morels are not like those of most ascomycetes. Rather than being cup- or saucer-shaped, the apothecium appears to be turned inside out, with the asci borne in depressions on the surface of a cap. Morels have not been successfully grown commercially.

Truffles produce fruiting bodies underground. While these mushrooms are a highly regarded delicacy, special skills are required to locate them. The unmatched experts for locating them are dogs and pigs trained to detect the mushrooms by odor. When unearthed, the master must quickly step forward. No part of this fungal treasure can go to the pig—not at $145.00 per ounce! The reward, instead, are acorns carried in the master's pocket. There are some people who have well-trained noses and thus do not need to rely on trained animals. Truffles are found in Oregon and California; but the best

❈ **Notes** ❈

## ❋ Notes ❋

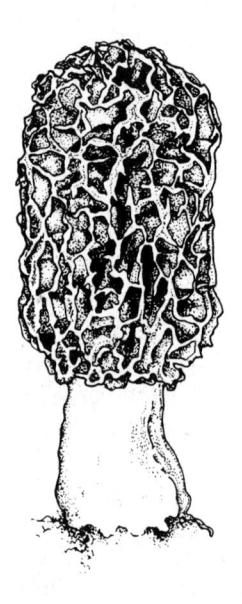

**Figure 19-9** *Morchella esculenta* (Illustration by Laurette Richin)

known sites are in the south of France, where oak forests are preserved primarily for truffle hunting. The truffle ascocarp is entirely enclosed. The only apparent means of spore dispersal is consumption of the fruiting bodies on the part of certain animals (mostly rats). The spores pass through the animal's body unaltered.

## Ergot

*Claviceps purpurea* is an important ascomycete that grows on the grains of rye grass cereals. It produces the drug **ergot**, which is both a severe poison and a useful medicinal drug. It affects the central nervous system as well as causes the contraction of involuntary muscle.

**Ergotism**, caused by eating infected rye bread, was common in the Middle Ages, when it was known as "St. Anthony's Fire." Epidemics of ergotism swept over France nine times in the seventeenth century and seven times in the eighteenth century. Even more episodes occurred in Germany. During an epidemic in the year 994 A.D., 40,000 people died from eating infected bread. It is recorded that when Czar Peter the Great was preparing to launch a military campaign against Turkey in 1722, his troops were felled

**Figure 19-10** Rye cereal showing sclerotia of *Claviceps purpurea*.

### ❋ Notes ❋

by ergotism. The military action did not take place and a war was thus lost because of a fungus.

Because ergot causes contraction of involuntary muscle, it has been useful in helping contraction of the uterus during labor. It has also been used to help retard bleeding and has proved valuable in treating migraine headaches.

Dried ergoty grains are also used for less noble purposes—specifically, in the manufacture of lysergic acid diethylamide (LSD), an hallucinogenic drug. Ergot is dangerous in that an overdose can cause gangrene and death. In affecting the nervous system, ergot can cause hysteria and temporary insanity. Could ergotism have been involved in the witchcraft that occurred in Salem, Massachusetts in 1692?

## Questions for Review

1. Describe ascus formation in detail.
2. Define the terms *paraphysis*, *apothecium*, and *cleistothecium*.
3. What is a crozier?
4. Asexual reproduction in ascomycetes occurs by formation of _____, which may form at the ends of hyphae by the process of _____.
5. Distinguish between perfect reproduction and imperfect reproduction.
6. The floor of the fruiting body, which bears asci, is called the _____.
7. Yeasts, which are classified as ascomycetes, are commercially important because of the enzymes they produce. _____ is an enzyme that produces alcohol in fermentation, and _____ is an enzyme that converts sucrose to two simple sugars: _____ and _____.
8. Name a fungus that infects the leaves of lilac plants.
9. What fungus threatens the extinction of the American Elm tree?
10. Who discovered penicillin, and what is the name of the fungus from which it was first derived?

## Suggestions for Further Reading

Alexopoulos, Constantine. 1962. *Introductory Mycology*. New York: John Wiley and Sons, Inc.

Christensen, Clyde M. 1965. *The Molds and Man*. Minneapolis: University of Minnesota Press.

Cooke, R. 1977. *The Biology of Symbiotic Fungi*. New York: John Wiley and Sons, Inc.

Hawker, L.E. 1967. *Fungi*. New York: Hillary House Publishers.

Rose, H. 1960. Yeasts. *Scientific American* 202(2):136–146.

Smith, A.H. 1963. *The Mushroom Hunter's Field Guide*. Ann Arbor: University of Michigan Press.

# 20

# The Basidiomycetes

The Basidiomycetes are the "club fungi," so-called because the sexual spores, called **basidiospores**, are borne on a club-shaped hypha, called the basidium. The typical number of basidiospores is four. Basidiomycetes includes mushrooms, toadstools, mildews, molds, **smuts**, **rusts**, puffballs, earth stars, **bracket fungi**, jelly fungi, bird's nest fungi, and stinkhorns. Approximately 100,000 species have been described and perhaps twice that number await discovery. The class is divided into two subclasses, one of which includes the rusts and smuts, and the other of which includes the mushrooms, bracket fungi, and puffballs. The hyphae are septate but with a remaining central pore, which makes a continuous channel for cytoplasm between adjacent cells. The mycelium, which forms after the sexual cycle, is typically binucleated. Nuclear fusion occurs just prior to the formation of basidiospores, which are produced by meiotic divisions.

## Rusts

*Puccinia graminis* fungus has taken a large toll on wheat yields throughout history. An epidemic in 1916 caused the loss of 14 million bushels in the United States and 100 million bushels in Canada. In 1935, 60 percent of the crop was lost in Minnesota and the Dakotas; in other areas, the loss was near total. In 1953 to 1954, the loss was more than 80 percent in the same area. In Roman times, it was thought that damaged wheat was a visitation upon wicked persons. For hundreds of years it was known or suspected that the common barberry plant was in some way involved; yet not until 1800 did it become known that **wheat rust** is caused by a fungus. German mycologist Heinrich DeBary (1831–1888) is credited with working out the details of

❋ Notes ❋

### ✱ Notes ✱

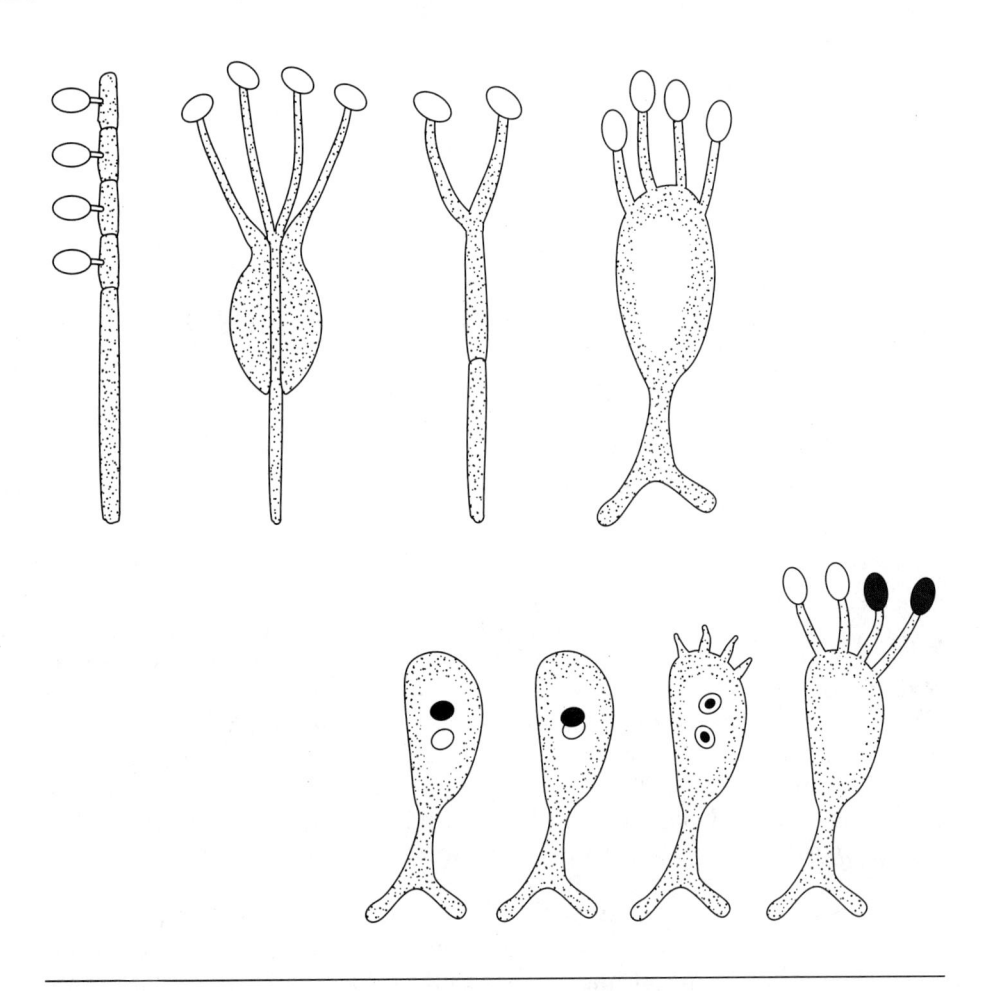

**Figure 20-1** Basidiospores are most commonly formed on a club-shaped hypha, but also can take the form shown at left. In the lower right-hand corner, the fusion of nuclei to create the diploid condition, which is then followed by meiosis to produce haploid basidiospores.

*P. graminis'* life cycle in 1865. The details are complicated, with five different kinds of spores being produced in the complete cycle (see figure 20-5).

The infection of the wheat plant in spring shows a rust-red zone, which, on examination with the microscope, reveals a cluster of sporangia. The cluster is called a sorus, and the spores that are produced here have two nuclei. These are called **uredospores**. Air currents carry the uredospores, thus spreading the infection to other wheat plants. In late summer, the nuclei fuse to create a diploid cell. These nuclei then divide, and a cross wall forms, thereby creating two cells at the end of a stalk. In this stage, the sorus becomes black, and the spores are called **teliospores**. Teliospores are thick walled and generally do not germinate until the following spring. They remain

*The Basidiomycetes* ♦ 197

❦ **Notes** ❦

**Figure 20-2** *Dictyophora*, a stinkhorn. Its odor makes it attractive to flies, which perch on the top. The fungus was formerly used to make a salve for the treatment of rheumatism. (Illustration by Laurette Richin)

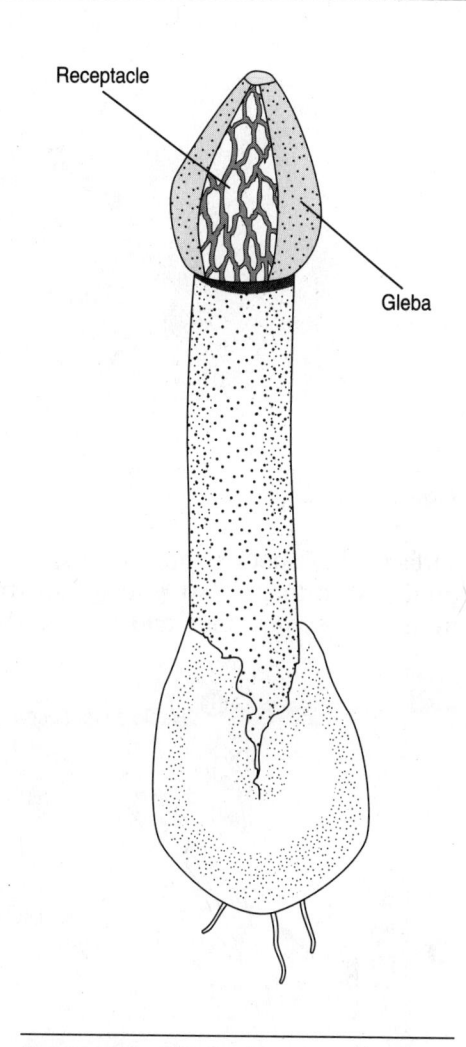

**Figure 20-3** *Phallus impudicus*, another stinkhorn.

on the straw and may germinate in the field or, perhaps, the manure pile. Germination produces an elongated basidium partitioned by walls. The process involves meiosis yielding four haploid basidiospores. As shown in figure 20-5, the basidiospores are suspended on the ends of pointed projections called the sterigmata (**sterigma**). There are two types of basidiospores: plus (+) and minus (−).

There is only one place that a basidiospore can germinate: on the surface of a leaf of the common barberry, *Berberis vulgaris*. The infection is highly host specific; there is no other place that it can grow. It is so specific, in fact, that it is able to grow only on the upper surface of the leaf. A mycelium is produced in the tissue of the leaf. After a time, pockets are formed on the upper

## ❃ Notes ❃

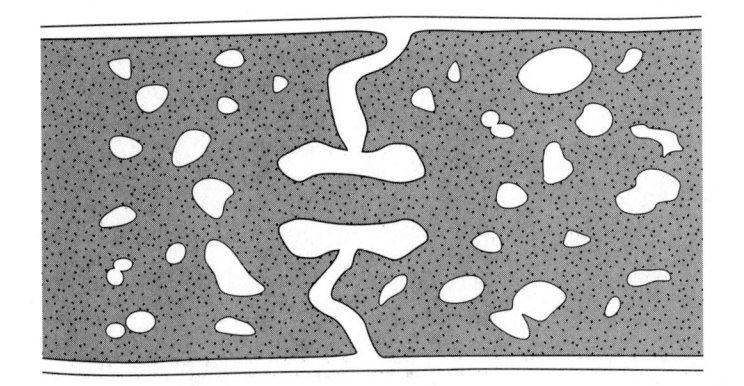

**Figure 20-4** End wall perforation

surface. These pockets are called **spermogonia** (or **pycnia**), and spermatia are produced in them. Two kinds of spermatia are produced from the plus and minus teliospores: plus and minus. The spermogonia secrete a sugary fluid

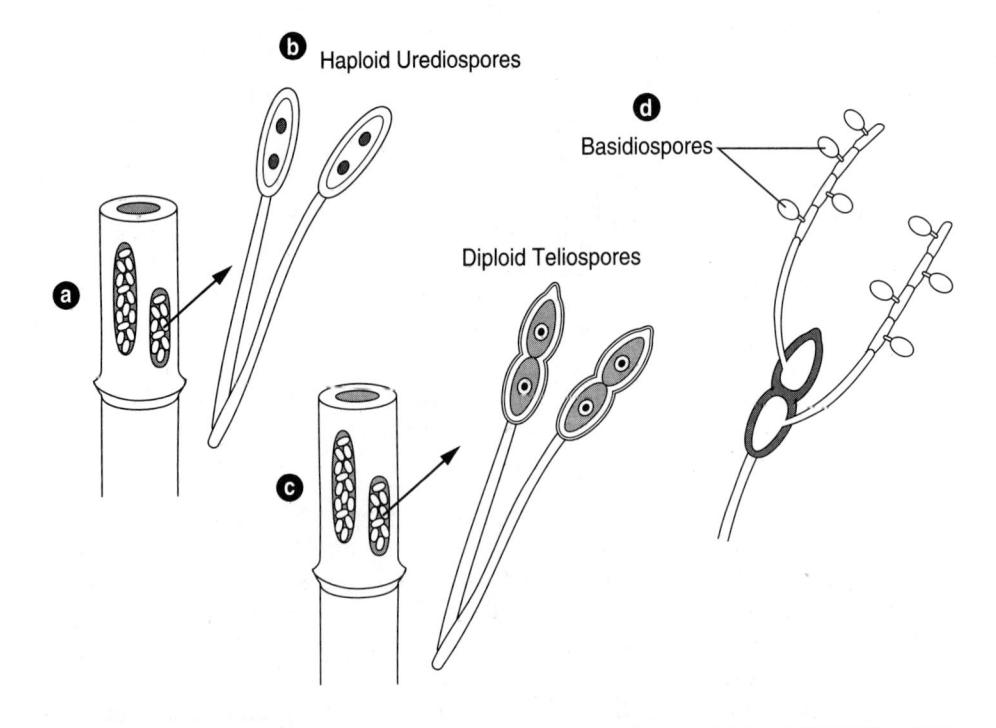

**Figure 20-5** *Puccinia graminis* (wheat rust) infection on the wheat plant. At left, (a) red-stage sorus occurring in spring. (b) The binucleated spores are haploid uredospores. At right of the red-stage sorus, the black-stage sorus (c) occurring in the fall. Diploid, 2N, teliospores are formed. At far right, a teliospore germinates to produce basidiospores, (d).

attractive to insects, which by seeking this nectar, bring mating types of spermatia together. Although plus and minus spermatia are thereby paired, nuclear fusion does not occur. The cells fuse, but the nuclei do not; a dikaryotic condition thus results. The dikaryotic cells then produce cup-shaped aecia (**aecium**) on the underside of the barberry leaves. Binucleated aeciospores are then produced in these aecia chains. These aeciospores are the agents able to infect the wheat plant, and they mark the end of the life cycle.

Recall now that the early-season infection of the wheat produces binucleated uredospores, and that nuclear fusion is long delayed. The nuclei do not fuse until late summer, when the black sorus stage occurs and teliospores are produced. Note also that there is an alternation of generations, an alternation of haploid cells and diploid cells, and an alternation of hosts, called **heteroecism**.

Although some species of rust fungi are able to complete the life cycles by parasitizing a single host, large numbers of rusts require two hosts. The white pine blister rust *Cronartium ribicola*, responsible for the loss of

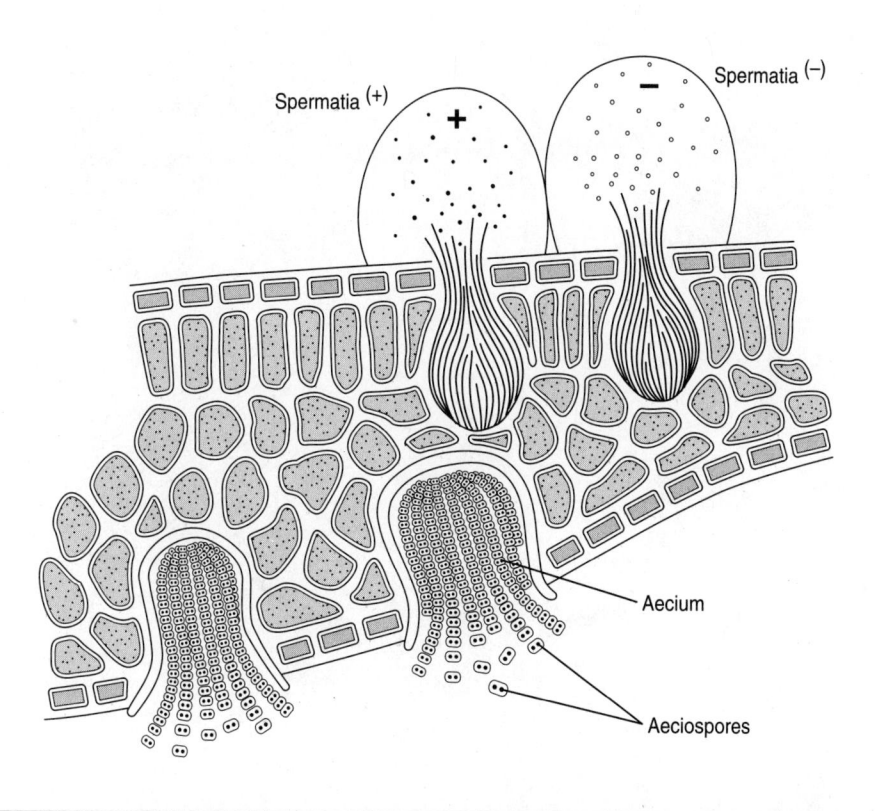

**Figure 20-6** *Puccinia graminis* infection on barberry leaf. On the upper surface of the leaf, spermogonia with spermatia, which grow from teliospores. Uninucleated spermatia become paired without nuclear fusion, producing aecia on the lower surface of the leaf. The aeciospores are binucleated.

### ✽ Notes ✽

millions of trees, infects white pine, with the alternate host being gooseberries or currents. Some measure of control has been attained by eradicating gooseberry and current plants in regions where white pine is an important lumber. In the same way, barberry eradication has met with modest success in controlling wheat rust.

## Smuts

*Ustilago zeae* (*U. maydis*) is an example of a smut. It infects maize, and its life cycle is much simpler than that of rusts. There are no spermatia, no

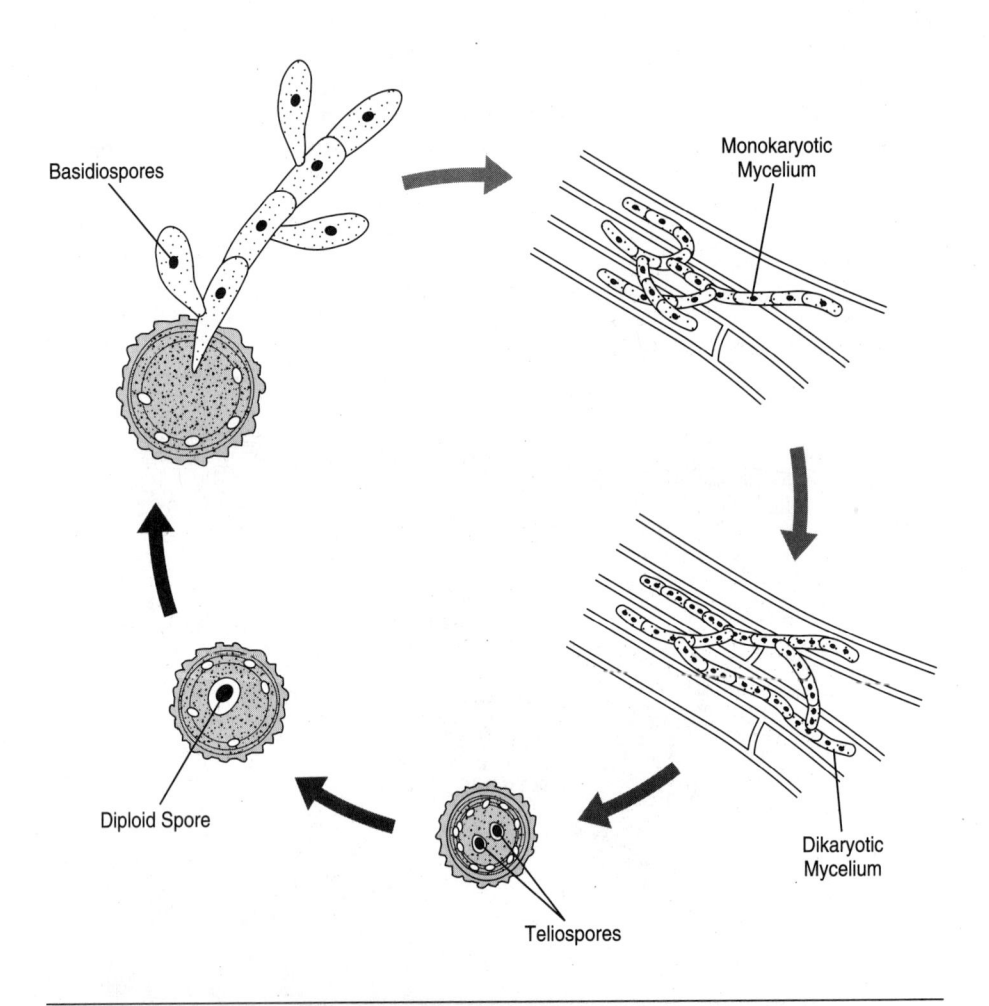

**Figure 20-7** *Ustilago maydis*, corn smut. Basidiospores infect the cells of the corn plant, producing a monokaryotic mycelium, which later becomes dikaryotic. The dikaryotic cells produce teliospores. The nuclei fuse, producing a diploid spore, which produces a basidium by meiosis.

aeciospores, no uredospores, and no alternation of hosts. Corn smut is a costly infection that commonly causes an annual loss of nearly fifty-five million bushels of grain in the United States.

Basidiospores invade the floral organs and the adjacent tissues of the corn and produce a mycelium, which is at first mononucleated. Later and as a result of nuclear division without cell division the mycelium becomes binucleated. The mycelium spreads and, in maturity, releases binucleated teliospores. The nuclei of the teliospores then fuse to create a diploid condition. The 2N spores can then germinate, forming a basidium and, by meiosis, four basidiospores. The basidiospores are budded from the sides of the basidium rather than being borne on the tips of sterigmata (as in the rusts).

A description of a typical basidium was given in the introductory comments of this chapter; but neither the rusts nor the smuts generate the typical form of basidium. Rusts and smuts belong to a subclass called Heterobasidiomycetidae, in which basidia commonly arise from thick-walled resting spores and divide into cells. The subclass Homobasidiomycetidae includes mushrooms, puffballs, and bracket fungi, which produce club-shaped, nonseptate basidia. *Coprinus*, a common gill fungus, is representative of this subclass. It has two kinds of spores: plus and minus. A spore resting on an appropriate soil will produce an extensive, monokaryotic mycelium. When this mycelium makes contact with another monokaryotic mycelium of a mating type (for example, a plus mycelium makes contact with a minus mycelium), a union will occur that results in a dikaryotic condition. The plus and minus nuclei are associated in the same cell but are not fused at this time.

The mycelium produces fruiting bodies called mushrooms, or toadstools, entirely composed of compacted hyphae. Figure 20-9 illustrates mushroom development. In the earliest stage, the mushroom takes the form of a **button**.

※ **Notes** ※

**Figure 20-8** *Coprinus*. At maturity, the pileus breaks down into an inky mass. (Illustration by Laurette Richin)

※ **Notes** ※

As the button expands, the covering, called the **universal veil**, breaks, revealing an elongated stem, the **stipe**, and, at the tip, the **pileus**. On the underside of the pileus, gills develop. They are made of compacted parallel hyphae that form a hymenium. The basidia grow out perpendicularly from the hymenium. In the early stages of basidium development, the two nuclei of the dikaryotic cell fuse to create a 2N zygotic nucleus. This nucleus then undergoes meiosis, yielding four haploid basidiospores, which come to lie at

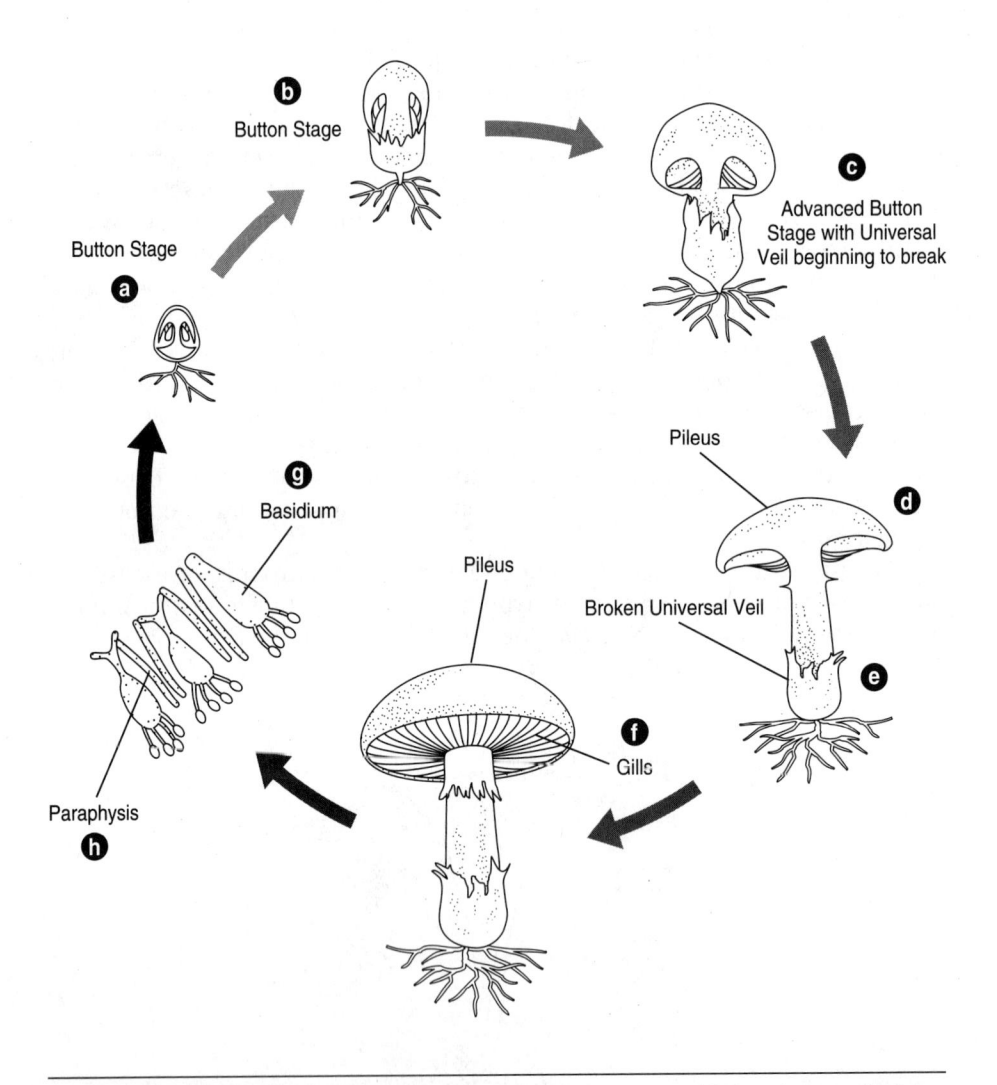

**Figure 20-9** Mushroom development. (a) and (b) Button stage. (c) Development further advanced, with the universal veil beginning to break. (d) The pileus. (e) The universal veil broken. (f) The gills in the pileus. (g) A basidium, which grows from the surface of the gill. (h) Paraphysis.

the tips of projections (sterigmata). Paraphyses (sterile filaments) also grow out from the hymenium and under microscopic examination can be seen resting among the basidia. *Coprinus* being a heterothallic form, the union of hyphae take place between mycelia of mating types. Some smut forms are homothallic and thus allow a union of cells from the same mycelium.

## Puffballs

Puffballs are another form of Basidiomycetes. They grow to varying sizes. While they often measure approximately one inch across, the diameter can sometimes be measured in feet. A very large puffball is estimated to produce trillions of spores, a somewhat smaller one, billions of spores. If two of a number of billions of spores should successfully grow new puffballs, the population of them would double, and one would conclude from this that the chances of their reproduction is extremely slight. Yet, they are common. They are also edible.

Remember that of the two nuclei in a dikaryotic cell, one is plus and the other minus. An elegant mechanism called a **clamp connection** ensures that when cells divide, the same arrangement of nuclei is maintained (see figure 20-10). A small lateral branch forms on the side of a terminal cell. This branch curves back to the hypha, makes contact with it, and constructs a passageway through which a nucleus can migrate. A new cell wall then forms. Nuclear migration ensures that daughter cells continue to have the plus and minus arrangement of nuclei. The formation of the side branch is suggestive of crozier formation in ascomycetes.

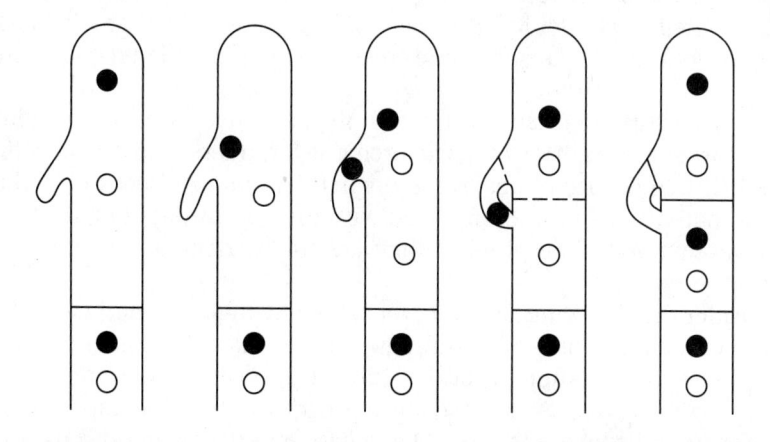

**Figure 20-10** Clamp connections, a mechanism that maintains the arrangement of plus and minus nuclei in each cell.

❋ Notes ❋

## Mycorrhiza

Many plant diseases are caused by fungi, too many, in fact, to catalog here. Silkworm disease, brown rot of plums, early blight of potatoes, *Fusarium* wilt of cotton, a smut disease of sugar cane are but a few. Certain other fungi, however, are worthy of praise.

Fungi assist in the germination of seeds. The mycelia of various basidiomycetes and ascomycetes are associated with the roots of flowering plants in symbiotic relationships. The name for such an association is **mycorrhiza** (*myco* meaning "fungus," *rhiza* meaning "root"). The fungi involved in such relationships are present in the soil. The fungal hyphae connect to the younger portions of roots and form a sheath of compact mycelium around the roots. Pines, heaths, and orchids lack root hairs and can grow successfully only when they associate in this way with fungal hyphae that furnish the absorbing mechanism. When pines are planted in an area where they have not grown before, the forester mixes the proper fungi with the soil to encourage successful growth.

Two kinds of union are recognized: an **ectotrophic** association, wherein the hyphae of the fungus grow between the cortical cells of the root but do not, or only rarely, penetrate the root cells; and an **endotrophic** association characterized by the absence of the fungal sheath and wherein the hyphae penetrate the cortex of the root and enter the cells. In endotrophic mycorrhizae, the hyphae that penetrate the cells disintegrate over time and are then digested by the host cells. This liberates certain products that can be absorbed by the roots. Perhaps one hundred species of fungi are capable of forming mycorrhizae with forest trees.

The mycorrhizal fungi may have special nutritional requirements, such as carbohydrates and vitamins, and it is presumed that they derive these benefits from tree roots. The reverse condition also occurs. A conifer grown in conditions wherein it is denied association with fungal hyphae often fails to develop normally but can sometimes recover when inoculated with a suitable fungus.

The mycorrhiza of orchids is endotrophic, the mycelium being widely distributed through the cortex or, if the orchid is a rootless form, penetrating the chlorophyll-free tissues of absorbing organs. The minute seeds of orchids do not germinate unless they are infected with the appropriate fungus, and until they have emerged aboveground, they are totally dependent on the fungal infection.

Consider the following scenario. A crayfish living in the darkness of a cave loses the capacity to form eyes. A fungus that ensnares worms (refer to chapter 18, figure 18-5) cannot produce the mechanism of entrapment in the absence of the worms. A tree having mycorrhiza loses the capacity to produce root hairs. In chapter 5 on Mendelian genetics, some discussion centered on Lamarck's concept of the inheritance of acquired characteristics, wherein organisms achieve traits to fulfill needs. Would the preceding examples, then, illustrate the disinheritance of unneeded characteristics? It sounds like reverse Lamarckism. What, then, causes the disappearance of genes for unused traits?

Insects are involved in some instances. For example, certain scale insects may press closely against the smooth bark of a tree, where they both suck the juices of the plant and become infected with a fungus. The fungus may derive benefits from the plant indirectly by absorbing plant nutrients from the insects.

## Questions for Review

1. In what way might the body of a fungus be compared to an iceberg?
2. Describe the structure of a basidium.
3. In the wheat rust fungus *Puccinia graminis*, a red _____ is formed in the spring; later in the season, this body changes color, to _____.
4. In the early season, red-stage form of *P. graminis*, the uredospores are _____; the spores formed during the later, black stage are _____ because of the fusion of the _____.
5. What does meiosis yield in *P. graminis*? _____ What is the only kind of plant on which these products of meiosis can grow? There they produce _____ bearing _____ (the name of the spore). How many nuclei do these spores have?
6. _____ is the name of a basidiomycete destructive to white pine. One alternate host of this infection is _____.
7. _____ _____ is the name of a fungus that infects the corn plant.
8. Where are basidiospores found on a mushroom?
9. Define the terms *mycorrhiza, clamp connections,* and *ectotrophic*.

## Suggestions for Further Reading

Alexopoulos, Constantine. 1962. *Introductory Mycology.* New York: John Wiley and Sons, Inc.

Christensen, Clyde M. 1965. *The Molds and Man.* Minneapolis: University of Minnesota Press.

Cooke, R. 1977. *The Biology of Symbiotic Fungi.* New York: John Wiley and Sons, Inc.

Hawker, L.E. 1967. *Fungi.* New York: Hillary House Publishers.

Rose, H. 1960. Yeasts. *Scientific American* 202(2):136–146.

Smith, A.H. 1963. *The Mushroom Hunter's Field Guide.* Ann Arbor: University of Michigan Press.

## Notes

# 21

# Fungi Imperfecti

Members of Phycomycetes, Ascomycetes, and Basidiomycetes all exhibit perfect stages in their reproductive cycles, producing zygospores, oospores, ascospores, or basidiospores. They also reproduce asexually by conidia. More than 10,000 species of fungus, however, do not reveal any perfect stage and reproduce only by conidia. Because the accepted method of classifying fungi is based on the fruiting structures exhibited in the perfect phase and such forms show evidence of only an imperfect phase, they are placed in a taxonomic catchall called the Fungi Imperfecti. Each year, studies of species relegated to Fungi Imperfecti uncover perfect stages; this makes possible removing such forms from Fungi Imperfecti and placing them in their proper classes. Some forms classified as Fungi Imperfecti for many years have eventually been properly reclassified. Reclassification is most often made to Ascomycetes and sometimes to Basidiomycetes. Does this mean a gradual demise of Fungi Imperfecti? The answer is "no," because many newly discovered species are added to this group. Also, many forms of fungi have permanently lost (or never had) the capacity for perfect reproduction; these will remain in Fungi Imperfecti. (Many of these fungi are of great importance because they are parasites that cause diseases of plants, animals, and human beings.) The conidial stages of most of these fungi are similar to conidial stages of Ascomycetes.

❋ **Notes** ❋

### Notes

The hyphae of this group are septate, having cross walls at the ends of cells. The end walls each have a central perforation (also seen in Ascomycetes) that allows the movement of cytoplasm and nuclei from one cell to another. The cells are usually multinucleated.

## Problems in Classification

In fungi exhibiting perfect phases in the reproductive cycle, classification is determined in part by the fruiting bodies produced in the sexual process. Because fruiting bodies are absent in Fungi Imperfecti, it is necessary to use the conidia for purposes of identification. The problems with doing so are legion, however, primarily because conidia are prone to alteration. Slight variations in culture conditions can change the form of conidia, as can nutritional changes and various methods of inoculation. One consequence of using this method is that the same organism has sometimes been classified in different groups and given different names. It is little wonder that workers in this field are led to complain that they are "working in the dark." The need to rely on pycnidia (flask-shaped structures that produce conidiospores) and conidia for classification defeats the hope of creating a phylogenetic arrangement reflecting genetic kinship.

## Moniliales

Fungi Imperfecti is also called Deuteromycetes. The largest group within Deuteromycetes is Moniliales, which includes more than 10,000 species. Many of these cause serious diseases. One example is *Cercospora*. It causes several plant diseases including leaf spot of beets and tobacco, early blight of celery, and infections of several other crops. Approximately 3,800 species of *Cercospora* have been described.

Injury to the mycelia of moniliales can result in the formation of pycnidia in the region of injury. In a laboratory culture of fungus growing on agar, a cut made across the surface of the growth will often stimulate the production of reproductive bodies along the line of the cut. This again illustrates an observation made earlier: unfavorable living conditions appear to encourage reproduction. This particular event may be related to the production of a hormone.

Conidiophores in Moniliales sometimes unite at the base and part way up toward the tip to form a structure called a **synnema**. A synnema can become much branched where conidia arise at the tips. This type of fruiting body is said to resemble a long-handled feather duster (see figure 21-1).

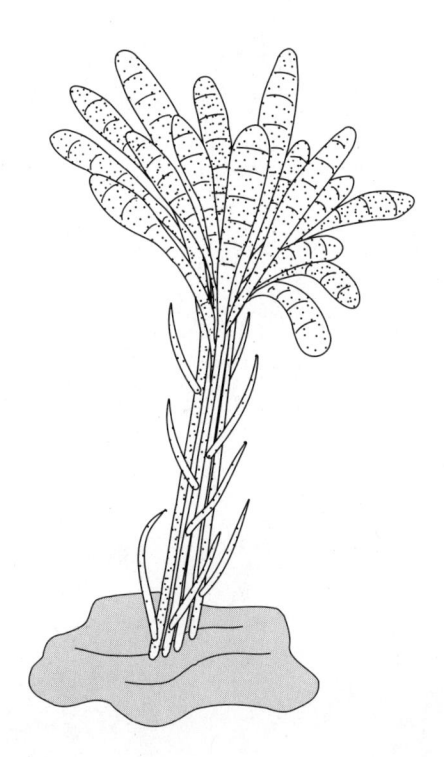

**Figure 21-1** A synnema: tightly compacted hyphae bearing conidiospores.

## Questions for Review

1. What is the meaning of the term *imperfect* in reference to fungi?

2. Why is such a category as Fungi Imperfecti needed in the classification of fungi?

3. Define *synnema* and tell in what group it is found.

✼ Notes ✼

## Suggestions for Further Reading

Alexopoulos, Constantine. 1962. *Introductory Mycology.* New York: John Wiley and Sons, Inc.

Christensen, Clyde M. 1965. *The Molds and Man.* Minneapolis: University of Minnesota Press.

Cooke, R. 1977. *The Biology of Symbiotic Fungi.* New York: John Wiley and Sons, Inc.

Hawker, L.E. 1967. *Fungi.* New York: Hillary House Publishers.

Rose, H. 1960. Yeasts. *Scientific American* 202(2):136–146.

Smith, A.H. 1963. *The Mushroom Hunter's Field Guide.* Ann Arbor: University of Michigan Press.

# 22

# Lichens

A lichen is created by the union of a fungus and an alga. As in the case of a successful marriage, these partners live together in harmony. Although some say that the fungus is parasitic on the alga, thereby distorting the alga, such an association is sufficiently successful to enable a single lichen to survive for several thousand years.

Lichens have scientific names, as do their components; thus, the fungus has a name, the alga has a name, and the association has a name.

Lichens grow in harsh environments: on rocks and the trunks of trees; in salt spray, the desert, the arctic, and tropical forests. Yet lichens are sensitive, and their presence or absence can be used as an indicator of pollution. For example, lichens often do not grow well on the trunks of trees in the city.

Lichens may be crustose, foliose, or gelatinous. Some crustose lichens seem to be painted on the substrate. *Verrucaria maura* is a lichen that forms a distinct black band just above high tide along the New England coast.

❈ Notes ❈

## Reproduction in Lichens

The algal cells of a lichen may be scattered among the hyphae of the fungus or, more frequently, may form a layer near the surface. In reproduction, both parts increase. Reproduction can be accomplished by an asexual process wherein **soredia** (several algal cells surrounded by fungal hyphae) are formed; or by the lichen being torn apart by wind. Reproduction can be accomplished by both parts independently; the alga divides by fission, whereas the fungus produces spores (perhaps ascospores).

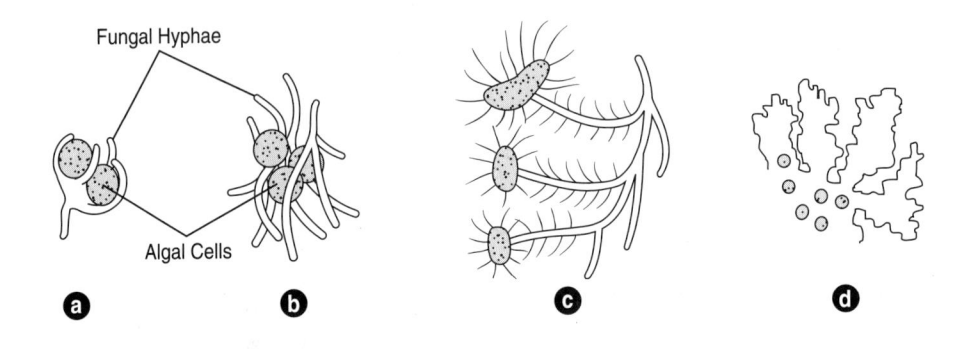

**Figure 22-1** (a) A soredium. Reproduction in lichens is vegetative, requiring that both components of the symbionts go together. The soredium, a minute fragment of lichen, contains both algal and fungal components. It may break away from the parent lichen, be carried by wind, and become newly established elsewhere. (b), (c), and (d) The hyphae of the fungus is closely associated with algal cells.

## The Members of a Lichen

In the laboratory, it is possible to tease apart the two members of a lichen, grow them independently in culture, and bring them back together again to re-establish the lichen. This, however, is strictly a laboratory feat. And while there may be a disassociation in nature that allows the alga to grow alone, the fungal component of a lichen does not survive alone in nature.

The fungal host is most commonly an ascomycete. In several species of lichen, however, it is a basidiomycete. When associated with algae as lichens, fungi are able to perform the perfect type of reproduction to produce asci or basidia. Attempts to bring this about in fungi that have been separated from their algal companions in the laboratory have not been successful, however.

Blue-green algae such as *Croococcus, Gloeocapsa,* and *Nostoc* often serve as symbionts in lichens. Among the green algae, *Chlorococcum* and *Trebouxia* often fulfill this role. Each lichen generally encompasses a single algal species; some, however, encompass two or even three.

## Growth of Lichens

Starvation is a condition necessary for the growth of lichens. When moisture is overly abundant, the alga outgrows the fungus. When nutrients are in generous supply, the fungus outgrows the alga. Either of the preceding two conditions is ruinous to the association. Many species grow very slowly. In the laboratory, it can take one year to obtain a lichen one millimeter in diameter. There are forms that grow much more rapidly, however.

## Products from Lichens

Several forms of lichen are used to produce dyes. *Ochrolechia tartarea* is one example. Litmus is derived from lichens such as *Roccella tinctoria*, *R. fuciformis*, and *Lecanora tartarea*. *Evernia prunastri* is used in making perfume. *Cladonia mitis*, *C. sylvatica*, and *Cetraria islandica* are also valuable forage for reindeer, caribou, and moose.

## Lichens and the Doctrine of Signatures

Long ago, it was reasoned that the universe was made for human beings, and that God, who designed everything for his "chosen ones," wanted to provide clues to help people learn the purposes of various things. This concept was called the *doctrine of signatures*. In accordance with this doctrine, *Lobaria pulmonaria*, a foliaceous lichen having a surface suggestive of a lung, was at the time used to treat tuberculosis and pneumonia. Likewise, *Peltigera canina*, a foliaceous lichen having the appearance of a row of teeth, was used to treat rabies. According to the doctrine of signatures, then, were a lichen to grow on a skull, that lichen would be good for the treatment of epilepsy; and were it to grow on the skull of a hanged criminal, its efficacy would be greatly enhanced.

The doctrine of signatures applied not only to lichen. For example, eyebright (*Euphrasia officinalis*), being marked with a spot like an eye, was considered an excellent treatment for diseases of the eye. Bugloss (*Anchusa officinalis* and several other forms), resembling a snake's head, was considered good for snake bites. And *Celandine*, having yellow juice, was used to treat jaundice. The age of Shakespeare, Francis Bacon, and Galileo was one of fetish cures and quack remedies—a time when it was said "things bad for the heart are beans, pease, sadness, onions, anger, evile tidings, and the loss of friends." With at least some of this, you must agree.

## Questions for Review

1. A lichen is created by the union of a _____ and an _____.

2. This union can be considered an example of _____, whereby each form benefits from the other.

3. An asexual form of reproduction in lichens occurs by the formation of _____.

4. Lichens may take several forms: _____ _____ or _____.

5. Some lichens are of commercial value in making _____ or _____.

❊ **Notes** ❊

❊ Notes ❊

## Suggestions for Further Reading

Ahmadjian, V. 1965. *The Lichen Symbiosis*. Waltham, MA: Blaisdell Publishing Co.

Ahmadjian, V., and Mason Hale (eds.). 1973. *The Lichens*. New York: Academic Press, Inc.

# 23

# Plant Classification

There are a number of different ways to classify plants: according to whether they are useful in medicine; are appropriate for use in religious ceremony; or are deemed harmful.

## Early Efforts at Plant Classification

Theophrastus (370–288 B.C.) may have been the first to publish on the subject of plant classification in the work called *Enquiry into Plants*. Here, he classified plants into four groups. In addition to botany, Theophrastus had a great many interests, including the causes of wind, scents, and sensations; physics; ethics; politics; and history. Another of his writings was entitled *On the Causes of Plants*. Dioscorides (fl ca A.D. 5) compiled *Materia Medica*, which illustrated 500 plants of supposed medicinal value. At one time, a plant name was actually a description of the characteristics of the plant. This often required the equivalent of an entire sentence and was thus a burdensome system.

## Carolus Linnaeus

Carolus Linnaeus (1707–1778) is considered the father of our modern system of naming plants. He is assigned credit for devising the binomial system of naming plants and animals. The scientific name of a plant is its genus name followed by its species name, and the binomial system requires that both names be either printed in italics or underlined. As an example, consider the shrub *Euonymus europaeus*. The value of the scientific name is quickly appreciated when common names are known. This one plant, for instance, is variously known as arrow-beam, prick-timber, prickwood, cat-tree, pegwood, pincushion-shrub, skiver-wood, witchwood, louseberry, butcher's prick tree, gatten tree, and spindle tree. Obviously, it would be well to abide by scientific names.

❋ Notes ❋

## ✻ Notes ✻

**Figure 23-1** Carolus Linnaeus (1707–1778), creator of the binomial system of naming plants. (Illustration by Donna Mariano.)

Present day concerns are to classify plants on the basis of kinship, putting together those that have a common evolutionary origin. This was not Linnaeus's concern. His arrangement divided the plant kingdom into two groups: the Cryptogamia and the Phaenerogamia. *Cryptogamia* literally means "hidden marriage," Phaenerogamia, "visible marriage." The latter includes the flowering plants. Cryptogams, such as **mosses** and **ferns**, have inconspicuous sex cells. The system, while entirely artificial, was useful.

The organisms discussed thus far in this text live in water or moist places (algae) and are saprophytic or parasitic (fungi), or symbiotic (lichens). In tracing from the most primitive (the blue-green algae) to somewhat more advanced forms, a significant change from prokaryotic to eukaryotic was noted. In upcoming chapters, we shall discuss plants that have achieved another profound change: emergence onto land. Among the variables that contributed to this change was the development of vascular (conducting) tissues. Development of vascular tissues, however, was neither the deciding nor the only and certainly not the first requirement for life on land. The mosses and liverworts, for example, have achieved some level of change without developing vascular tissues. Land plants also evolved structural modifications against excessive water loss, an altered method of gas exchange, supporting tissues, and, over time, new methods of transferring sperm to egg. Yet even the most advanced plants have not completely resolved all the hazards of life on land.

## The Theory of Evolution

When Charles Darwin (1809–1882) published *On the Origin of Species*[1], he upset the entire world with his theory that human beings were descended from anthropoid ancestors. By introducing the idea of **evolution**, however, he compelled botanists to consider anew their heretofore accepted systems of classification. Methods of classification to this point had been, to a large degree, artificial. The theory of evolution led to efforts to devise a natural system of classification whereby those forms having a common ancestor would be grouped together.

## Problems in Classification

These efforts have seen a measure of success; some parts of the scheme still, however, remain artificial. In fact, the five kingdom system employed in this text is yet on trial. The problems are not so acute with regard to higher plants such as the flowering plants. Here, flower parts are the most reliable indicators in determining allied forms. The number of parts, the way they are arranged, and the presence or absence of stamens serve to accurately classify flowering plants. The difficulty arises, rather, when working with microscopic forms. Even the distinction between plants and animals becomes hazy. It may be said, for instance, that plants are distinguished from animals by the lack of contractile tissue or protoplasm; but this is not entirely true. It may also be said that plant cells have walls and animal cells do not. This is largely but not entirely true. It may be said that plants have the green pigment chlorophyll and, therefore, can carry on photosynthesis, whereas animals cannot. Some bacteria, however, are capable of photosynthesis. Using other microcharacteristics in an attempt to classify organisms is useful though also somewhat flawed. Chromosome number is an example. Some species have different chromosome numbers in different races. *Xanthisma texanum* has one less pair of chromosomes in the root than in the shoot. In the new field

## Notes

---

[1] *On the Origin of Species by Means of Natural Selection, or the Preservation of Favored Races in the Struggle for Life.*

### ✵ Notes ✵

of chemical taxonomy, plants are grouped according to common chemical substances. The plants of the Cruciferae family, for example, all have mustard oils. Even the presence of common chemicals, however, does not guarantee kinship. Nicotine, for example, occurs in some widely separated families and not just in tobacco. Substances called tropolones occur in fungi (*Penicillium*) and also in certain conifers. The mention of these exceptions is not meant to entirely invalidate chemical taxonomy or other attempts at classifying organisms. Rather, it is meant to illustrate the importance of considering a *collection* of information when attempting to identify relationships between organisms. The computer is being used as a tool to this end.

## Monophyletic or Polyphyletic?

The theory that vascular plants are descended from green algae is generally accepted. Whether it happened once or several times, however, is unknown. If it happened only one time, the flowering plants are said to be of **monophyletic** origin; if several times, they are said to be of **polyphyletic** origin.

## System of Plant Classification

How are the land plants classified? This text adopts a five kingdom system. Monera, Protista (which includes the algae), and Fungi have already been considered. Land plants can be studied a number of different ways. The emphasis in this text is on simplicity, focusing on the subkingdom Embryophytes. Embryophytes are subdivided as follows:

> **bryophytes**—land plants lacking vascular tissue:
>    **liverworts** *(Marchantia)*
>    **hornworts** *(Anthoceros)*
>    mosses *(Mnium)*
> tracheophytes—land plants having vascular tissue:
>    **pteridophytes**
>      ferns *(Dryopteris)*
>      club mosses *(Lycopodium)*
>      horsetails *(Equisetum)*
>    spermatophytes
>      **gymnosperms** (coniferous trees)
>      **angiosperms** (flowering plants)

Bryophytes lack vascular tissue and, thus, take water in through **rhizoids** (an undifferentiated tissue) and by absorption from the air. Pteridophytes have vascular tissue, producing **xylem** and **phloem** in bundles. Spermatophytes are vascular plants that produce seeds.

Each of the primary subdivisions noted preceding is called a division, or **phylum** (phyla, plural). Phyla are divided into smaller groups as follows:

Phylum<br>
Class<br>
Order<br>
Family<br>
Genus<br>
Species

In writing the scientific name—that is, the genus name and the species name—proper protocol is to begin the genus name with an uppercase letter, the species name with a lowercase letter, and to italicize (or underline to indicate italics) both names.

## What Is a Species?

A common but somewhat difficult question to answer is, "What is a species?" The easy answer is "all of the same kind." This answer might be expanded to mention that the members of a species have mutual fertility and that a species is reproductively isolated from other species. These latter two claims do not entirely fit, however. Thus, the struggle to define the term *species* goes on. The claim that a species is all of the same kind, however, still holds.

## Questions for Review

1. Explain the differences between artificial and natural systems of classification.
2. What is meant by the binomial system of nomenclature?
3. There is a theory that vascular plants descended from _____ _____. If this happened only once, vascular plants are said to be of _____ origin.

## Suggestions for Further Reading

Bold, H.C. 1973. *Morphology of Plants*. Harper and Row Publishing, Inc.

Cronquist, A. 1968. *The Evolution and Classification of Flowering Plants*. Boston: Houghton Mifflin Co.

Porter, C.L. 1967. *Taxonomy of Flowering Plants*. San Francisco: W.H. Freeman Co.

Scagel, R.F., R.J. Bandoni, G.E. Rouse, W.B. Schofield, J.R. Stein, and T.M.C. Taylor. 1967. *An Evolutionary Survey of the Plant Kingdom*. Belmont, CA: Wadsworth Publishing Co.

❈ **Notes** ❈

# 24

# Bryophytes: The Liverworts, Hornworts, and Mosses

A long honored system of classification divides the plant kingdom into two subkingdoms: Thallophyta (which includes all plants lacking roots, stems, leaves, and vascular tissues) and Embryophyta (by definition, plants having embryos). The embryos in Embryophyta are contained within the female structure called the **archegonium**, which develops on the gametophyte plant. The embryo represents the early stages of the sporophyte generation. In this group, the reproductive organs (antheridia and archegonia) are multicellular. They are oogamous (occurring as egg and sperm) and exhibit alternation of generations. Embryophyta is divided into two phyla: **Bryophyta** and **Tracheophyta**. Whereas the tracheophytes are the vascular plants, the bryophytes are the liverworts, hornworts, and mosses. It is to the bryophytes that we now turn our attention.

Bryophytes include a group of very small plants having no conducting tissues. Whereas they have rhizoids (which penetrate the soil and are able to take up water), they lack vascular tissue such as xylem and phloem (which occur in higher plants). Bryophytes can also absorb water from the air. The gametophyte is the "conspicuous" generation (somewhat of a misnomer because the plant is actually quite inconspicuous and just more visible than the sporophyte). This distinction is important. (In ferns, the roles are reversed; the sporophyte is considered the conspicuous generation.)

## Liverworts

*Marchantia* is an example of a thallose liverwort. The gametophyte thallus is dichotymously branched and usually approximately two centimeters wide and four to six centimeters long. The sporophytes are nearly microscopically

❋ Notes ❋

## ✺ Notes ✺

sized. *Marchantia* is found in wet places, flattened on the soil and anchored by rhizoids extending from its underside. When one examines the thallus with a lens, small openings can be seen. These pores open to chambers immediately below, where columns of cells extend up from the floor of the chamber. These cells have chloroplasts and are active in photosynthesis. Carbon dioxide, which is used in the photosynthetic process, drifts into the chamber through the pore, and oxygen, a product of photosynthesis, drifts outward. The pores may be called pseudostomates, *pseudo* because they are not considered to be the same as or related to the stomates of more advanced plants.

Three kinds of appendages are found on the underside of the thallus: peg rhizoids, which have internal, peglike structures on the cell wall and function in absorption; smooth rhizoids, which also function in absorption as well as anchor the plant to the soil; and scales, which have no known function but may be protective.

The sexes are separate. On the female gametophyte, special stalks called **archegoniophores** rise up from the surface of the thallus. Upon maturity, the archegoniophores have fingerlike lobes at the upper end. Archegonia grow on the undersides of these fingerlike lobes. Each archegonium is a flask-shaped structure bearing an egg cell in its bottom portion, the **venter**. When the archegonium is quite young, there are cells in the neck portion. These cells must break down and dissolve away as the archegonium matures to allow a passageway for sperm cells.

The male gametophyte also produces stalks, called **antheridiophores**. Antheridiophores have a different appearance at the upper end than do archegoniophores. The upper end of an antheridiophore is a flat, disc-shaped head on the upper surface of which antheridia are borne. When the antheridia mature, they break open and release large numbers of biflagellated

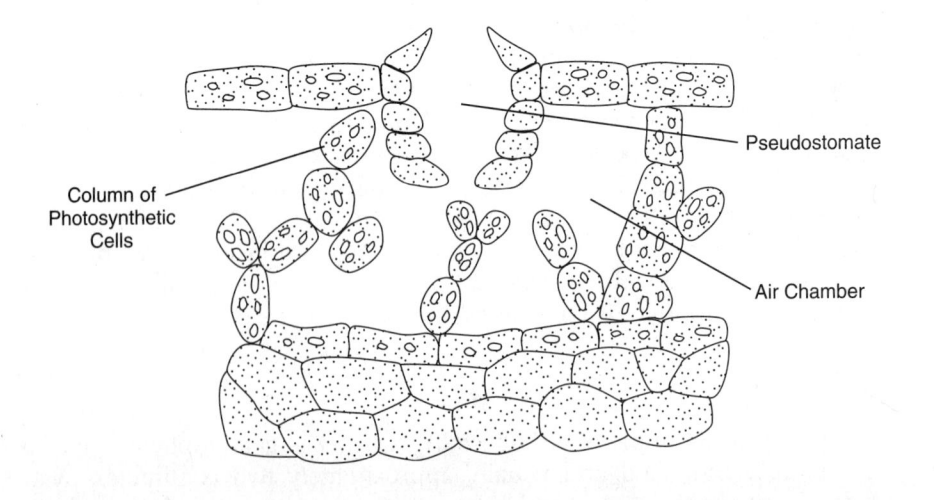

**Figure 24-1** *Marchantia* gametophyte thallus.

sperm cells. Because the sexes are separate and the gametes are produced at the tops of special stalks, the way whereby sperm reach the egg cells is perplexing. One supposition is that a rain drop striking the tip of an antheridiophore causes the sperm to be catapulted to their destinations. Another is that they are carried by insects. Yet another is that the sperm have great enough endurance to swim down the length of the antheridiophore, across a certain expanse of thallus, up the length of a neighboring archegoniophore, and arrive at the archegonia to fulfill their purpose.

❋ **Notes** ❋

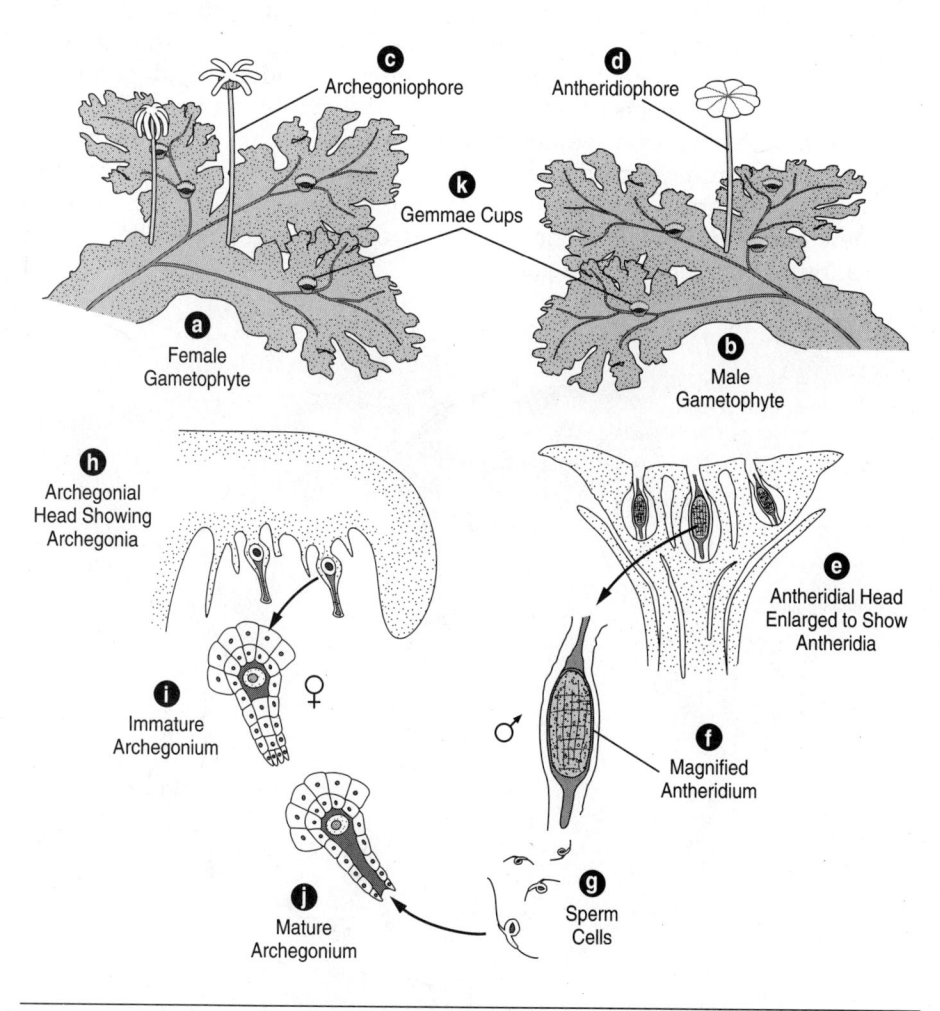

**Figure 24-2** *Marchantia* gametophyte generation. (a) The female gametophyte. (b) The male gametophyte. (c) An archegoniophore at the tip of which are archegonia under the lobes. (d) An antheridiophore bearing antheridia at the tip. (e) An enlarged antheridial head, showing antheridia. (f) A magnified antheridium. (g) A sperm cell. (h) Archegonial head, showing archegonia on the under side of the lobes. (i) An immature archegonium. (j) A mature archegonium. (k) A gemma cup bearing gemmae (agents of asexual reproduction).

### Notes

The egg cell is fertilized in the venter of the archegonium, where the sporophyte develops. The sporophyte remains attached to the gametophyte (in a sense, parasitic upon it) and microscopically sized. The sporophyte generation is diploid. It develops a **foot**, which grows through the base of the archegonium and anchors the sporophyte to the tissue of the gametophyte. A short **seta** and a **capsule** in which spores and **elaters** develop also form. Spore mother cells (diploid cells destined to undergo meiosis and produce haploid spores) are in the capsule. Upon maturity, the capsule splits open, and the spores and elaters are liberated. The elaters are **hygroscopic** filaments, coiling and uncoiling under the influence of varying conditions of humidity. These movements aid in the dispersal of spores.

Asexual reproduction is achieved either by the thallus fragmenting through decay or an enchanting evolutionary wonder called **gemmae**. Gemmae are small clusters of cells that are scarcely visible without a lens and rest in a minute basket on the upper surface of the thallus. This little basket is called the **gemma cup**, and the little, egglike clusters of cells are the gemmae. The gemmae can be dislodged and, falling on favorable soil, grow into new *Marchantia* plants. The method is entirely vegetative. There is no alternation of generations with gemmae.

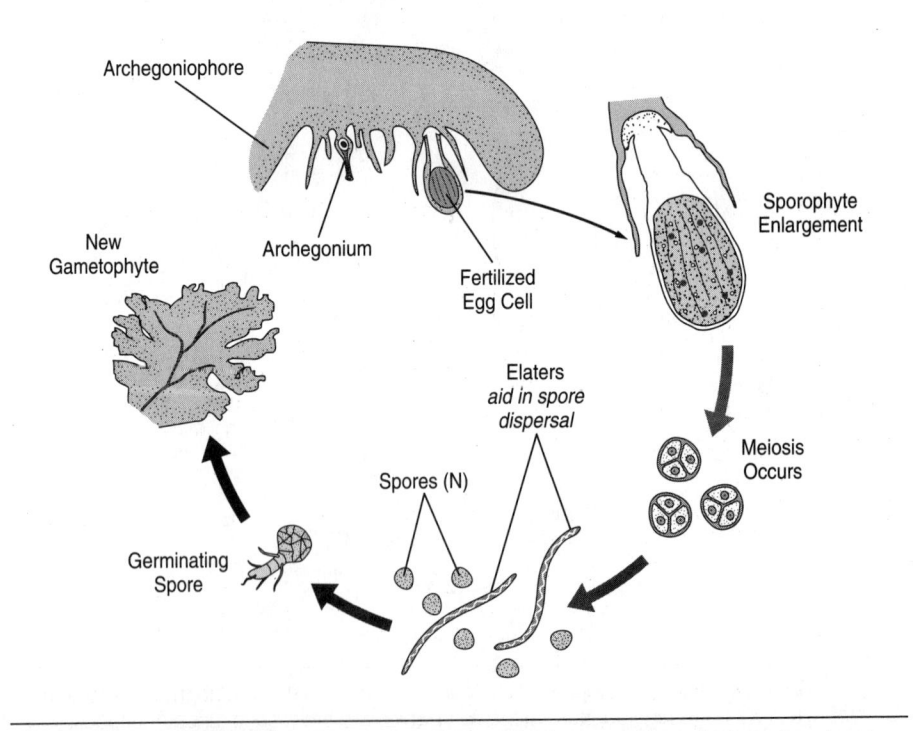

**Figure 24-3** *Marchantia* sporophyte generation. At left, an archegoniophore bearing an archegonium. The egg cell has been fertilized, and the young sporophyte is growing. At upper right, an enlargement of the sporophyte, showing spore mother cells in the capsule. Just below, meiosis occurs. At bottom, a spore germinates to produce a new ametophyte.

## Hornworts

*Anthoceros* is an example of a hornwort. The life cycle with regard to alternation of generations is much like that of *Marchantia*. One interesting difference, however, is that each cell of the gametophyte has only one chloroplast, and each chloroplast possesses pyrenoids similar to those of green algae. Again, the sporophyte grows on the gametophyte, the gametophyte being the conspicuous generation. But then a special development takes place. The spore-producing tissue forms a long cylinder parallel to the axis, and the spores mature from the top down. The foot of the sporophyte, which is anchored to the tissue of the gametophyte, grows downwardly and through the gametophyte tissue to penetrate the soil beneath. While one may infer that this represents a harbinger of independence in the sporophyte generation and, thus, a step toward the higher plants, this inference is probably without sufficient basis. It is more likely that *Anthoceros*, rather than being an ancestor of higher forms, is a dead end.

## Mosses

Mosses are low plants that produce dense mats. They occur mostly in moist places, but some forms can be found in the desert. Approximately 14,500 species have been described. Mosses provide an elegant example of alternation

❊ **Notes** ❊

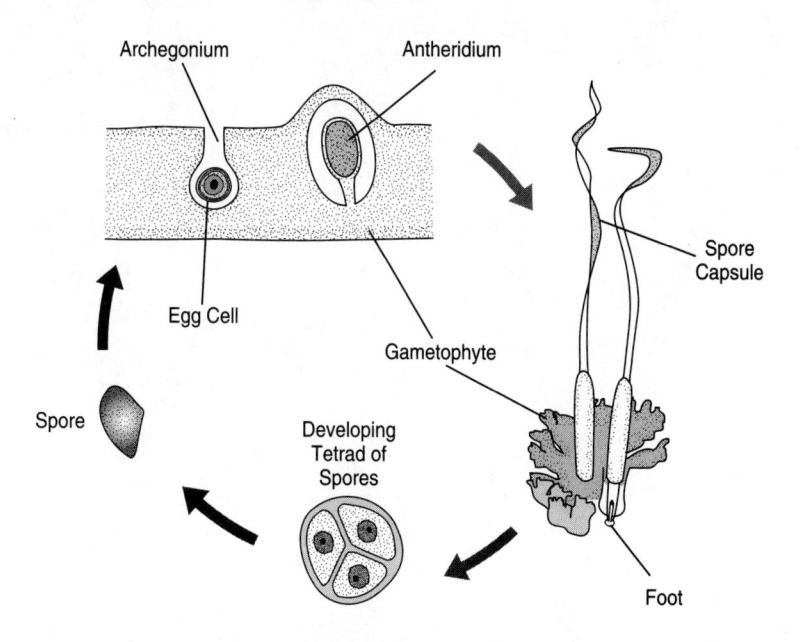

**Figure 24-4** *Anthoceros*. At left, an egg cell in the center of the archegonium and to its right an antheridium. At right, the sporophyte generation growing in the remnant of the archegonium. It produces a foot, which grows into gametophyte tissue and may continue into the soil below.

### ❋ Notes ❋

of generations. The gamete-forming generation begins with a meiospore. The meiospore produces a chlorophyll-bearing delicate thread called the **protonema**, rather than growing directly into the gametophyte. The protonema soon produces rhizoids, which grow down into the soil, and buds, which represent the first cells of the gametophyte plant. As the buds continue to grow, they each produce an elongated leafy stalk at the apex of which the reproductive organs are produced. In *Polytrichum*, the hairy cap moss, the sexes are separate. The upper end of the female gametophyte bears a cluster of archegonia among paraphyses, and the male plant bears the antheridia. Both antheridia and archegonia are on short stalks. Canal cells in the neck of the archegonium

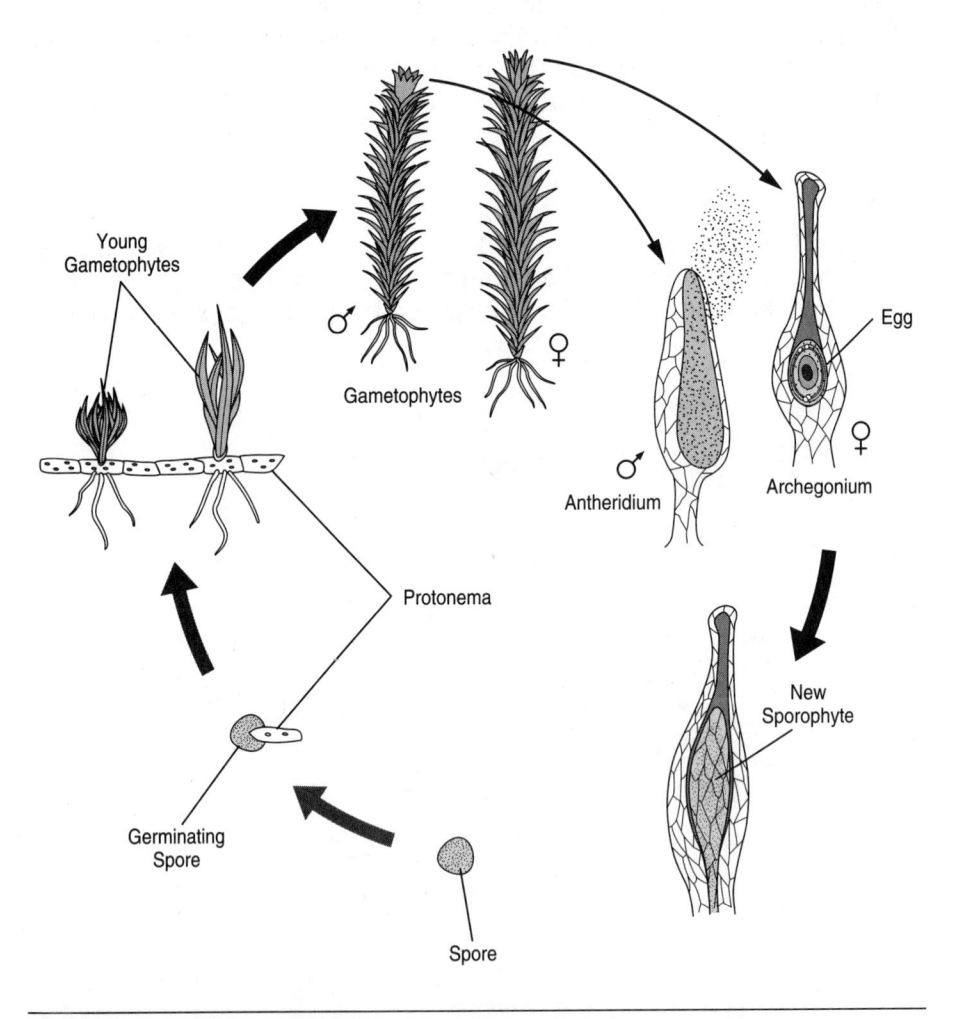

**Figure 24-5** Moss gametophyte generation. At left a protonema with the beginnings of two young gametophytes. At center, the male gametophyte is distinguishable from the female gametophyte by the star-shaped expansion of the upper leaves. At right, the egg cell fertilized in the venter of the archegonium grows into a new sporophyte.

must break down to allow a passageway for sperm cells. Biflagellated sperm cells are released from the antheridium and attracted to the archegonium by chemical substances released there.

The egg cell, being fertilized in the venter of the archegonium, grows into a sporophyte. The sporophyte remains attached to the upper end of the gametophyte. As the sporophyte grows, it produces a foot, which grows downward into the gametophyte tissue, and an elongated seta, which bears the spore capsule at its upper end. The archegonium also continues to grow. As the sporophyte elongates, a portion of the archegonium breaks away and is carried aloft as a cap on the spore capsule. This is called the **calyptra**, or hairy cap. When the sporophyte matures, the tip of the seta bends, causing the capsule to hang downward. The calyptra then falls off, and the covering of the capsule, called the **operculum**, breaks away. The operculum is **dehiscent**; it thus breaks away along specific lines. At this point, the capsule is not yet open to release the spores, however. The **peristome**, a membrane below the operculum, dehisces along radial lines to produce a circle of membranous points, called the **peristome teeth**. The hygroscopic points move in and out with changes in humidity and aid in the dispersal of spores. The spores, landing on favorable soil, can then produce protonemata.

※ **Notes** ※

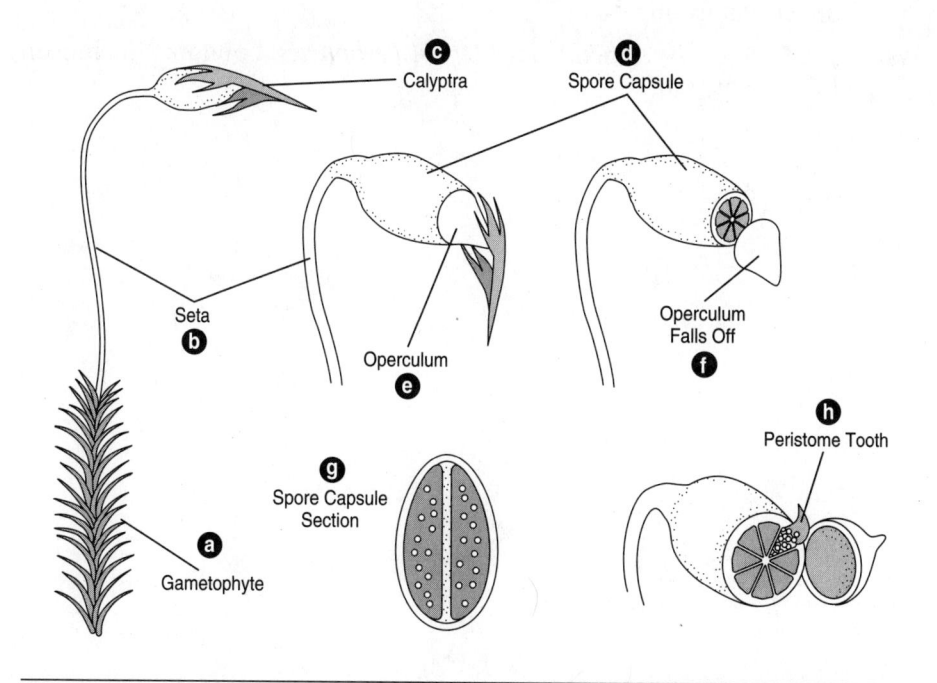

**Figure 24-6** Moss sporophyte generation. (a) The leafy gametophyte. (b) The stalk, or seta, rising up from the gametophyte. (c) The calyptra. The stalk bends, and the calyptra falls away, revealing the spore capsule, (d), capped by the operculum, (e). (f) The operculum falls off. (g) A cross section of the spore capsule revealing a central column of tissue. (h) A head-on view of the spore capsule showing a peristome tooth extended.

## ✹ Notes ✹

## Questions for Review

1. What are some of the significant traits of Bryophytes?
2. What is a protonema?
3. What stage in the life cycle of a bryophyte is haploid?
4. Define the terms *calyptra, operculum, peristome,* and *seta.*
5. Describe the life cycle of *Marchantia.*
6. Hornworts possess a trait that some have suggested connects them with higher plants. What is this trait?
7. How does *Marchantia* obtain water?
8. What is a pseudostomate?
9. Describe an asexual form of reproduction in *Marchantia.*
10. What is the dominant generation in all bryophytes?

## Suggestions for Further Reading

Conard, H.S. 1956. *How to Know the Mosses and Liverworts.* Dubuque, IA: Wm C. Brown Publishing Co.

Watson, E.V. 1971. *The Structure and Life of Bryophytes.* London: Hutchinson and Company, Ltd.

# 25

# Pteridophytes: The Ferns, Club Mosses, and Horsetails

In Bryophytes, the conspicuous generation is the gametophyte. In Pteridophytes, we find a reversal of roles. The conspicuous generation is the sporophyte, and the gametophyte is small and often unseen. This is true of all higher plants. The members of Pteridophytes have vascular tissue, but they do not form seeds.

※ Notes ※

## Ferns

There are approximately 10,000 species of ferns. Ferns grow in varied habitats. *Marsilia*, a rooted fern, grows in water. *Salvinia* floats on water. The woodland walking fern (*Camptosorus*) forms new plants wherever a leaf tip touches the soil; it thus can go (or grow) from one site to another. The shield fern (*Dryopteris*), maidenhair fern (*Adiantum*), adder's tongue (*Ophioglossum*), and interrupted fern (*Osmunda*) are woodland species. The shoestring fern (*Vittoria*) is an epiphyte on tree trunks.

The familiar leafy plant is the sporophyte. In many species of fern, the sporangia are found on the undersides of leaves. Sporangia occur in clusters called sori, and in many species, the sori are covered by a membrane called the **indusium**. In the sporangium, spore mother cells go through meiosis to produce meiospores. When the time comes for the spores to be released, the indusium becomes dry and splits. The sporangia open to release the spores, which then grow into gametophytes. The sporangium has a collar of cells with thickened walls, which tend to straighten and bend as the humidity changes. The sporangium dehisces where the lip cells are located.

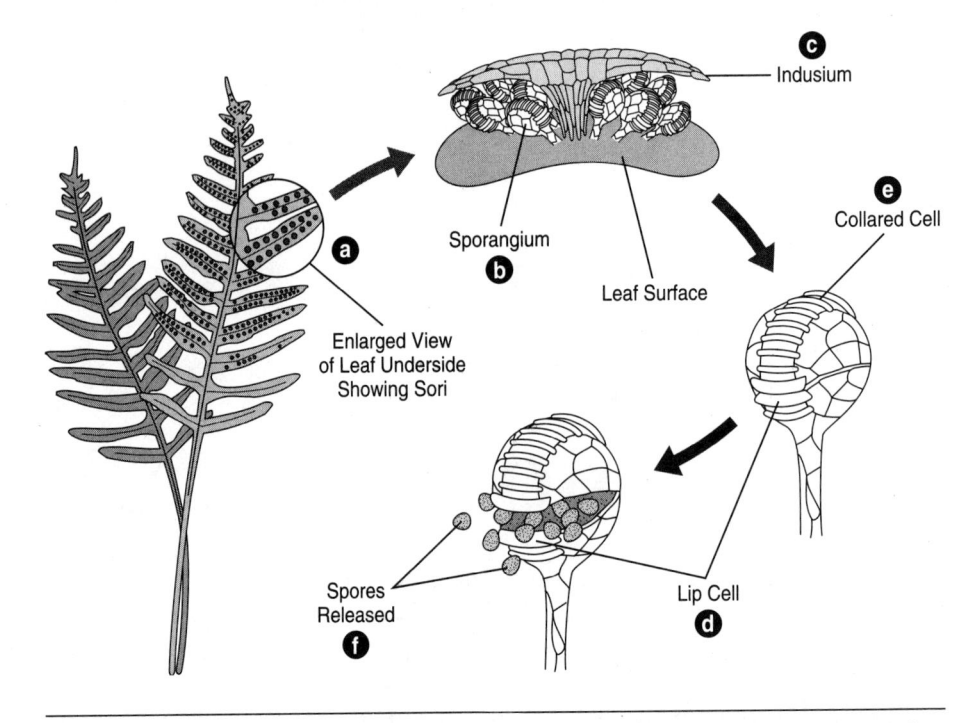

**Figure 25-1** The sporophyte generation in the life cycle of a fern. (a) The underside of a small portion of fern leaf, showing sori. An enlarged drawing of a sorus shows a sporangium, (b), and the indusium, (c). A lip cell, (d), and a collared cell, (e), in a yet to open sporangium. (f) The sporangium dehisces at the lip cells. (g) Spores are released.

The gametophyte is approximately one-quarter of an inch across. In some fern species, both antheridia and archegonia grow on the same thallus; in other species, the sexes are separate. When the plant is wet with rain or dew, multiflagellated sperm cells swim to the egg cell in the archegonium. Here, the new embryo sporophyte is briefly attached to and dependent on the gametophyte. Soon, however, a root penetrates the tissue of the gametophyte, and the sporophyte is established as an independent organism. Figure 25-2 shows that the first leaf of the young sporophyte is different from the leaves that come later.

## Club Mosses

*Lycopodium* is an example of club mosses. These plants are often seen in the woods and may be mistaken for pine seedlings. In fact, they go by the vernacular name "ground pines." The sporophytes produce upright stems, which bear minute leaves, called **microsporophylls**, each having a single, central vein without branches. Stomates can be found on both surfaces of the leaves. The leaves arise without forming leaf gaps (an advanced characteristic found

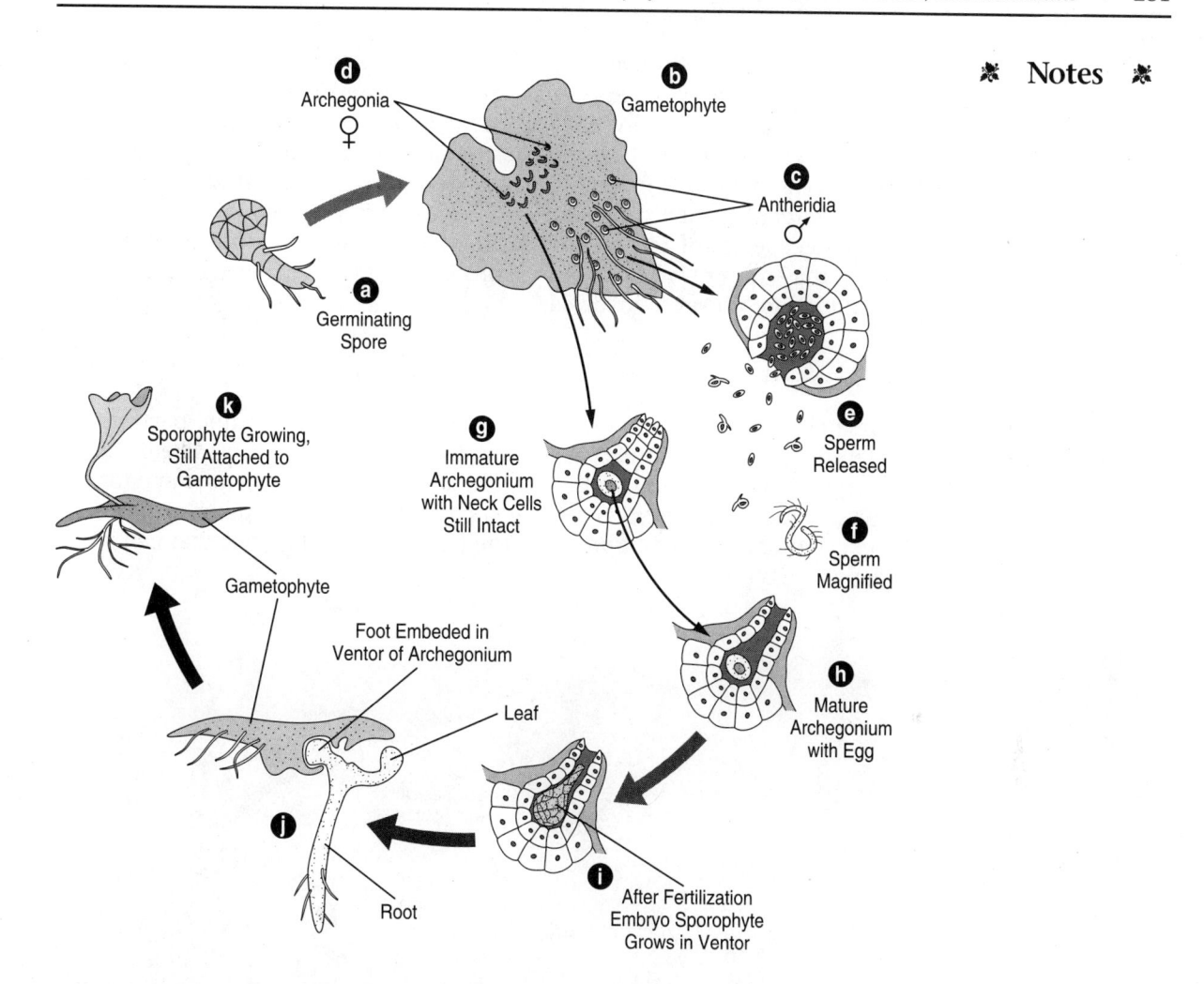

*Pteridophytes: The Ferns, Club Mosses, and Horsetails* ♦ 231

**Figure 25-2** The gametophyte generation in the life cycle of fern. (a) A spore germinates to produce the first few cells of the gametophyte. (b) The heart-shaped, mature gametophyte. (c) Antheridia among the rhizoids. (d) Archegonia near the notch of the upper portion of the gametophyte. (e) An antheridium opens to release sperm. (f) A much magnified sperm cell, showing many cilia. (g) An immature archegonium with the neck canal cells still intact. (h) A mature archegonium with an egg. (i) The embryo sporophyte growing in the venter of the archegonium. (j) The young sporophyte sends a root downward and a shoot upward with the primordium its first leaf. (k) The sporophyte continues, still attached to the gametophyte.

in higher plants and defined as an interruption of the vascular tissue immediately above the leaf trace).

The vascular tissue is similar to that in higher plants. The xylem and phloem alternate, producing a pattern similar to that found in the roots of higher plants. The stems and roots grow up from horizontal rhizomes (stems).

### ❋ Notes ❋

Such roots are called **adventitious**. Sporangia are borne in clusters, forming club-shaped strobili (**strobulus**). Spore mother cells undergo meiotic divisions, and the resulting meiospores grow into minute gametophytes.

The gametophytes bear both antheridia and archegonia. Biflagellated sperm cells swim to the venter of the archegonium to fertilize the egg cell, which is then the first cell of the new sporophyte generation.

Some species of *Lycopodium* form small masses of tissue called *bulbils*, which may drop off to form new sporophytes. This is a vegetative means of reproduction; no sex is involved.

## Horsetails

Horsetails were large marks on the landscape in the latter part of the **Paleozoic** era, more than 300 million years ago. Then, they were like big trees, reaching heights of 100 feet. Now, they are but remnants of their former selves, limited in height to several inches. The sole exception is found in tropical America; *Equisetum gigantium* has a weak, waxy stalk that can reach a height of 30 feet.

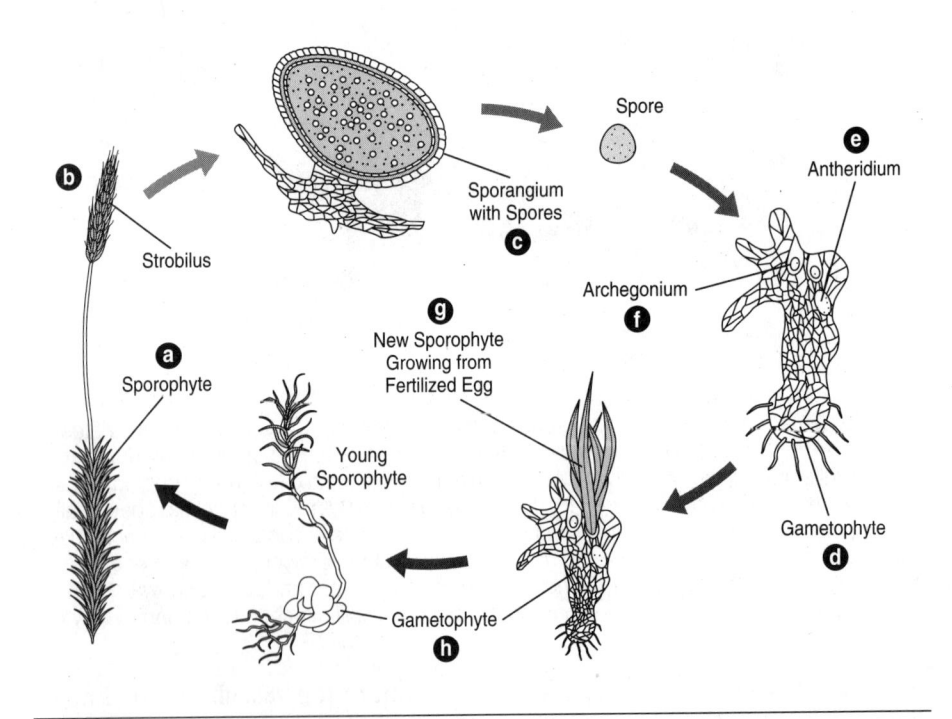

**Figure 25-3** *Lycopodium*. (a) The sporophyte. (b) A strobilus. (c) A longitudinal section through a sporangium with spores. (d) A gametophyte. (e) The antheridium. (f) The archegonium. (g) A new sporophyte beginning to grow up from the fertilized egg in the venter of the archegonium. (h) A gametophyte.

*Equisetum arvense* serves as a good example of horsetails. It produces both vegetative and sporogenous shoots. The stems are ribbed and hollow. A cross-sectional view shows three kinds of canals. The central canal is called the **pith** canal because when first formed, it is filled with pith, which in maturity breaks down to leave an open canal. The two other kinds of canals are the vallecular canals and the carinal canals. A *vallecula* is a groove, and the vallecular canals are aligned with the grooves of the stem. A *carina* is a ridge, and the smaller, carinal canals are aligned with the ridges of the stem.

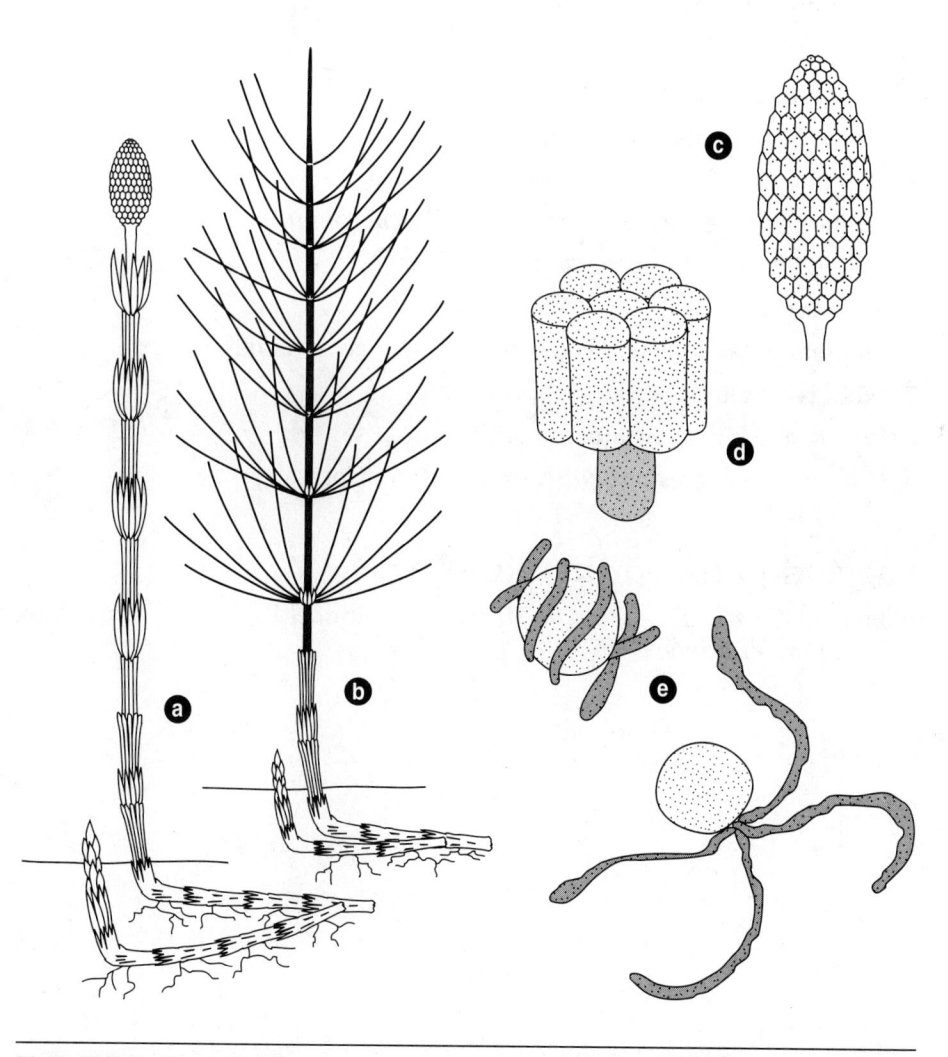

**Figure 25-4** *Equisetum arvense*, a horsetail. (a) A reproductive shoot bearing a strobilus at the tip. (b) A vegetative shoot. The shoots arise from horizontal rhizomes. (c) An enlargement of the strobilus. (d) A sporangiophore bearing six sporangia. (e) Spores, one with elaters coiled about the spore, and the other with elaters extended.

### ✿ Notes ✿

At the tip of the spore-producing shoot, a strobilus develops where the sporangia occur. Upon maturity, the sporangia release spores entwined with elaters. In drying, the elaters uncoil as the humidity changes and thus aid in the dispersal of the spores. Spores germinate within several hours to produce minute gametophytes, which look like protonemata at first. The antheridia produce multiciliated sperm cells. The female gametophytes produce archegonia with egg cells unless conditions are impoverished, in which case they undergo sex reversal and produce antheridia. When fertilization occurs, a new sporophyte is produced.

## Questions for Review

1. Where is an indusium found?
2. Describe the life cycle of a fern.
3. Define the terms *sorus*, *indusium*, and *archegonium*.
4. Define the term *hygroscopic*.
5. What are sori?
6. *Equisetum* has three kinds of canals in the stem; name these canals.
7. What is a bulbil?
8. When were horsetails a dominant form?
9. How are the spores of *Equisetum* aided in dispersal?

## Suggestion for Further Reading

Billington, Cecil. 1952. *Ferns of Michigan*. Bloomfield Hills, MI: Cranbrook Institute of Science.

# 26

# Tissues

**※ Notes ※**

Plant tissues are of two kinds: meristematic and **permanent**. Meristematic tissues are those that remain capable of dividing; permanent tissues are those that have lost this ability. Sometimes, permanent cells recover the capacity to divide (and, thus, are only temporarily permanent!).

## Meristematic Tissue

Meristematic tissues are further divided into two groups: **apical meristem** and **cambium**. Apical meristems are found at the tips of **stems** and **roots**. These cells increase the length of shoot and root, and are responsible for primary (first) growth. The cambium is of two kinds: **vascular cambium** and **cork cambium**. Vascular cambium produces two kinds of tissues: xylem, which is deposited central to the cambium, and phloem, which is deposited peripheral to the cambium. Cork cambium also produces two kinds of tissues: **cork**, which lies outside of the cork cambium, and **phelloderm**, which lies internal to it.

## Simple Tissues and Complex Tissues

Another way whereby tissues can be classified is again into two kinds: simple and complex. **Simple tissue** is composed of only one kind of cell, and **complex tissue** is composed of two or more kinds of cells.

### Simple Tissues

There are five types of simple tissue: **epidermis**, **parenchyma**, **collenchyma**, **sclerenchyma**, and **cork**. Epidermis covers the outer surfaces of leaves, young roots, and stems. As plants (especially woody plants) age, the epidermis (a primary tissue) on stems and roots is replaced by bark (a secondary tissue). Stomates are found with epidermal cells on leaves and stems but not on roots. Epidermal cells are living. They are almost entirely covered on the outer surface by a waxy **cuticle**, which retards water passage.

### ✽ Notes ✽

Parenchyma cells are commonly **isodiametric** (as broad as they are long) and have thin walls, living protoplasts, and conspicuous intercellular spaces. Parenchyma is regarded as an unspecialized tissue. If the cells have chloroplasts, the tissue is called **chlorenchyma**. In potatoes and fruits, the parenchyma cells store starch. Parenchyma is the main tissue of apples, pears, plums, and tomatoes. One of the most significant features of parenchyma is that it retains living protoplasts at maturity.

Collenchyma is the first of the supporting tissues found in young stems and leaves. These cells retain living protoplasts in maturity, as do parenchyma cells. Wall thickenings are conspicuous at the corners of cells. While these thickenings contribute to the strength of cells, they also remain plastic so the cells can stretch. The walls are composed primarily of cellulose but also contain some pectic compounds. These cells become elongated (often as many

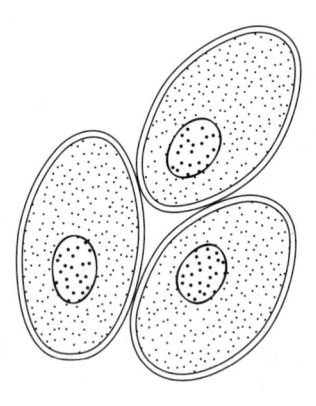

**Figure 26-1** Parenchyma, an unspecialized living tissue having conspicuous intercellular spaces.

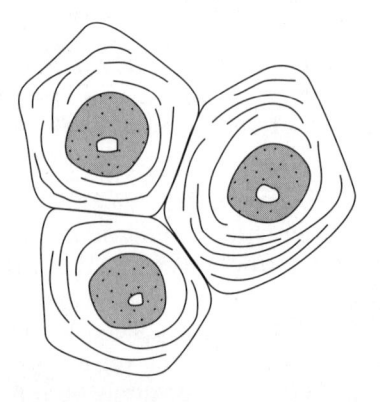

**Figure 26-2** Collenchyma

as two millimeters long) and frequently have chloroplasts. Intercellular spaces are sometimes noted.

There are two kinds of sclerenchyma, another supporting tissue: **stone cells** and **fibers**. Stone cells (also called **sclereids**) are short, nearly isodiametric, and have thick, lignified walls with tubular **pits**. They occur in small clusters in the tissue of pears. They also compose the hard shells of walnuts. In maturity, these cells do not have any living contents. In this sense, they can be distinguished from collenchyma.

✤ Notes ✤

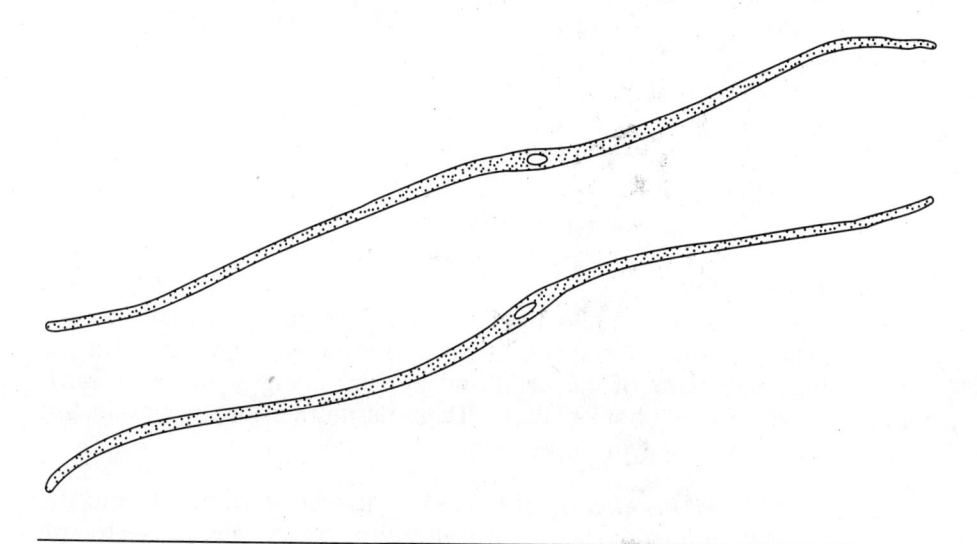

**Figure 26-3**  Fibers. Note the small lumen in the center of the cell.

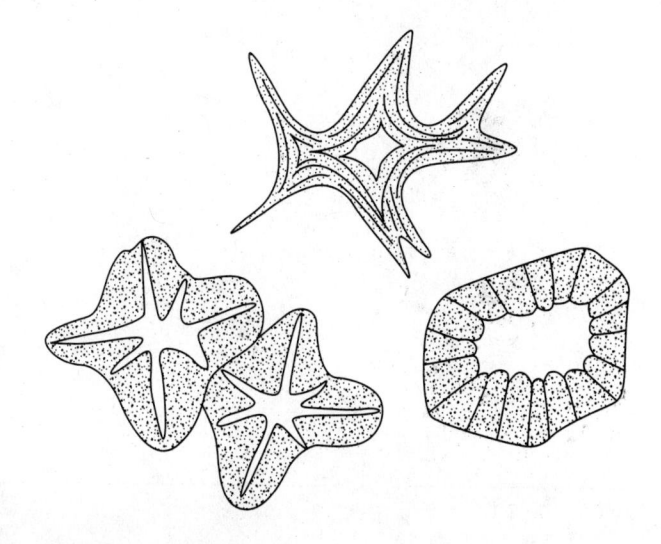

**Figure 26-4**  Sclereids (stone cells)

### Notes

Fibers are much elongated and have greatly thickened secondary walls, which may be of cellulose and lignin. The walls are often so thick that the lumen of the cell nearly disappears. There are no living protoplasts. Fibers are of two kinds: those that occur in wood and those that occur elsewhere. The latter are called **bast** fibers, and they are also of two kinds: those having lignified cell walls and those having cellulose cell walls. The latter are valued commercially. Bast fibers are often associated with the phloem, or inner bark, and may reach lengths of several feet—that is, an individual cell reaches a length of several feet. Examples include sisal, jute, hemp, ramie, hennekan, sansievaria, maguey, palma, formium, caroa, coir, latona, esparto, amuritius, and cabuya.

The last of the simple tissues is cork. Cork cells are found on the surfaces of stems and roots of the older parts of plants. Cork cells are flattened, and their walls are impregnated with the fatty substance suberin. There are no intercellular spaces, and the protoplasts soon die.

### Complex Tissues

There are two kinds of complex tissues: xylem and phloem. Both are conducting tissues, xylem being involved in the movement of water from the roots to the upper parts of the plant and phloem being involved in transporting the products of photosynthesis. These tissues are somewhat different in angiosperms than in gymnosperms.

*Xylem and Phloem in Angiosperms.* Several tissues are derived from the apical meristem; one of these is **procambium**, which gives rise to the

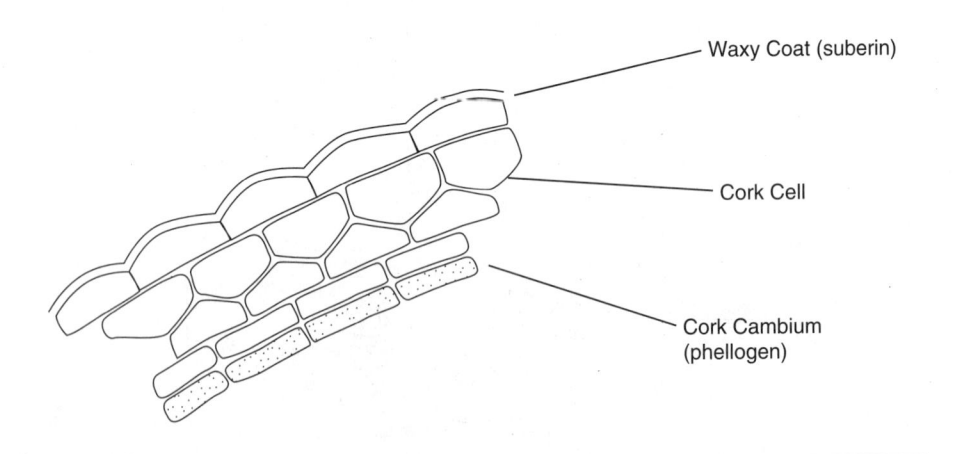

**Figure 26-5** Cork

cambium. In a woody plant, the cambium is a cylinder of meristematic cells one cell in thickness and lying just below the bark. Its divisions produce xylem toward the center and phloem toward the outside. The xylem has two kinds of conducting cells, **tracheids** and **vessels**, and two kinds of non-conducting cells, fibers and parenchyma.

The tracheids, vessels, and fibers are nonliving when functional. Tracheids are tapered at their ends and make contact with other tracheids. The secondary wall thickenings in tracheids produce a variety of patterns. Some forms show rings, some continuous spirals. In others, the wall thickenings produce a reticulate pattern, or a pitted appearance. A variety of names can therefore be applied to tracheids: annular tracheids, spiral tracheids, scalariform tracheids, reticulate tracheids, and pitted tracheids (see figure 26-6).

Vessels are composed of vessel elements, individual cells with the end walls digested away. Vessels formed in this way can be up to one meter long. The ends of vessels have walls at top and bottom. These end walls are perforated; vessels are thus able to conduct water more readily than are tracheids.

❈ **Notes** ❈

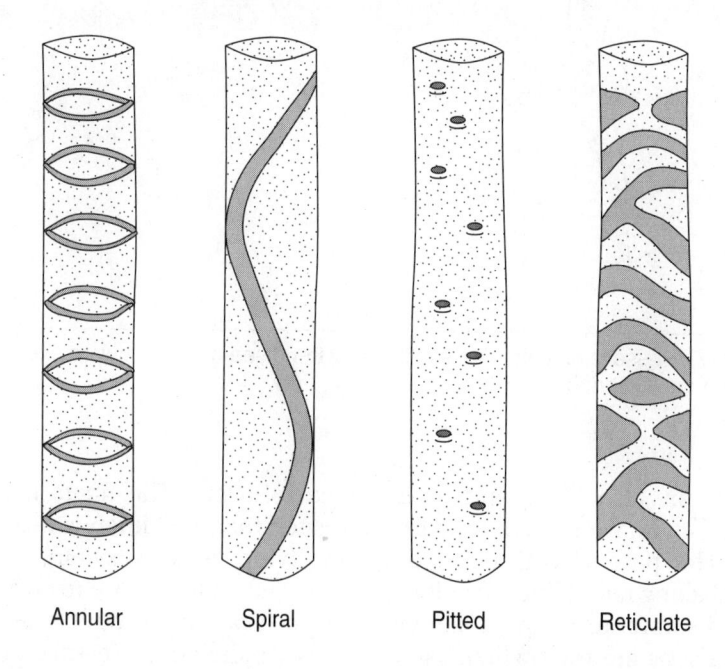

Annular    Spiral    Pitted    Reticulate

**Figure 26-6** Types of tracheids

## ❋ Notes ❋

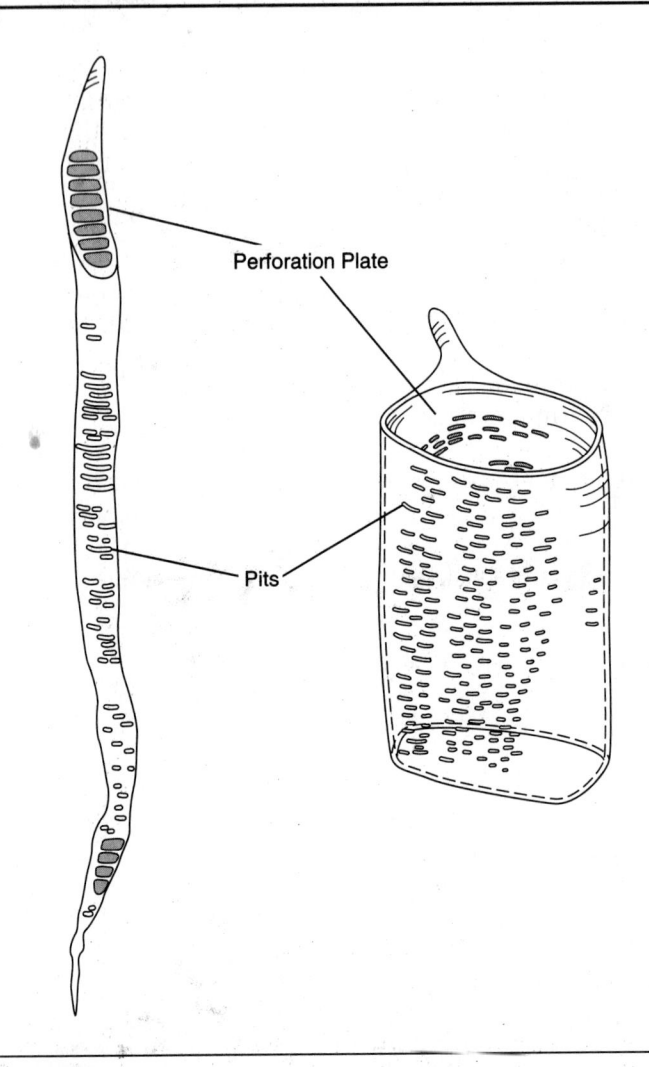

**Figure 26-7** Vessel elements. The evolutionary trend seems to be toward shorter and wider elements as at right.

There are also fibers and parenchyma in xylem. The fibers are greatly elongated and have thick secondary walls such that the lumen is nearly obliterated. The parenchyma cells, the only living cells in xylem, are often arranged along radial lines, resulting in rays. They may be one to several cells wide and several to many cells high. These cells are the products of cambial activity just as are the tracheids and vessels. As the circumference of a stem increases, new rays are produced. (Further details are provided in chapter 32 on stems.)

In angiosperms, the phloem (which is involved in transporting the products of photosynthesis from the leaves to the other parts of the plant) is distinguished by one particular feature: the **sieve tubes**. These are composed of elongated slender cells, the **sieve-tube elements**, having perforations in the end walls and positioned in a manner to produce continuous channels. Secondary wall thickenings do not occur, and the cells retain living cytoplasm although the nucleus is absent. Cytoplasm runs through the perforations in the cells to connect with adjacent sieve-tube elements. A central vacuole persists and extends through the end-wall perforations. The vacuolar membrane is not apparent.

Another feature of angiosperm phloem is the **companion cell**, which lies beside each sieve-tube element. This is a parenchymatous, living cell. The sieve-tube element and companion cell are derived from the division of a parent cell, which can be called the sieve-tube mother cell. The cytoplasm of these two cells remains in contact by delicate strands that run through microholes in the walls. These are plasmodesmata. There may be a single companion cell lying beside each sieve-tube element or several cells, which arise by subsequent divisions of the companion cell.

A complex carbohydrate composed of glucose units and called **callose** tends to be deposited at the site where cell perforations are most developed. This deposition eventually results in closing off the holes, thereby rendering the sieve tubes nonfunctional.

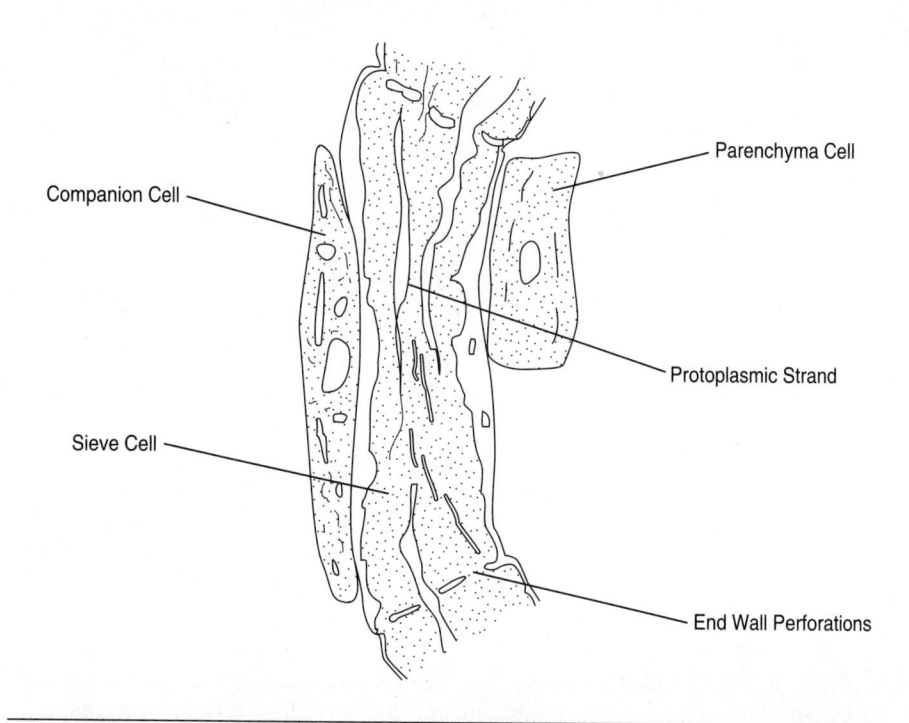

**Figure 26-8** Sieve cell with companion cell. The protoplasmic strand runs through the end-wall perforations.

### ※ Notes ※

In addition to the sieve tubes and companion cells the phloem also contains fibers and ordinary parenchyma.

The movement of water and its dissolved materials through the phloem is not thoroughly understood. The solutions can move in opposite directions at different times, and different solutes can move in opposite directions through the same tubes at the same time. This transport is achieved by living cells, and it is clear that energy is expended in the process. If phloem cells are killed, the movement ceases.

*Xylem and Phloem in Gymnosperms.* When the xylem of gymnosperms is compared to that of angiosperms, the most conspicuous difference is the lack of vessels. The only conducting cells in the wood of gymnosperms are the tracheids. It is yet possible to discern **annual rings** because the tracheids formed in the spring are wider in diameter than those produced later in the season.

The pits formed in the secondary walls of the tracheids are **bordered pits** quite unlike those of angiosperms (see figure 26-9). The primary wall in the center of the pit is thickened; it is called a **torus**. The secondary wall produces borders. The pits of adjacent cells are directly opposite, and the pits are sometimes collectively called a pit pair. It has been observed that the torus can be deflected and press closely to the mouth of the pit, thereby obstructing entry into the pit cavity. This seems to be without significance, however.

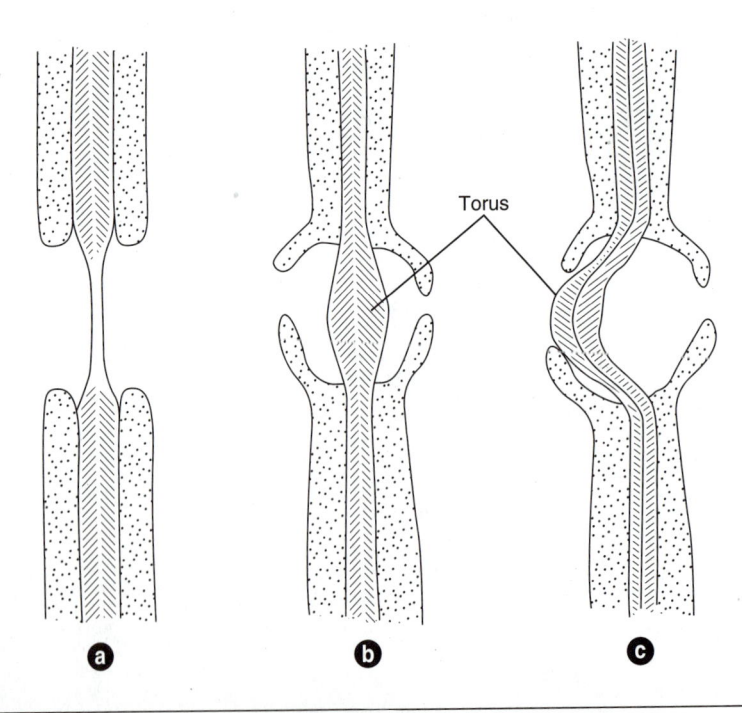

**Figure 26-9** (a) A pit as it occurs in angiosperms. (b) A bordered pit as seen in gymnosperms. (c) The torus, here deflected and appearing to close the pit mouth on one side.

The phloem of gymnosperms is different from that of angiosperms in that sieve tubes are not formed. **Sieve cells** occur, but the end walls are not perforated as they are in angiosperms. Furthermore, there are no companion cells associated with the sieve cells in gymnosperms.

## Questions for Review

1. Characterize parenchyma, collenchyma, and sclerenchyma.
2. What are pits? Do they have any significance to the welfare of plants?
3. Define the terms *meristematic* and *permanent*.
4. The two recognized types of meristematic tissue are _____ and _____.
5. If a tissue is composed of only one kind of cell, it is called a _____ tissue; if it is composed of more than one kind of cell, it is called a _____ tissue.
6. Define the terms *bast*, *lignin*, and *suberin*.
7. Define the terms *xylem*, *phloem*, and *companion cell*.

## Suggestions for Further Reading

Cutter, E.G. 1978. *Plant Anatomy*, Part I, "Cells and Tissues." Reading, MA: Addison-Wesley Publishing Co.

Esau, K. 1977. *Anatomy of Seed Plants*. New York: John Wiley and Sons.

O'Brien, T.P., and M.E. McCully. 1969. *Plant Structure and Development: A Pictorial and Physiological Approach*. New York: The MacMillan Co.

※ Notes ※

# 27

# Gymnosperms

Before considering the two great groups of seed plants, the gymnosperms and the angiosperms, take a moment to speculate on their origin. Think back to a time before woody gymnosperms appeared. In the life-forms studied thus far, liberated spores grow into free-living gametophytes. In many cases, the spores that produce the female gametophytes and those that produce the male gametophytes appear to be the same; they are, thus, called homosporous. The club moss *Selaginella*, a pteridophyte, is heterosporous, however. Its **megaspores** give rise to female gametophytes, which remain within their spore walls even after fertilization; and its **microspores** give rise to male gametophytes, which also remain within their spore walls. They do not become free-living. Yet, the female gametophyte is not surrounded by sporophytic tissue, and thus cannot receive any nourishment from this tissue. Water is still required for the sperm to swim to the egg. Now, take this a step further; envision that the megaspore and resultant female gametophyte remain in the megasporangium on the sporophyte plant and that wind carries sperm to the plant, where the sperm fulfills its mission of fertilization. This sets the stage for the origin of seed plants. As part of this evolutionary trend, microspores and male gametophytes evolve into pollen grains in which germination and sperm development are delayed until transported by wind or insects to the stigma of a flower (that is, to a sporophyte plant). The female gametophyte develops within the **ovary** of a flower. In this way, the seed plant stage of evolution was accomplished.

❈ Notes ❈

## The First Seed Plants

The first seed plants were seed ferns and some primitive **conifers**. All of these can be called gymnosperms, forms wherein the seeds are not enclosed within ovaries. Seed ferns that bore seeds on their leaves appeared during the **Carboniferous** period of the Paleozoic era, approximately 340 to 280 million years ago. Although these ferns became extinct during the Mesozoic era, the plants that evolved into modern seed plants probably trace their origin to them.

## Classifying the Gymnosperms

An often used system of classifying plants is to place all the seed-bearing plants in the phylum **Spermatophyta** (*sperm* meaning "seed"). The spermatophytes are then divided into the gymnosperms (naked-seeded plants) and the angiosperms (enclosed seeds). An angiosperm has its seeds enclosed in an ovary, in maturity called the **fruit**. In many forms of gymnosperm, the seeds are not enclosed but, rather, lie on the surface of a bract, or cone scale. Gymnosperms are classified into four orders: Cordaitales, Coniferales, Ginkgoales, and Gnetales. The most important of these is Coniferales, which includes the largest of the terrestrial plants as well as plants of great economic importance.

## Coniferales

Consider the pine as an example of Coniferales. Pines are monoecious, producing both male and female cones (strobili) on the same tree. Thus, they produce both microspores and megaspores. Spore formation begins when the trees are approximately ten years old. The male cones, perhaps one-half inch long, are borne in clusters on side branches. The scales of the cone are microsporophylls, and two **microsporangia** are found on the underside of each scale. Microspore mother cells in the microsporangium undergo meiosis

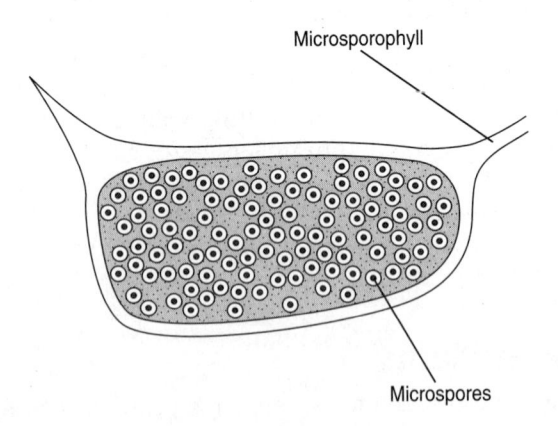

**Figure 27-1** Microsporangium with microspores. The microsporangium is borne on the lower surface of the microsporophyll.

to produce haploid microspores. The microspore is an immature pollen grain with a two-layered wall: an outer **exine** and an inner **intine**. The exine bulges away from the intine to produce air sacs, which aids wind dispersal of the pollen. Several nuclear divisions occur before the pollen is mature and released. These divisions produce two prothallial cells, a generative cell, and a tube cell. At this stage, the microsporangium splits open to release the pollen grains.

In the female cones, **megasporangia** are on the upper surfaces of the **megasporophylls**. In each megasporangium, a **megaspore mother cell** undergoes meiosis to produce a tetrad of megaspores. Three of these megaspores degenerate and disappear. The remaining megaspore resides in a tissue produced by cells of the megasporangium, the **nucellus**, the whole being surrounded by an **integument**. The integument is not complete, but, rather, leaves a passageway, the micropyle, through which the male gamete can gain

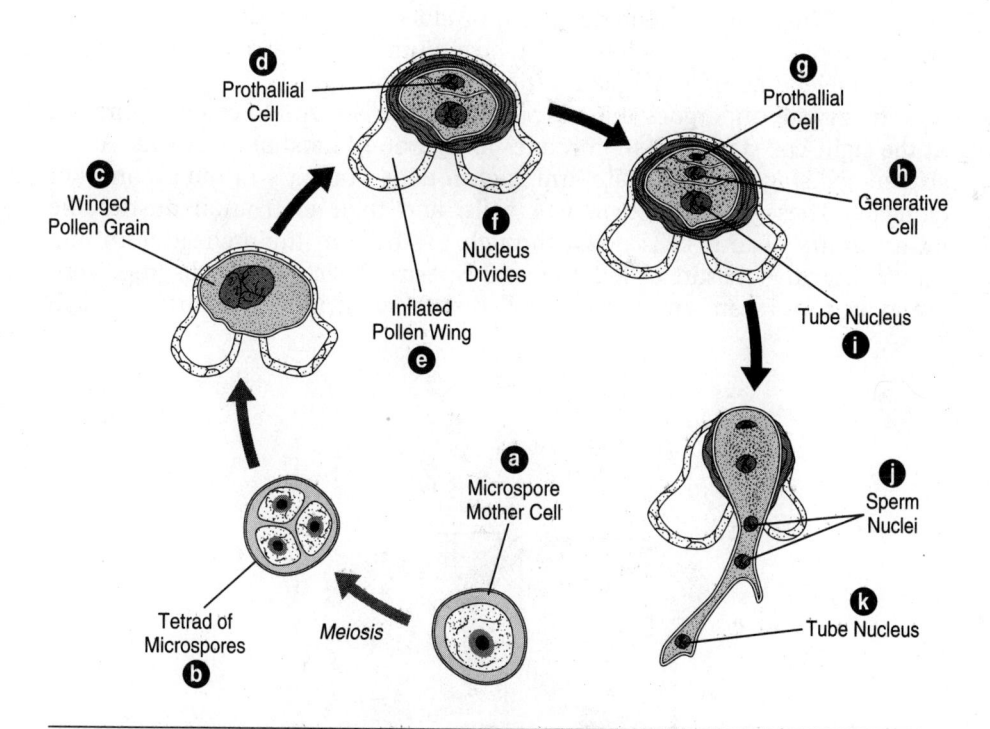

**Figure 27-2** Development of the male gametophyte from a microspore. (a) Microspore mother cell. (b) Microspore mother cell undergoes meiosis to produce a tetrad of microspores. (Only three microspores are shown here.) (c) Winged pollen grain. (d) The prothallial cell. (e) The inflated wing of the pollen. (f) The nucleus divides. (g) The prothallial cell. (h) The generative cell. (i) The tube nucleus. The generative cell divides to form a stalk cell and a body cell. (j) The body cell then divides to form two sperm nuclei. It is then a mature male gametophyte. (k) The tube nucleus.

※ **Notes** ※

access to the egg. The position of the micropyle is toward the central axis of the cone.

Some cells of the nucellus produce an exudate. This exudate, called a **pollination drop**, flows out to the mouth of the micropyle, where pollen grains adhere to the exudate. The pollination drop then shrinks and draws the pollen grains in through the micropyle to bring them into contact with the nucellus. The pollen is thus positioned an extremely short distance from where the egg will reside.

At the time of pollination, the female gametophyte has not yet grown to maturity; yet the **pollen tube** begins its slow growth through the nucellus. Digestive enzymes are secreted to create a channel through the nucellar tissue. The megaspore has produced the female gametophyte by numerous divisions. Toward the micropylar end of the gametophyte are two or more archegonia, each of which bears an egg cell. The growth of the pollen tube is very slow, taking as long as one year to reach the egg, which matures during this passage of time. The generative cell divides to form a stalk cell and a body cell. The body cell then divides to produce two nonmotile sperm nuclei. When the pollen tube reaches the archegonium, it ruptures and releases the sperm nuclei, one of which can fertilize the egg.

The zygote undergoes mitoses without initially forming cell membranes. At the eight-cell stage, cell membranes form around most of the nuclei. At the sixteen-cell stage, the cells are arranged in tiers. The cells in the second tier elongate. These are the **suspensor cells**, and their elongation pushes the lowest group of four cells down through the base of the archegonium and into a nutritive tissue called the **endosperm**. These four cells represent embryos; thus, there are four one-celled embryos at this moment. Although

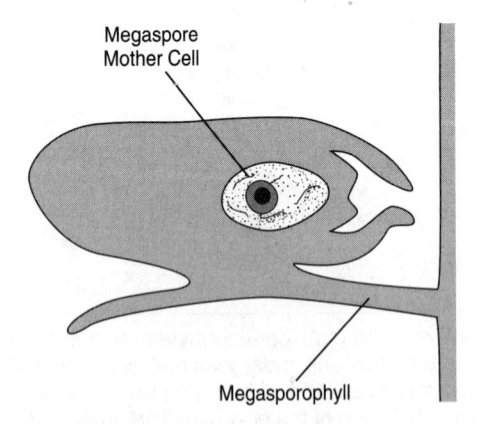

**Figure 27-3** A megasporangium, showing the megaspore mother cell before meiosis. The megasporangium is borne upon the upper surface of the megasporophyll.

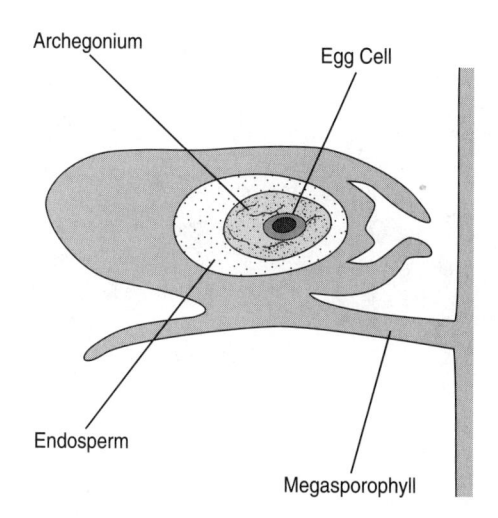

**Figure 27-4** The fully developed female gametophyte.

all of them may begin growth, one will gain over the others, which will be unable to continue development.

The growth of the embryo continues slowly, and the seeds may not be shed until after two or three years of development. Once shed, they may remain dormant for an extended period. The pine embryo is composed of a central **hypocotyl**, a **radicle**, a **plumule**, and several **cotyledons**.

Pines are described as **evergreen**, meaning that the leaves are not shed all at one time. Leaves remain on the tree varying lengths of time, depending on the species. Whereas *Pinus strobus*, the white pine, holds its leaves for two years, for example, *Pinus aristata* retains its leaves for as many as fourteen years. The leaves of *P. aristata* are long and slender (needle-like) and produced in bundles. The number of leaves in a bundle is usually three to five. Each leaf has a thick, waxy cuticle and sunken stomates, characteristics valuable in reducing water loss. Because leaves remain on the tree during winter, when water uptake is much reduced, the cuticle and sunken stomates thus represent important adaptations. A cross-sectional view of a pine leaf (figure 27-6) reveals a hypodermis two or three cells in thickness and immediately beneath the epidermis. The hypodermis also helps prevent water loss. A **vascular bundle** or two is located at the center of the leaf. The bundle is surrounded by a parenchyma, which is bounded by the endodermis. Outside the endodermis is a zone of mesophyll, through which resin ducts may be found passing.

Before discussing the tissues of a pine stem, two terms need to be introduced: **primary growth** and **secondary growth**. Primary growth derives from cells at the tip of a stem, whereas secondary growth derives from a ring of

## ❉ Notes ❉

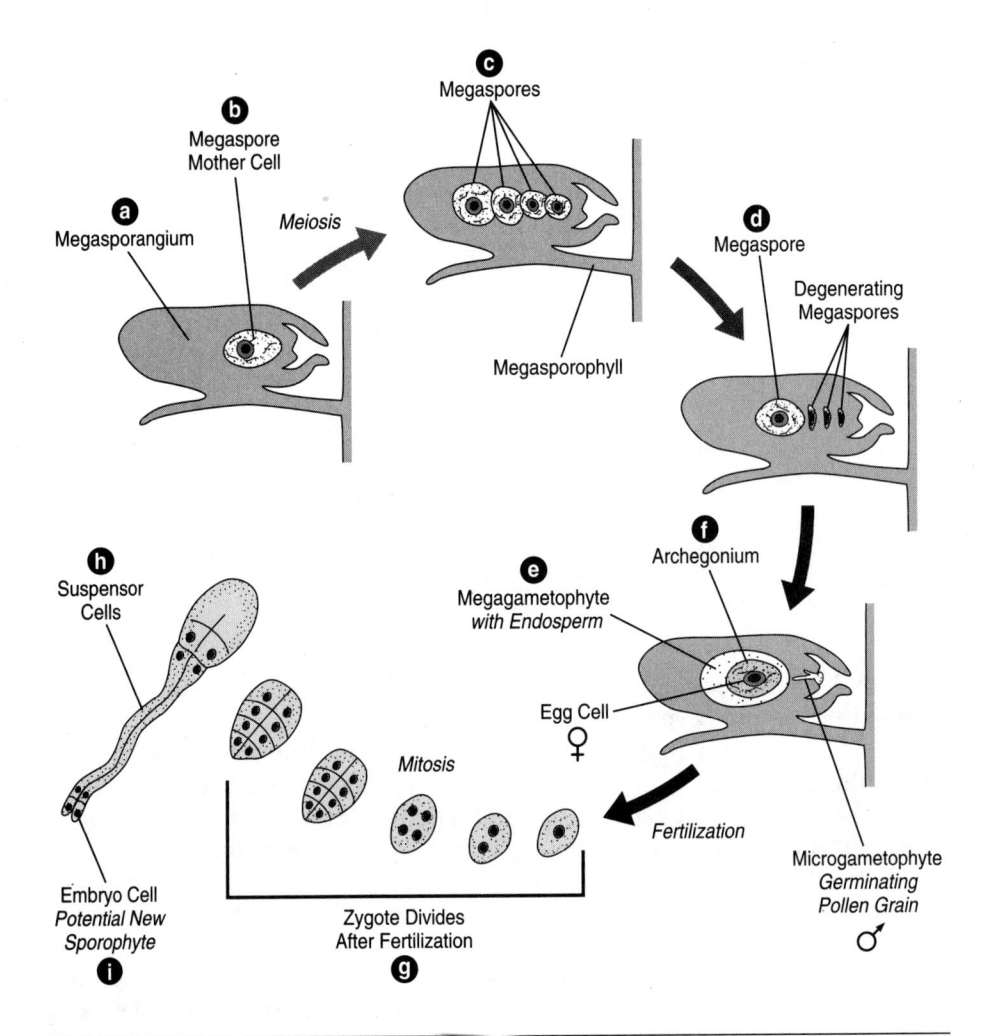

**Figure 27-5** Female gametophyte development and early embryo formation. (a) Megasporangium. (b) Megaspore mother cell. (c) Four megaspores are produced by meiosis. (d) Three of the megaspores degenerate, leaving only one. The megaspore grows into the female gametophyte having endosperm, (e), and archegonia, (f), each with an egg cell. (g) After fertilization, the zygote undergoes nuclear divisions. Here shown is the sixteen-cell stage. (h) The second tier of cells, the suspensor cells, elongate. (i) Below the suspensor cells are embryo cells. Each embryo cell is a potential new sporophyte.

cells a short distance below the tip (the cambium). A cross section of a stem near the apex, where no secondary growth has yet occurred, reveals a centrally located pith (parenchyma tissue) immediately outside of which is a circle of vascular strands and then a **cortex** bounded by an epidermis. A cross section of stem slightly below this reveals secondary tissues. A layer of cells

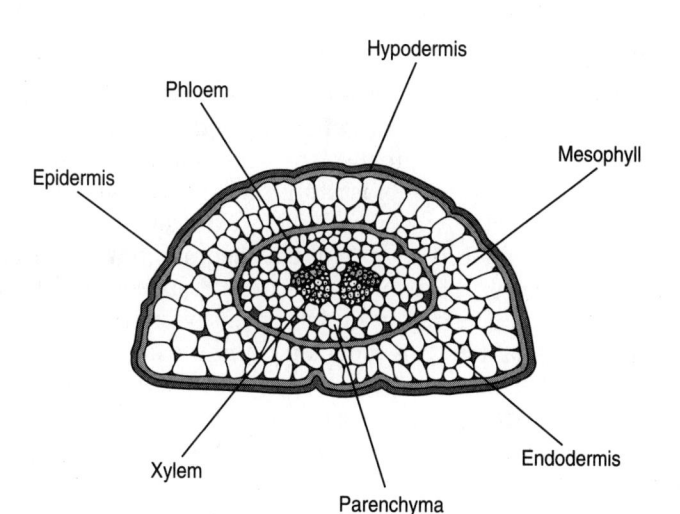

**Figure 27-6** Cross section of pine leaf.

between the primary xylem and primary phloem of the vascular strands as well as cells between the vascular strands collectively constitute the cambium, or actively dividing cells. Cambial cell divisions produce the secondary vascular tissues, xylem and phloem. The xylem is deposited centrally to the cambium; the phloem, peripherally. The secondary xylem is composed of two kinds of cells: tracheids and fibers. In cross-sectional view, the tracheids formed during spring are relatively large compared to the smaller tracheids formed during autumn. This results in a clear delineation of annual rings in the wood. Numbers of resin ducts can be seen scattered throughout the wood, or secondary xylem.

As the secondary xylem increases, the cambium, which lies outside of this xylem, must move outward and increase its circumference. Cambium must therefore produce not only secondary xylem and secondary phloem, but also additional cambium cells. As the diameter of the stem increases by cambial activity, the cortex and epidermis (the outermost primary tissues) split and slough away.

A longitudinal section reveals rows of bordered pits in the walls of the tracheids. As indicated previously, tracheid cells have both primary and secondary cell walls. Whereas the pits are created by interruptions in the secondary wall, the primary wall is continuous through the pits. The primary wall in the region of the pit has a thickening, called the torus. The secondary wall overhangs the pit, thus making a bordered pit. The pits of adjacent cells are aligned, making a pit pair. The pit membrane is flexible and can be pressed one way or the other.

## Notes

### Ginkgo biloba

Gymnosperms are generally and correctly considered to be coniferous trees having needle-shaped leaves commonly retained on the plant for several years. They are thus called evergreens. An exception is *Ginkgo biloba*, which has many traits of a hardwood angiosperm. *Ginkgo biloba* is the only remaining survivor of the Ginkgophytes, which were extensive in the late Paleozoic period. Their ancestors are unknown, but they seem to have arisen from more than a single precursor. In past times, there were sixteen genera; now there is only one. The tree is commonly grown in cultivation, and probably does not grow in its native habitat. It is dioecious, and the staminate trees are preferred in cultivation because the "female" trees produce foul-smelling "fruits." Leaves appear in late spring and are shed in the fall. *Ginkgo biloba* is commonly called the maidenhair tree because of the resemblance of its leaves to the maidenhair fern, *Adiantum*.

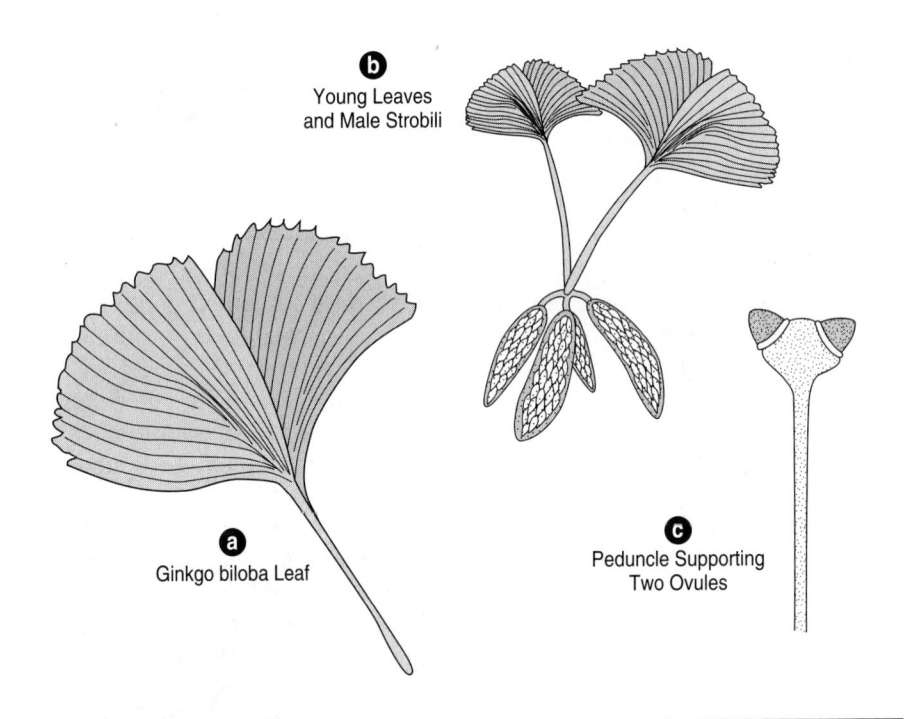

**Figure 27-7** (a) A *Ginkgo biloba* leaf. (b) Young leaves and catkinlike male strobili. (c) A peduncle supporting two young ovules.

Two types of branches are noted: long-shoot branches and short-shoot branches. Long shoots form the main branches of the tree, and short shoots, which form on the main branches, bear clusters of leaves at their tips.

The male strobili of microsporophylls hang down, giving the appearance of catkins (a kind of **inflorescence**); the female structure develops pairs of ovules on short spur shoots. When the pollen grains come in contact with the nucellus, a passageway is digested through the nucellus and discharges sperm into the archegonial chamber. Although the sperm cells do not swim to the egg cell, their flagellated structure is evidence of their primitive ancestry.

### ❀ Notes ❀

## Questions for Review

1. What is the meaning of the word *gymnosperm*? List several examples of gymnosperms.

2. In pine, the megasporophyll bears a _____ on its upper surface. Within this structure is a diploid cell, called the _____ _____ _____, which is destined to undergo meiosis to produce _____. One of these cells grows into the _____.

3. The egg cell of the pine gametophyte generation is borne in an _____. After fertilization, several nuclear divisions occur but do not all contribute to the formation of embryos, some of the cells being _____.

4. The cells immediately above the embryo cells elongate. They are called _____ cells, and their elongation pushes the embryo cells into the _____.

5. What is meant by the term *evergreen*?

6. Cambial activity produces secondary _____, secondary _____, as well as additional _____.

## Suggestions for Further Reading

Denison, W.C. 1973. Life in tall trees. *Scientific American* 228(6):74–80.

Sporne, K.R. 1965. *The Morphology of Gymnosperms*. London: Hutchinson and Company, Ltd.

# 28

# Angiosperms

In animal reproduction, a sperm cell unites with an egg cell to produce a zygote, which grows to the next generation. As we have thus far seen, the reproductive events in the life cycle of a moss or fern are different. They are also different in flowering plants, where an egg is fertilized by a sperm, and the new individual that grows from the zygote reproduces by asexual means (spores). The spore-producing plant can then produce a new individual that reproduces sexually. As previously described, this is alternation of generations: a sexually reproducing generation alternating with an asexually reproducing generation; a gametophyte producing a sporophyte and the sporophyte then producing a gametophyte.

When gametes (egg and sperm) are produced in animals, meiosis results in a reduction of the chromosome number. In plant reproduction, as has already been seen, meiosis does not occur at the time of gamete production, and, in fact, the gametophyte, which produces the gametes, is already haploid.

❊ Notes ❊

## Life Cycle

Now consider flowering plants. The conspicuous generation is the sporophyte. Flowers are a part of the sporophyte generation. Flowers are composed of modified leaves arranged in several whorls. The innermost whorl constitutes what is generally known as the female part. This raises an apparent contradiction, however. If the plant is a sporophyte, it seems contradictory to indicate it having female parts. As will be seen, not only is this contradictory, it is also acceptable.

The innermost whorl is called the **pistil**. It possesses the stigma (at its upper end); the long, slender **style**; and the ovary containing **ovules** (at the base). The ovules become seeds. Proceeding from the innermost whorl outward, the next whorl of flower parts are the **stamens**, composed of **filaments** and **anthers**. The anthers bear pollen grains. Next come the **petals**, which compose the **corolla**. The outermost whorl, the **sepals**, compose the **calyx**.[1]

---

[1] Flowers are further discussed in chapter 35.

## ✿ Notes ✿

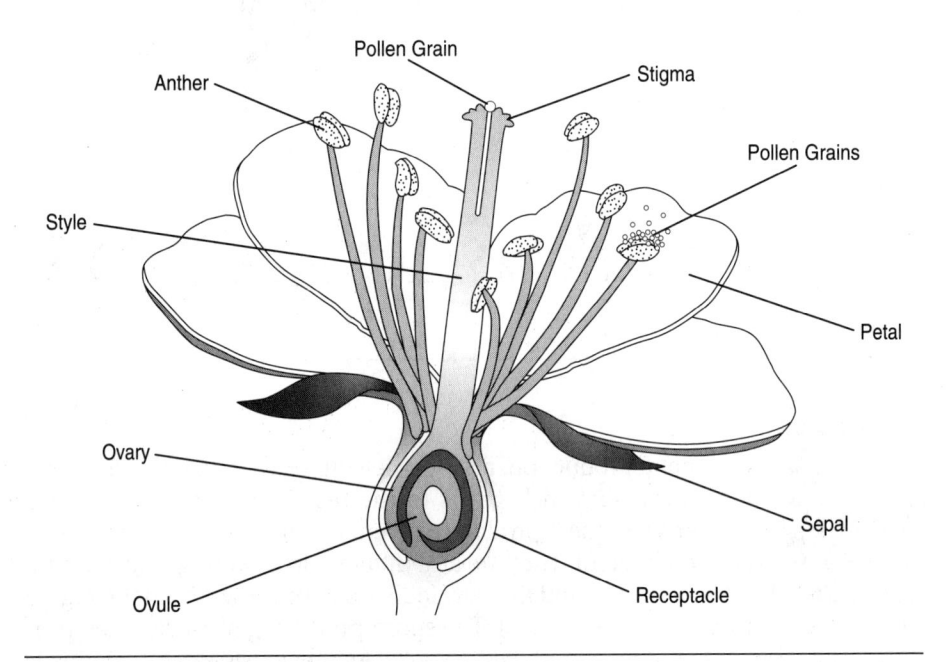

**Figure 28-1**  A generalized flower.

    Figure 28-2 shows a developing ovule. It is composed of many cells, one of which has a destiny different from that of the others. This cell is the megaspore mother cell, whose destiny is to undergo meiosis to produce a tetrad of haploid megaspores. Three of the four megaspores degenerate and disappear. The remaining spore grows into a female gametophyte, a microscopically sized generation. This megaspore undergoes three mitotic divisions without forming cell membranes. The resulting **embryo sac** has eight nuclei, which position themselves so that there are three nuclei at the upper end, three at the lower end (the micropyler end), and two in the middle. Cell membranes then form around the three upper nuclei, now called **antipodal cells**. Membranes also form around the lower nuclei, and here, an egg cell is bounded on each side by a **synergid**. The two nuclei in the center remain naked. These are **polar nuclei**. At this point, the embryo sac is properly called a mature female gametophyte.
    In the anther, the microspore mother cell undergoes meiosis to produce haploid microspores. The nucleus of each microspore then undergoes a single mitotic division to produce a binucleated pollen grain. It is in this condition that the pollen is shed. After landing on the stigma of the pistil, the pollen grain germinates, thus producing a tube that grows down through the style

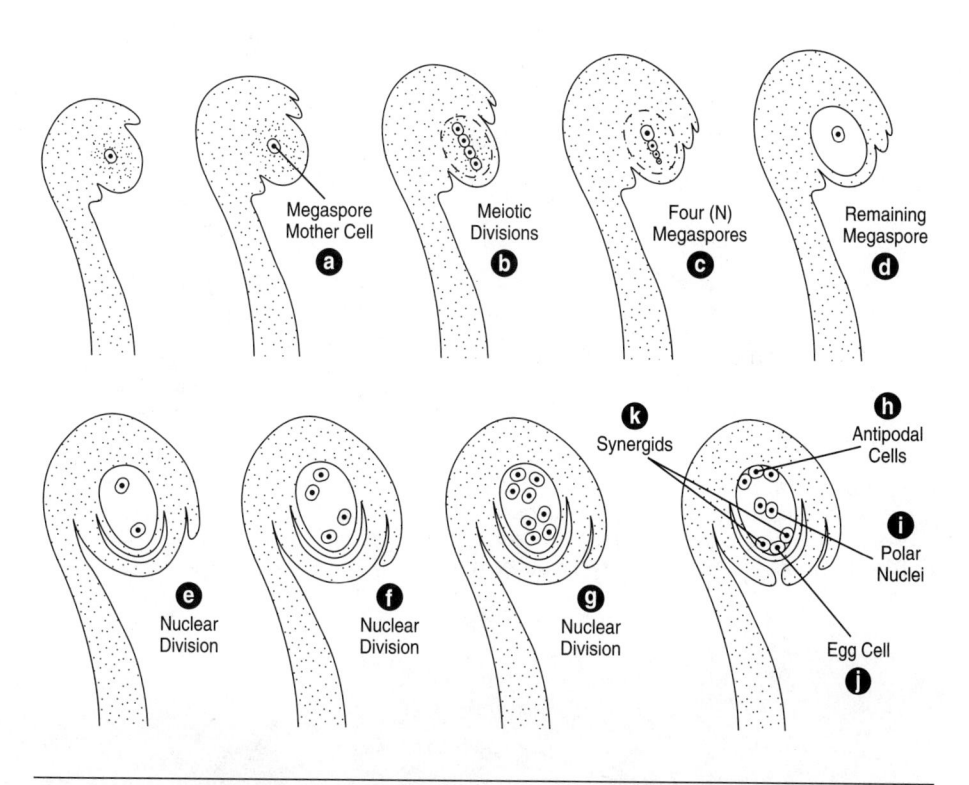

**Figure 28-2** Female gametophyte development in an ovule. (a) Ovule with a megaspore mother cell. (b) The first and second divisions of meiosis produce four haploid megaspores, (c), three of which degenerate and disappear, leaving only one megaspore, (d). (e), (f), and (g) The megaspore undergoes three nuclear divisions to produce an eight-nucleated embryo sac. (h) Three nuclei migrate to the upper end to form antipodal cells surrounded by cell membranes. (i) Two polar nuclei remain in the center of the embryo sac. Three nuclei migrate to the lower end (the micropylar end) and form cell membranes, resulting in an egg cell in the center, (i), bounded by two synergids, (j).

and toward the ovary. One of the nuclei leads the way near the end of the tube. This is called the **tube nucleus**. The other nucleus, called the **generative nucleus**, comes along behind and undergoes a division to produce two sperm nuclei. At this point, the pollen grain is properly considered a mature male gametophyte. The pollen tube grows to the ovule. In the ovule, the female gametophyte is largely but not completely surrounded by sporophytic tissue, the seed coats. A small opening called the **micropyle** remains in the tissue; and it is here that the pollen tube makes its entry. The end of the pollen tube secretes enzymes that digest a passageway through the outer membrane of the embryo sac. The tube nucleus then degenerates and disappears.

258 ♦ Chapter 28

※ **Notes** ※

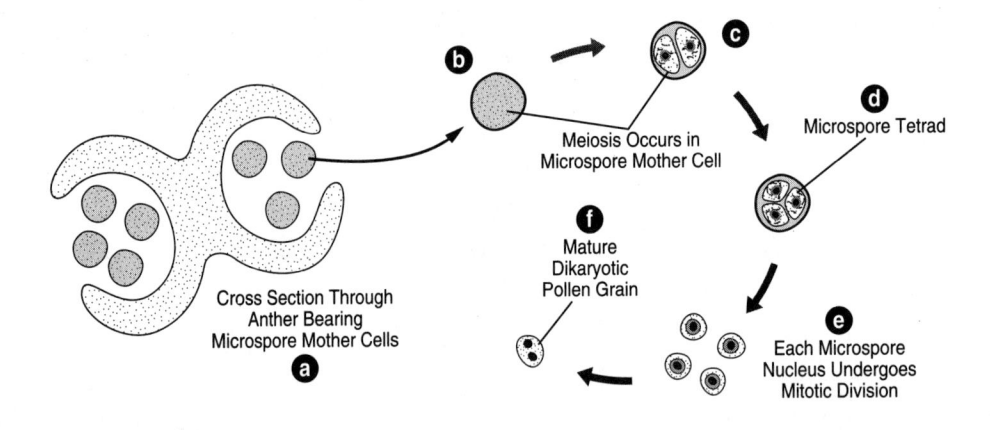

**Figure 28-3** (a) Cross section of an anther bearing microspore mother cells. (b) and (c) A microspore mother cell undergoes meiosis to produce a tetrad of microspores, (d). The microspores separate (e). The nucleus of each microspore undergoes a single mitotic division to produce a mature dikaryotic pollen grain (f).

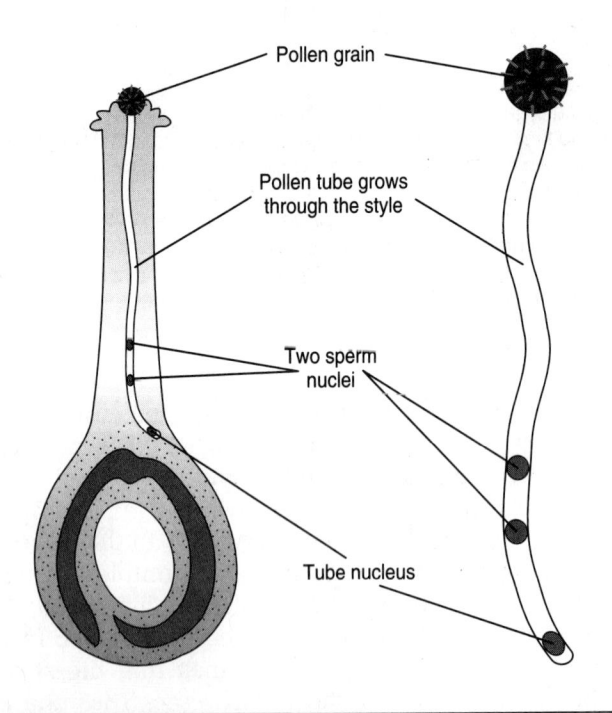

**Figure 28-4** At left, a male gametophyte in the pistil of a flower. At right, the male gametophyte alone.

The next event is termed **double fertilization**. Both sperm nuclei are involved in fusions. One sperm nucleus fertilizes the egg, and the other sperm nucleus migrates to and fuses with the two polar nuclei, yielding a triploid endosperm nucleus. The zygote divides numerous times to produce the embryo sporophyte, which resides in the developing seed. At the same time, the endosperm nucleus undergoes a number of divisions to produce a reserve of nutrients. The divisions of the endosperm nucleus are not initially accompanied by cell membrane formation; the endosperm is thus coenocytic. After the endosperm has formed, cell membranes are generated, and the tissue becomes cellular. Although the events vary in different species, figure 28-6 shows the endosperm being consumed by the embryo sporophyte as the sporophyte increases in size. When the seed is mature, nearly no endosperm remains.

The seed contains a 2N embryo sporophyte surrounded in early development by a 3N endosperm. The antipodal cells and synergids have degenerated. All, in turn, are surrounded by seed coats, the tissues of an older sporophyte generation. There are thus three generations of tissues present: two sporophyte generations with a gametophyte generation in between.

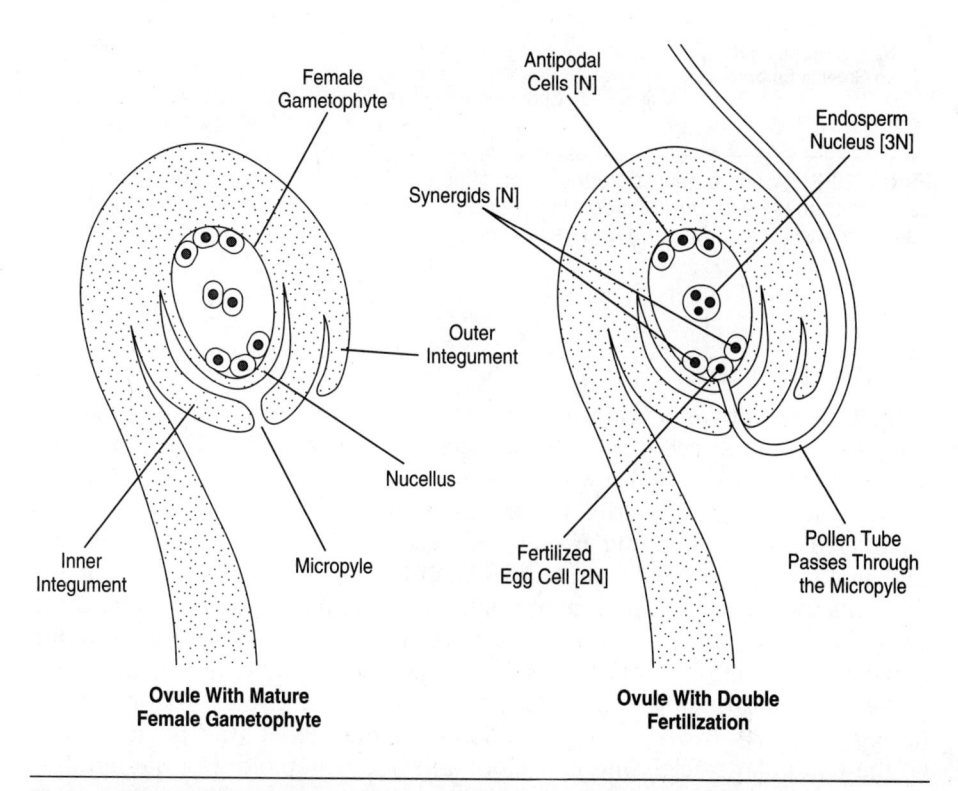

**Figure 28-5** At left, an ovule with a mature female gametophyte. At right, an ovule in which double fertilization has taken place.

### Notes

**New Sporophyte Begins to Grow in Embryo Sac**

- Coenocytic Endosperm
- Embryo
- Nucellus

**Cell Membranes Form Around the Endosperm Nuclei**

- Endosperm Nuclei
- Embryo Plantlet
- Nucellus

**The Embryo Plantlet has Consumed the Endosperm**

- Hypocotyl
- Cotyledons

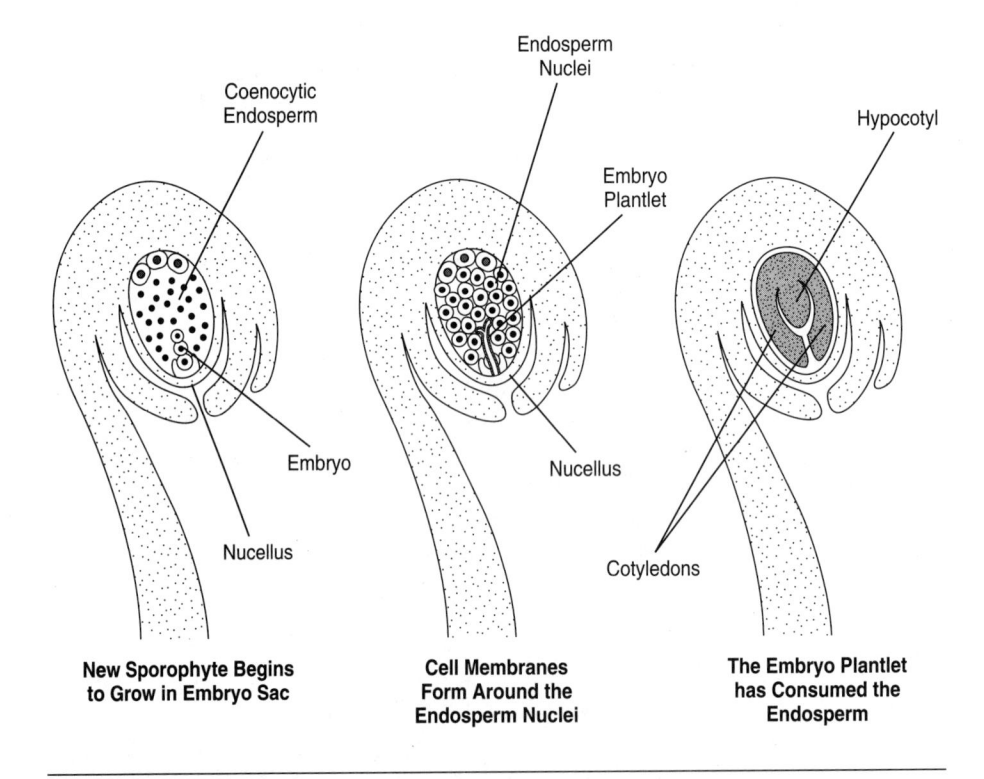

**Figure 28-6** At left, a new sporophyte has begun to grow in the embryo sac. At center, cell membranes have formed around the endosperm nuclei. At right, the embryo plantlet has consumed the endosperm and developed two cotyledons and the hypocotyl.

## Lilies

Female gametophyte development in lilies takes a different course. When the megaspore mother cell divides to produce four megaspores, all of the megaspores remain functional and reside within the embryo sac. Three megaspores then migrate to the upper end, where they fuse to produce a 3N nucleus. The remaining megaspore rests at the lower end of the sac. The remaining two nuclei undergo two mitotic divisions. The first of the divisions increases the number of nuclei in the embryo sac to four, and the second division increases the number to eight. Four of the nuclei are triploid, and four are haploid. These nuclei then position themselves as described earlier: three at the upper end (the antipodal nuclei), three at the lower end (an egg cell and two synergids), and two in the center (the polar nuclei). Of the two polar nuclei, one is triploid and one is haploid. Double fertilization as earlier described then occurs. One sperm unites with the egg to produce the zygote, a one-celled sporophyte, and the other sperm unites with the polar nuclei to produce a 5N endosperm nucleus.

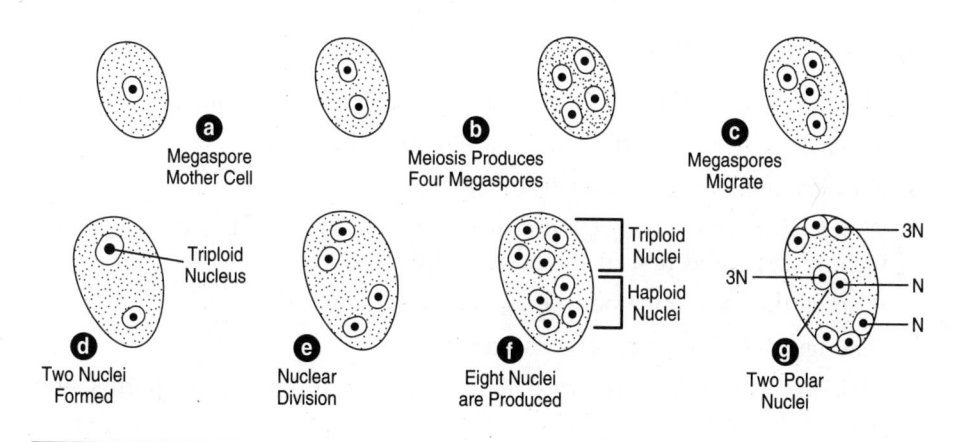

**Figure 28-7** Female gametophyte development in a lily. (a) Megaspore mother cell undergoes meiosis to produce four megaspores, (b). All four megaspores remain functional. (c) Three migrate to the upper end of the embryo sac, one to the lower end. (d) The three at the upper end fuse, forming a triploid nucleus. There are then two nuclei. (e) They divide twice, producing eight nuclei, four of which are triploid, and four haploid. (f) One of each cluster of four nuclei migrates toward the center of the embryo sac to form polar nuclei (g).

## Comparing Angiosperms to Gymnosperms

Having described both angiosperms and gymnosperms, it would be interesting to now compare the two. The following table lists characteristics of each for purposes of comparison.

| Angiosperms | Gymnosperms |
| --- | --- |
| Seeds are enclosed in a fruit. | Seeds are naked. |
| The dominant plants of the Earth, encompassing more than 250,000 species. | Encompass approximately 750 species. |
| Occupy both land and water. | Are land dwelling only. |
| Of more recent evolutionary origin. | Of more ancient origin. |
| Encompass both woody and herbaceous forms. | Encompass only woody forms. |
| Leaves are mostly broad and flat. | Leaves are mostly needle-like. |
| Have flowers. | Have strobili or cones. |
| Lack archegonia in embryo sac. | Archegonia are present. |
| Possess mostly fibrous roots. | **Taproots** predominate. |
| Have both tracheids and vessels in the xylem. | Have tracheids only. |
| Exhibit **deliquescent** branching. | Exhibit **excurrent** branching. |
| Many species have both microsporangia and megasporangia in the same flower. | Microsporangia and megasporangia are always separate. |
| Divided into **monocots** and **dicots**. | **Polycotyledonous**. |

Table 28-1 A Comparison of Angiosperms and Gymnosperms.

## ❈ Notes ❈

## Questions for Review

1. At what stage in the life cycle of a plant does haplosis take place?
2. At what point do haploid complements of chromosomes come together?
3. How do angiosperms differ from gymnosperms?
4. Which of the two groups, gymnosperms or angiosperms, are considered to have appeared first in the course of evolution?
5. What is the difference between a seed and a spore?
6. Distinguish between monocots and dicots.
7. How does the endosperm arise in angiosperms and in gymnosperms, and what is its value?
8. Explain double fertilization in angiosperms.
9. How do the numbers of angiosperms compare to the numbers of gymnosperms in living species?
10. The seed plants, called _____, are divided into two groups: the _____ in which the seeds are not enclosed, and the _____, each of which has seeds enclosed in an ovary. The structure that encloses the seeds is in maturity called a _____.

## Suggestions for Further Reading

Bierhorst, D.W. 1971. *Morphology of Vascular Plants*. New York: The MacMillan Co.

Esau, K. 1977. *The Anatomy of Seed Plants*. New York: John Wiley and Sons, Inc.

Ray, P.M. 1972. *The Living Plant*. New York: Holt, Rinehart, and Winston, Inc.

# 29

# Hormones

**Notes**

Many people are more familiar with hormones being produced in animals than in plants. For example, insulin is produced in the pancreas and is involved in the metabolism of sugar. The thyroid gland produces the iodine-containing hormone called thyroxine. The adrenal glands produce adrenalin. With regard to animals, hormones are defined in part as being produced by ductless glands. With regard to plants, hormones are defined as substances produced in minute amounts at one site that have a physiological effect at another site.

## In Search of Auxin

It has long been known that stems bend toward the light; and one might infer that a plant bends toward the light because it needs light. But how does it do this?

While working with germinating grass seedlings, Charles Darwin (famous for his theory of evolution) observed that the stem bent toward the light when the tips were exposed, and that curvature did not take place if the tips of the seedlings were shielded from light.

Figure 29-1 shows only the tip being shielded from light, and the curvature itself occurring a short distance below the tip. The name for this phenomenon is **heliotropism** (*helio* meaning "sun"), and, as Darwin observed, curvature occurs even when light is extremely dim.

If germinating grass seedlings are covered with a vial painted black and a slight scratch is made in the paint, the seedlings will curve toward the slight amount of light able to enter through the scratch. Likewise, if a tin can having a pin hole is placed over several seedlings and a light is placed outside the tin can, the seedlings will curve toward the pinhole.

Years after Darwin made his observations, Peter Boysen-Jensen performed experiments on oat seedlings. He observed that when he cut off the oat seedlings' tips and subsequently exposed the seedlings to light, no curvature took place. When he put the tip back on again, however, curvature did occur.

## ✽ Notes ✽

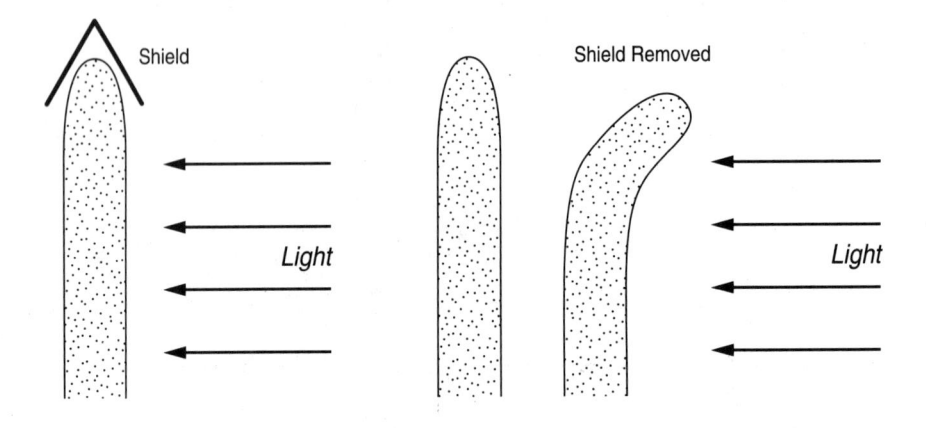

**Figure 29-1** When the growing tip is shielded from light, no bending occurs. When the shield is removed, bending takes place.

Boysen-Jensen performed another experiment that demonstrated that something produced in the tip of the seedling moved down to have an effect a short distance below the tip. First, he cut the tip off the seedling, placed a thin wedge of gelatin on the cut surface, and replaced the tip on top of the gelatin. In this case, curvature did occur. Next, he performed the same steps but substituted a piece of mica for the gelatin. In this case, no curvature occurred. It seemed, then, that the gelatin allowed some diffusible substance to pass through, while the mica instead acted as an obstruction.

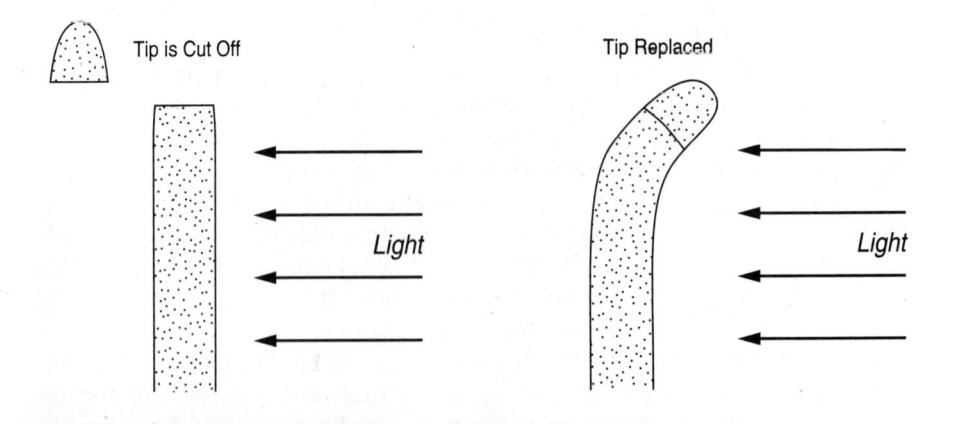

**Figure 29-2** When the growing tip is cut off, no curvature takes place. If the tip is then replaced, curvature will again occur.

Hormones ◆ 265

❧ **Notes** ❧

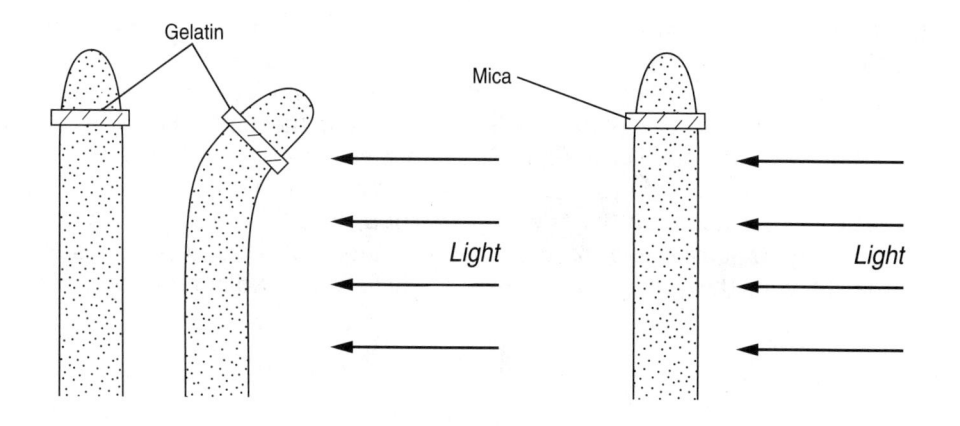

**Figure 29-3** At left, when a thin wedge of gelatin is placed between the tip and the region of curvature, curvature occurs. At right, when a thin piece of mica is substituted for the gelatin, no curvature occurs.

Another interesting thing was observed by Arpad Paal in Hungary. When he cut off the tip and then replaced it off center, curvature occurred without light being involved.

It was thus learned that a diffusible substance called **auxin** is produced in the growing tip and diffuses to the cells below. Here, it acts as a cell-wall softener, allowing elongation of the cells. This substance is apparently inactivated by light such that the concentration of it is greater on the side away from the light. Note that when a cell stretches more on one side than on the other, it curves. Thus, the uneven distribution of auxin, the substance being

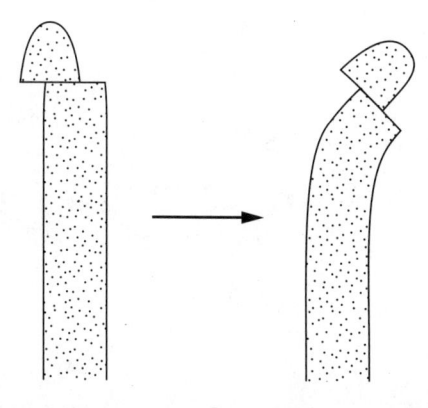

**Figure 29-4** When the growing tip is replaced off center, auxin will diffuse into the cells on the left and allows curvature.

### Notes

greater on the side away from the light, causes the stem tip to turn toward the light. Auxin, then, fits the definition of a plant hormone.

If a stem is placed in a horizontal position for a time, it will turn upward. This is not because the plant needs light. Rather, the auxin is apparently pulled down by gravity. Having a higher concentration of auxin on the lower side, the cells on this side elongate, thereby causing the stem to turn upward.

If a root is placed in a horizontal position, just the opposite response takes place—that is, the root curves downward. Although this seems inconsistent with the preceding discussion, there is an explanation. Auxin is

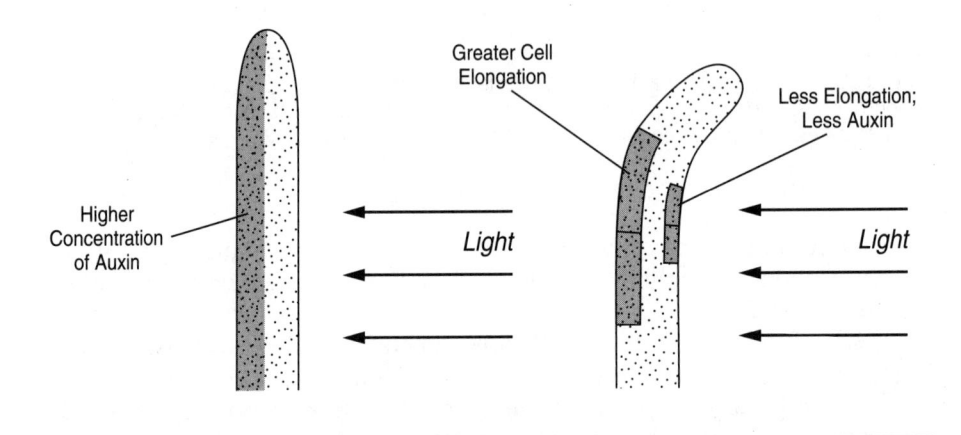

**Figure 29-5** Curvature is explained by the elongation of cells on the side opposite the light.

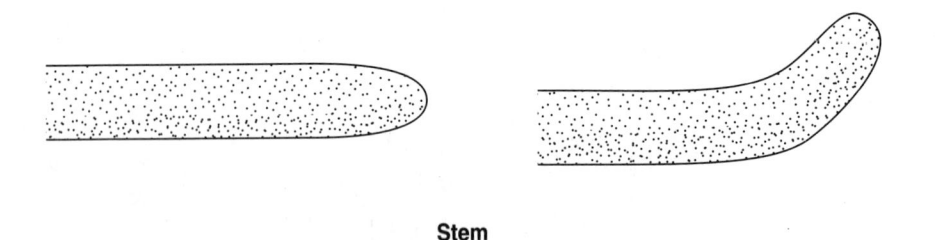

Stem

**Figure 29-6** A stem tip placed horizontally will curve upwards.

pulled to the lower side of the root just as it is in stems. But root cells are sensitive to a lower concentration of auxin. The higher concentration of auxin on the lower side of a root thus has a toxic effect, inhibiting cell elongation. The cells therefore stretch more on the upper side.

※ **Notes** ※

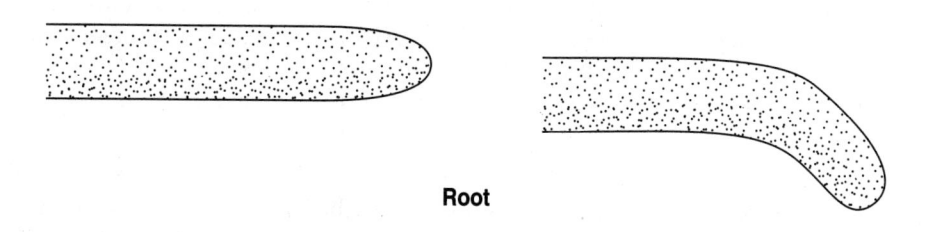

**Root**

**Figure 29-7** A root tip placed horizontally will curve downwards.

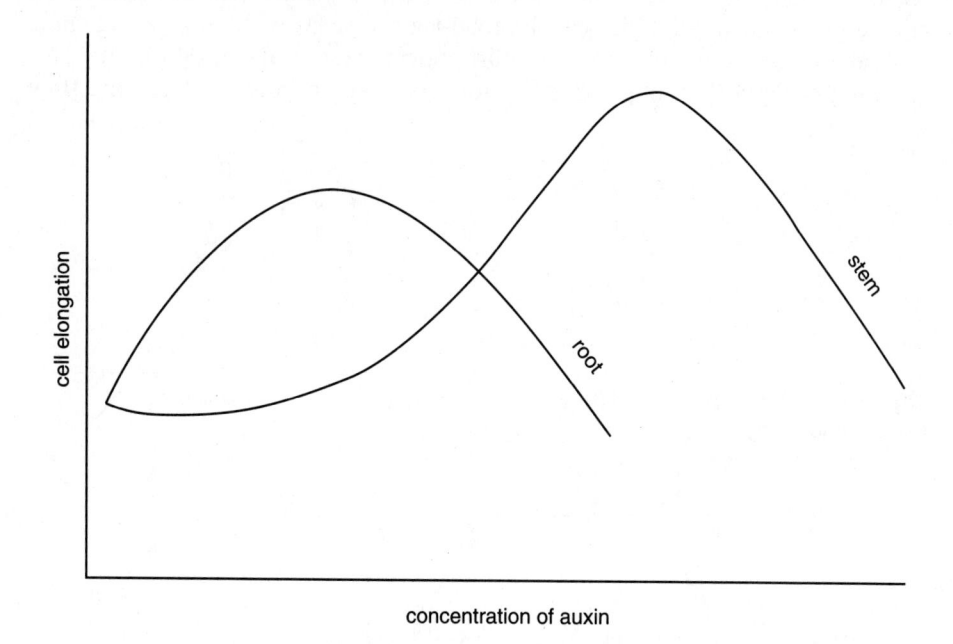

**Figure 29-8** The sensitivity of root cells to auxin is far greater than that of stem cells.

### ❊ Notes ❊

It should be evident, then, that light, rather than being a promoter of growth, has a destructive influence; more growth takes place on the unlighted side, resulting in curvature toward the light.

In 1928 Fritz Went isolated auxin. F. Kogl determined its molecular structure: indole acetic acid, or **IAA**. As shown in figure 29-9, the molecular structure of IAA is remarkably similar to that of the amino acid tryptophane. It is therefore likely that tryptophane is the precursor of IAA.

The degree of bending can be used as a method of measuring IAA concentration. Using a living organism to detect the presence of something is called **bioassay**. It has been shown that curvature occurs in response to an extremely small concentration of IAA. An amount as small as eight hundred-millionths of a milligram ($8 \times 10^{-8}$ mg) will, in fact, cause curvature.

## Uses for Auxins

Upon learning the molecular structure of IAA, chemists set about to manufacture related compounds in an effort to determine what biological effects such related compounds might have. One of these compounds is the widely used commercial preparation indole butyric acid.

When such substances are used at a certain concentration, they act as growth promoters; when used at a higher concentration, however, they kill plants. In this respect, then, auxins can be used as weed killers, as is 2-4 dichlorophenoxyacetic acid (2-4-D). This compound is used as a weed killer on lawns. Used at a certain concentration, it can kill broad-leaved weeds without harming grass.

How auxins are used and in what concentrations depends on the presenting problem. If an apple tree has too many apples and the fruits are thus

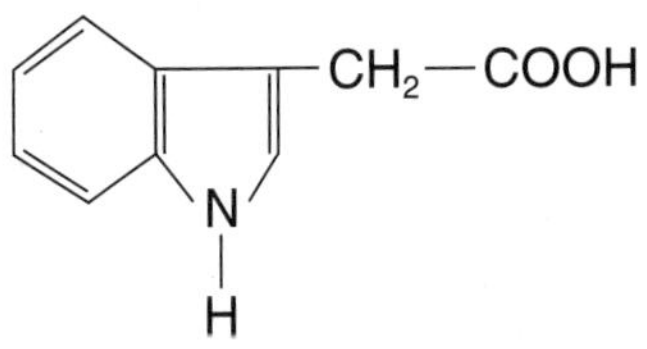

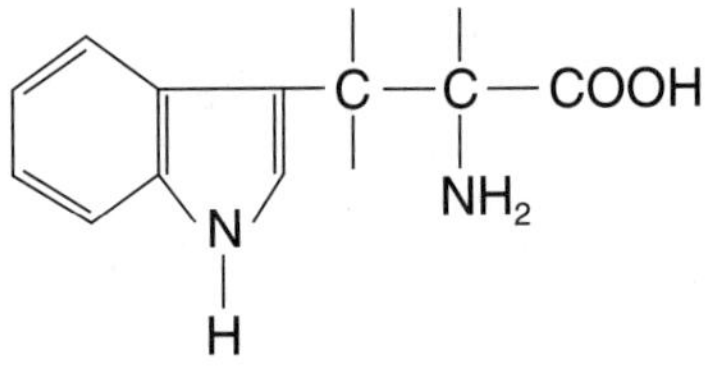

**Figure 29-9** Indole acetic acid (IAA), at left, is closely related to the amino acid tryptophane, at right.

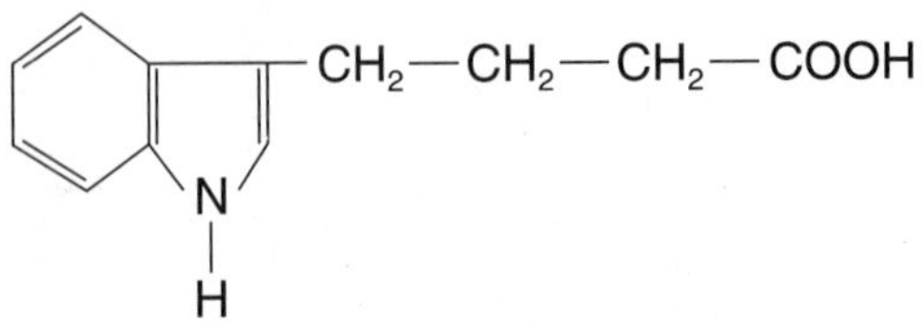

**Figure 29-10** Indole butyric acid

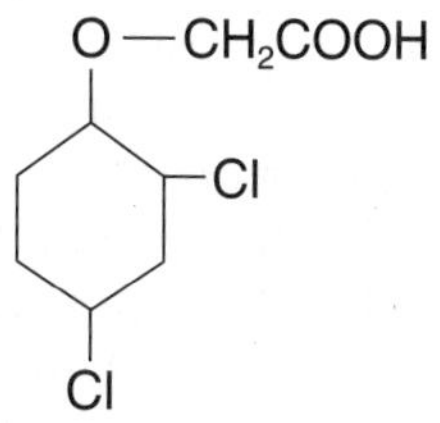

**Figure 29-11** 2-4 dichlorophenoxyacetic acid

unable to develop to full size, for example, a particular auxin at a given concentration can be used to induce some of the fruits to be cast off. If the apples instead tend to drop prematurely, a different concentration of the same auxin can be used to induce retention of the fruits.

Pollination and fertilization are generally necessary for fruits to develop. In certain circumstances, such as the absence of bees, for example, auxins can be used to encourage fruit development without pollination. This is a way that seedless fruits are produced.

### *Obtaining Auxins*
Auxins can be extracted from the medium used in the culture of mushrooms. Auxin for use in experimental work can be obtained by germinating a large number of oat seedlings, cutting off their tips, and extracting the auxin. There is, however, an easier way. Not only is the medium used in the culture of mushrooms a convenient source, but so is urine. Urine actually contains larger quantities of this growth-promoting substance than do plants.

## Other Plant Hormones
Since the discovery of auxin, a number of other plant-growth regulators have been uncovered; ethylene, **gibberellins**, **cytokinins**, **abscisic acid**, florigen (and **photoperiodism**), **phytochrome**, and **wound hormone** are examples.

### *Ethylene*
While ethylene certainly exhibits hormonal properties in its effects on plants, its simple molecular structure and the fact that it is a gas seem incompatible with our concept of hormones. Thus, it may be better to refer to ethylene as a hormone-like substance. Ethylene's effect on plants and fruits was observed long before it was known to be produced in plants. For instance, it had long been known that unripened bananas ripened rapidly when stored near a gas leak in a warehouse. Further, shade trees located near leaking street lamps took on altered forms.

### ❋ Notes ❋

Ripe fruits produce ethylene, and ethylene hastens ripening. (The latter fact became known in 1934.) A ripened apple placed in a bowl with green apples will, by its release of ethylene, hasten the ripening of its neighbors, which, in turn, increase their own ethylene production. Ethylene plays various roles in plant behavior. It releases potato tuber buds from dormancy, stimulates leaf **abscission**, influences the flowering of pineapple plants, induces the formation of adventitious roots in stem cuttings, helps germinating seeds push up through the soil, and helps bring about **senescence**.

Ethylene is produced not only by ripening fruit but also by flowers, seeds, leaves, and roots. If fruits are bruised or cut, ethylene seems to be produced in a surge. Application of IAA stimulates ethylene production sometimes as much as tenfold. When pea seeds germinate, an increase of ethylene causes the cells of the stem tip to produce a tighter bend, thus aiding the seedling in pushing through to the surface.

Ethylene is used commercially to control the ripening of fruit, as during the shipping of green bananas. Because ethylene hastens ripening, there is often the need to slow the process by careful ventilation to remove the gas. The same effect can be achieved by pumping in carbon dioxide, which counteracts the effects of ethylene.

Ethylene is also theorized to hasten senescence, perhaps by either contributing to a decline of metabolic rate or interfering with RNA synthesis or protein manufacture. The argument regarding senescence in both plants and animals is that organisms, or the cells of organisms, are "born to die." In plants, this process seems to be advanced by ethylene and delayed by auxins, gibberellins, or a combination of both. It is interesting to note that woody perennials seem to challenge the concept of the aging process being normal. These trees seem to have no mechanism to bring growth to a close. Growth occurs in cambium and presumably can continue without limit. Of course, trees do not live forever. They fall prey to infection or fire; or if the bark is so thick as to protect them from infection and fire (as is the case with the giant Sequoias of California), perhaps they grow until they eventually fall under their own weight.

### Gibberellins

Gibberellin is a growth-promoting substance first isolated from the fungus *Gibberella fujikuroi*. This fungus infects rice seedlings, causing them to grow to great height. The infection is called the "foolish seedling disease."

Scientists in the Western nations did not learn about gibberellin until after the end of World War II, in approximately 1950. As many as fifty gibberellins have thus far been isolated from fungi and from the healthy tissues of higher plants (buds, leaf primordia, immature seeds, fruit tissue, and roots). Gibberellin production in roots is generous, and gibberellin is translocated to other parts of the plant.

The way whereby gibberellins accelerate growth is not the same as with auxins. Whereas auxins are involved in cell elongation, gibberellins affect the rate of mitosis. Gibberellins, therefore, do not affect curvature of the stem. In some kinds of plants, flowering is induced by the application of a gibberellin. Dormancy of buds and seeds can be broken, and some gibberellins make possible the growth of plants at lower than normal temperatures. A lawn can be induced to grow two to three weeks earlier than is typical by the use of gibberellins. Gibberellins can also be used to increase the size of seedless grapes. A drop of a gibberellin extract placed on an unfolding leaf will cause the leaf to increase in length by as much as 400 percent. Gibberellins are designated as A1, A2, and so on. Gibberellic acid, extracted from *Gibberella fujikuroi*, is designated as A3.

## Cytokinins

Growing cells in a tissue culture medium composed in part of coconut milk led to the realization that some substance in coconut milk promotes cell division. The "milk" of the coconut is actually a liquid endosperm containing large numbers of nuclei. It was from kernels of corn, however, that the substance was first isolated in 1964, twenty years after its presence in coconut milk was known. The substance obtained from corn is called **zeatin**, and it is one of many cytokinins.

While cytokinin acts mainly to promote cell division, this hormone also serves other functions: cell enlargement in young leaves, tissue differentiation, flowering, fruiting, and delay of aging in leaves.

There are now known to be as many as 100 cytokinins, some of which occur naturally and others of which are manufactured. Zeatin is the most active of the natural cytokinins and less active than some of the synthetic ones.

When several living cells are isolated from a living plant and cultured in a medium containing both cytokinin and auxin, cell division proceeds, forming a mass of undifferentiated cells, called a **callus**. The callus is then able to differentiate and produce both shoot and root, thus developing an entire new

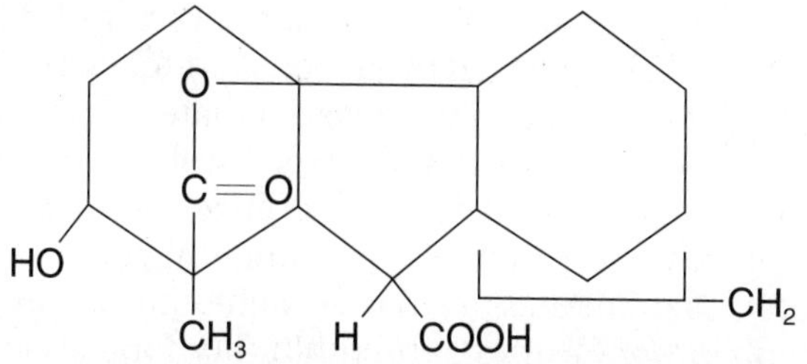

**Figure 29-12** Gibberellic acid

※ **Notes** ※

plant. Auxin favors root formation, and cytokinin favors the growth of the shoot. This means of **vegetative propagation** makes possible the perpetuation of superior forms.

## *Abscisic Acid*

Abscisic acid (ABA) is a powerful hormone that inhibits the action of auxins and gibberellins. It is called a stress hormone because it is formed when the plant is subjected to an unfavorable environment. When water is in short supply, abscisic acid contributes to closing the stomates. The presence of the acid in guard cells causes potassium ions to leave the cells; the guard cells thus lose water, and the stomates close. Abscisic acid affects seed dormancy as well as fruit and leaf abscission. Because the action of abscisic acid appears to oppose that of auxins and gibberellins, these hormones are said to be antagonistic.

## *Photoperiodism and Florigen*

Much evidence suggests the existence of a hormone that induces a plant to produce flowers. The discovery of such a flowering hormone has proved elusive, however.

What causes a plant to produce flowers? Some plants produce flowers in the spring, some in the summer, some in the fall. In order for a plant to flower, it must reach a certain maturity. Temperature can also be significant. In many species, the number of hours of daylight or of uninterrupted darkness is the controlling factor. The cocklebur, for example, requires not less than 8½ hours of an uninterrupted darkness in order to produce flowers. If this period of darkness is interrupted by the merest flash of light, flowering will not occur. The wavelength of the interrupting light is significant, however. If the light is of the **far-red** end of the spectrum, exposure will not inhibit flowering. In fact, exposure to far-red light can reverse the adverse effect of a flash of ordinary light. Thus, if a cocklebur plant is left in darkness for 8½ hours and then brought into ordinary light, flowering will occur. If the dark period is interrupted by a flash of ordinary light, flowering will not occur. If the dark period is interrupted by ordinary light followed by a brief flash of far-red light, however, flowering will occur. The far-red light nullifies the adverse effect of ordinary light and can do so a number of times. If the plant is exposed to ordinary light followed by far-red light followed by ordinary light followed by far-red light, flowering will occur. Whatever light is used last in a series, then, controls whether or not the plant will flower. Light can alter a chemical reaction only if absorbed, and the effect of absorbed far-red light offsets the damage inflicted by absorbed ordinary light.

It is postulated that darkness or far-red light changes an unknown substance from an inhibitory form to a noninhibitory form. It is clear that such a substance is produced in the leaves of the plant because when the leaves are removed after 8½ hours of darkness, flowering does not occur. The substance, then, must be transported from the leaves to the site of flowering. Flowers are modified leaves; thus, the so-called flowering hormone causes the plant to stop producing ordinary leaves and instead produce the modified leaves called flowers.

Although the flowering hormone has not been isolated, it has been given a name: florigen. Several steps are involved in the synthesis and action of the flowering hormone:

1. a buildup of florigen achieved by photosynthesis
2. conversion of florigen to a noninhibitory form
3. the synthesis of florigen, also achieved in the dark
4. possibly, a chemical step requiring light
5. movement of florigen from the leaves to the site of flowering.

## Phytochrome

Phytochrome also plays a role in flowering. Its actions are similar to those of florigen, but the interconversions of phytochrome do not alone account for flowering.

Phytochrome is a pale-blue, proteinaceous pigment found in all higher plants. Only minute amounts are produced. Given that phytochrome is essentially invisible, it is not surprising that it remained hidden for so long. Phytochrome was isolated and identified in 1959. A special pigment-analysis instrument had to be utilized to detect it. Phytochrome occurs mostly in meristematic tissue and takes two forms: phytochrome-red (Pr) and phytochrome-far red (Pfr). As mentioned earlier, far red lies at the nearly invisible, long-wave end of the spectrum. Either form of phytochrome can be converted to the other. Phytochrome-far red becomes phytochrome-red when it absorbs far-red light. Conversely, Pr becomes Pfr when it absorbs red light. In nature, Pr becomes Pfr far more than the reverse. Phytochrome-far red converts back to phytochrome-red in the dark. The conversion process in the light is instantaneous. (This should be no surprise given the previous observations regarding the flowering of cocklebur.)

## Wound Hormone

Wound hormone is another yet to be isolated hormone. While there is evidence of its existence, the wound hormone remains hidden. Most vascular plants respond to wounding by resuming growth and forming a region of callus (undifferentiated parenchyma cells) covering the wounded area. Wounding triggers the capacity of cells that have ceased dividing to begin growth again. Experiments indicate that the damaged cells produce substances that stimulate cells to resume mitoses. If a wound is immediately washed with water, callus formation does not take place; if wound juice is instead applied, growth will renew.

❋ **Notes** ❋

## Questions for Review

1. How is ethylene gas used commercially?

2. How do hormones differ from enzymes?

3. Define the term *photoperiodism*. Give some examples of plants whose flowering is dependent on photoperiod.

4. Describe an experiment performed on grass seedlings that showed the bending of the stem tip toward light. Give evidence that this behavior is traceable to a hormone.

5. What is a general name for the hormone referred to in question number 4? Where is it produced? When it was isolated and its chemical structure determined, it was shown to be _____ _____ _____.

6. The hormone referred to in questions number 4 and 5 has its effect by causing the _____ _____ to soften.

7. Ripening fruits produce ethylene gas. If a fruit is injured, what happens to the ethylene production?

8. Gibberellin was first isolated from a fungus called _____ _____.

9. In tissue culture, a mass of undifferentiated cells is called a _____; this tissue may undergo differentiation to produce an entire plant.

## Suggestions for Further Reading

Galston, A.W., P.J. Davies, and R.L. Satter. 1980. *The Life of the Green Plant*. Engelwood Cliffs, NJ: Prentice-Hall, Inc.

Ray, P.M. 1972. *The Living Plant*. New York: Holt, Rinehart, and Winston, Inc.

Thimann, K.V. 1977. *Hormone Action in the Whole Life of Plants*. Amherst, MA: University of Massachusetts Press.

# 30

# Biological Clocks

**B**ean plants regularly extend their leaves in the daytime so as to take in the maximum amount of light. At night, they lift their leaves parallel to the side of the stem in what may be described as a "sleeping" position. These leaf movements occur day after day throughout the entire life of the plant. It might be inferred that these responses are triggered by the coming of daylight in the morning and the coming of darkness at night. However, when bean plants are placed in a controlled situation wherein the light is constant twenty-four hours a day, the same responses occur at the same times noted earlier (that is, the plants extend their leaves at morning and lift them up parallel to their stems at night). Likewise, when the plants are placed in darkness for twenty-four hours a day, the responses are again the same. How does this happen? Why does it happen? The answers to both questions are unknown (although the temptation exists to say that they extend their leaves because they need light).

❋ Notes ❋

## Circadian Rhythms

Because such plant movements occur at approximately the same time every day, and because they appear to be based on a twenty-four hour day, they are called *circadian rhythms*. They are said to be based on "biological clocks." The French investigator Jean Jacques de Mairan noted the diurnal movements of bean plants in 1729, observing that the movements occurred in the same rhythm when the plants were kept in a constant environment. Similar responses occur in other flowering plants. The flowers of some plants open at the approach of dusk; the flowers of others, at the approach of light. In fact, the flowers of different species open at various times throughout the night and day. Carolus Linnaeus recognized this fact. He constructed a circular garden having pie-shaped plantings of species whose flowers opened at

### ❦ Notes ❦

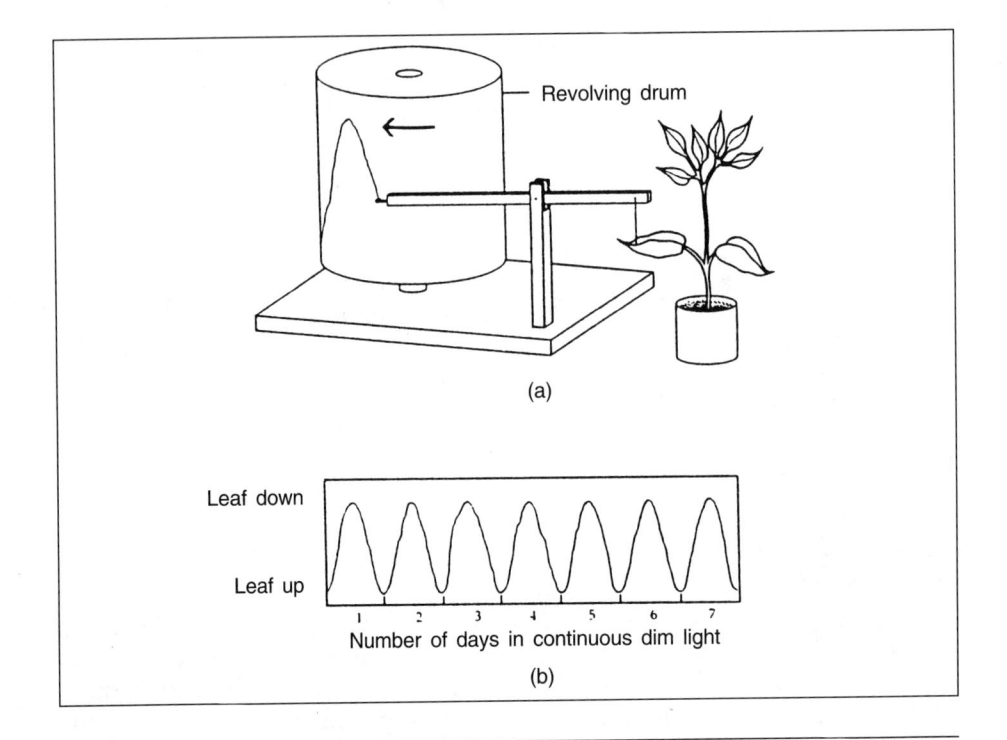

**Figure 30-1** (a) A very sensitive apparatus designed to record the movements of a leaf. (b) Graph showing the diurnal rhythm of leaf movements under constant light conditions.

differing times. By careful arrangement, he was able to devise a "floral" clock that indicated the time of day by which flowers were opened or which leaves were folded upward.

When plants are placed in an environment where it is dark during the day and light during the night, they reverse their responses. The leaves open during the new hours of illumination and assume the sleeping position during the new hours of darkness. In contrast with the observations regarding keeping plants in continuous light (that is, that external factors do not appear to influence plant responses), these observations seem to indicate that external factors such as light do influence plant responses.

The rate of a chemical reaction increases when temperature increases and decreases when temperature decreases. If plant rhythms of behavior are related to chemical events, they should thus also vary with change of temperature. In fact, the type and timing of plant responses are unaffected by temperature changes; the magnitude of such responses, however, is affected as anticipated.

A number of other features of plant life have been observed to demonstrate circadian rhythm. Among these are the rate of cell division and the rate of photosynthesis; both are maximized at approximately the same time every day regardless of external conditions. Likewise, certain flowers produce nectar at a certain time of day—the same time of day that bees, with their own biological clocks, visit the flowers. The causes of these respective behaviors lie within the organisms.

## *Gonyaulax polyedra*

*Gonyaulax polyedra*, a single-celled marine alga belonging to the dinoflagellates, is a close relative of *G. catanella*, which causes poisoning in shellfish. *Gonyaulax polyedra* is **bioluminescent**, the light reaching a peak in the middle of the night. Photosynthesis, conversely, peaks in the middle of the day, and cell division is restricted to the hours before dawn. If *G. polyedra* is kept in constant conditions, it will reproduce numerous times, making available cells that have never before existed in conditions of both day and night. These organisms exhibit the same circadian behaviors for scores of generations, that is, producing maximum light in the middle of the night, demonstrating peak photosynthesis in the middle of the day, and reproducing just prior to dawn.

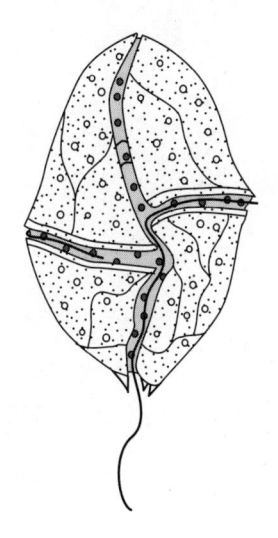

**Figure 30-2** *Gonyaulax polyedra*

❈ **Notes** ❈

## Questions for Review

1. Define the terms *circadian* and *endogenous*.
2. What sort of biological clocks seem to be exhibited by the dinoflagellate, *Gonyaulax polyedra*?

## Suggestion for Further Reading

Brown, F.A., Jr. 1962. *The Biological Clock.* Boston: D.C. Heath Co.

# 31

# Plant Nutrition

**P**lants are **autotrophic**, meaning they make their own nutritional requirements by photosynthesis. Actually, this statement is only partially true. Plants need water, carbon dioxide, and sunlight to manufacture by photosynthesis. The primary nutrients required by a plant are the three elements carbon, hydrogen, and oxygen, which they gain through photosynthesis.

## Required Minerals

In addition to these three elements, a number of minerals—twelve to be exact—are essential to the welfare of plants. The majority of these are found dissolved in water in the soil: phosphorus, sulphur, calcium, potassium, magnesium, iron, zinc, manganese, copper, boron, and molybdenum are all taken into the plant through the roots. Nitrogen, the twelfth mineral, may be supplied by the atmosphere, yet it is taken in through the roots in a changed form. Because plants must take in minerals through the roots, nitrogen as an element is not available to a plant. In order to be used by the plant, nitrogen must therefore be converted to **nitrate**.

Some plants also require cobalt. Calcium is not required by certain fungi and algae.

## Determining Mineral Needs

How do we determine which minerals are required for plant welfare? This is done in the same way we determine which vitamins are necessary for human welfare. Namely, we observe what happens when given minerals are absent. The challenge is to arrange a scenario wherein every required mineral but one is present. This is not easy to do. Very precise methods must be employed to ensure that the mineral being studied is not present, even in minute amounts. Further, some of the required minerals are needed in only micro

✺ Notes ✺

### ✽ Notes ✽

amounts. Should any of these minerals be present as impurities in a chemical, float into the test on dust from the atmosphere, or be present on the seed when it is planted, testing to determine the consequences of that mineral's absence could be compromised. Sachs[1] and Knop, two early investigators, learned this lesson. They prepared a nutrient solution containing calcium nitrate, potassium nitrate, potassium di-acid phosphate, magnesium sulfate, and ferric phosphate. They then supplied it to plants growing in solution culture. In seeing no provision of zinc, manganese, copper, or molybdenum in this nutrient solution you might suspect that the plants would show a lack of something. This, however, was not noted. Rather, the plants thrived in good order. Sachs and Knop were thus misled to believe that the minerals provided in the solution were adequate to the need. In actuality, however, other minerals were present as impurities in the chemicals they used. This example illustrates the importance of using the most meticulous methods to ensure that the only elements present are those intended. The water should be twice distilled. Water that has been stored in a glass bottle for a time is not desirable because part of the bottle has dissolved in the water, yielding boron. Water that is distilled but allowed to flow through a metal pipe is also unsatisfactory, having remnants of zinc and copper.

Zinc is needed in amounts of approximately three parts per million of the plant's dry weight. Molybdenum is likewise needed in only trace amounts.

## Symptoms of Improper Nutrition

What happens when an essential element is lacking? When nitrogen is lacking, the leaves become yellow and are unable to form chlorophyll. When phosphorus is lacking, the leaves turn dark green. One may infer, then, that elemental deficiency can be determined solely by looking at the plant and observing the symptoms. While this is true to a certain degree, experience is also required to accurately assess such deficiencies. For example, plants turn yellow when nitrogen is lacking. But they also turn yellow when potassium is lacking, when magnesium is lacking, and when iron is lacking.

Insufficient calcium results in degeneration of terminal root and shoot growth and malformation of young leaves. The deficiency itself, however, does not directly cause these problems. Lack of calcium leads to abnormally high uptake of magnesium; these symptoms, then, are actually indicative of magnesium toxicity.

In addition to suffering from deficiencies, plants can also suffer from an overabundance of elements. The enthusiastic gardener who supplies too much nitrate, for example, will reap plants having leaves that exceed the capacity of the roots to provide for them.

---

[1] Julius von Sachs (1832–1897) published *Keimungsgeschichten*, which laid the foundation for using microchemical methods in physiology studies.

## Questions for Review

1. What are the primary chemical elements required by plants?
2. Name some elements required in micro amounts for plant growth.
3. How can one determine that a plant requires a certain mineral nutrient?
4. What can happen when an essential mineral is lacking?

## Suggestions for Further Reading

Buckman, Harry O., and K.C. Brady. 1969. *The Nature and Properties of Soils.* New York: The MacMillan Co.

Epstein, E. 1972. *Mineral Nutrition in Plants: Principles and Perspectives.* New York: John Wiley and Sons, Inc.

Marschner, H. 1986. *Mineral Nutrition of Higher Plants.* London: Academic Press.

Mengel, K., and E.A. Kirby. 1979. *Principles of Plant Nutrition.* Bern, Switzerland: International Potash Institute.

❋ Notes ❋

# 32

# Stems

A stem may be defined as something that bears leaves. But this definition is too simplistic; it needs to be expanded so as to answer some pertinent questions. From where does a stem come? What part of the seed produces it? When one dissects a seed, it is found to bear a seed coat or two and, within the seed coat, an embryo. The embryo consists of cotyledons (the seed leaves, which are food sources), a radicle (which grows downward upon germination to produce the root), and a plumule (which grows upward to produce the stem and leaves). The cotyledons are attached to the plumule. The seedling thus has two parts: the **epicotyl** (that part above the cotyledons) and the hypocotyl (that part below the cotyledons). Thus, a stem can be defined as that part of a plant above the hypocotyl.

Gymnosperms are entirely woody, while both woody and herbaceous forms occur in angiosperms. Angiosperms include both monocots and dicots. These differences create a need for several descriptions of stem anatomy.

## The Woody Dicot Stem

First, consider the woody dicot stem. As explained earlier, cells are either meristematic or permanent. Meristematic cells are those that retain the ability to divide, whereas permanent cells are those that have lost this ability. Cells at the growing tip of a stem are meristematic. A cluster of young leaves is generally present at the apex of the stem so as to protect the young point. This apex is the terminal **bud**. Growth in length is accomplished by both divisions of the cells lying under the bud and the elongation of cells lying just below the region of cell division. The cells that continue to divide, called the **initials**, differentiate into regions, becoming the **protoderm**, **provascular tissue**, and **ground meristem**. The protoderm later becomes epidermis; the provascular tissue becomes primary xylem, primary phloem, and cambium; and the ground meristem produces the remaining primary tissues (that is,

❈ Notes ❈

### ❋ Notes ❋

the pith, [or cortex] and tissue separating the vascular bundles). Primary tissues are so-called because they are the first formed tissues.

As a cell at the apex divides, one of the two resulting cells may lose the ability to divide, thus becoming permanent. The other cell, oriented toward the tip, remains meristematic. In this way, the meristem continually progresses upward.

At left of figure 32-2 is a longitudinal section of a stem; at right, corresponding cross sections. No leaf primordia are included. The topmost cross section (figure 32-2a) is cut through a region of apical meristem, where there has been no differentiation of tissues. At the next section down (figure 32-2b), some differentiation of vascular bundles can be seen. The vascular bundles have a primary xylem directed inwardly, a primary phloem directed externally, and a cambium (a zone of meristematic cells), which has not yet produced any divisions. Hence, the only tissues present are primary tissues. In the lowermost section (figure 32-2c), the bundles have coalesced to produce a continuous cylinder of vascular tissue; but there is still no secondary growth. In the center is a zone of undifferentiated (parenchymatous) cells, called the pith. Lying immediately outside of the pith is the vascular cylinder having a primary xylem, a primary phloem, and a cambial layer one cell thick and between the vascular tissues. The cambium has not yet functioned to produce secondary xylem and phloem. Peripheral to the vascular cylinder is another zone of parenchyma (the cortex), and on the outside is the epidermis.

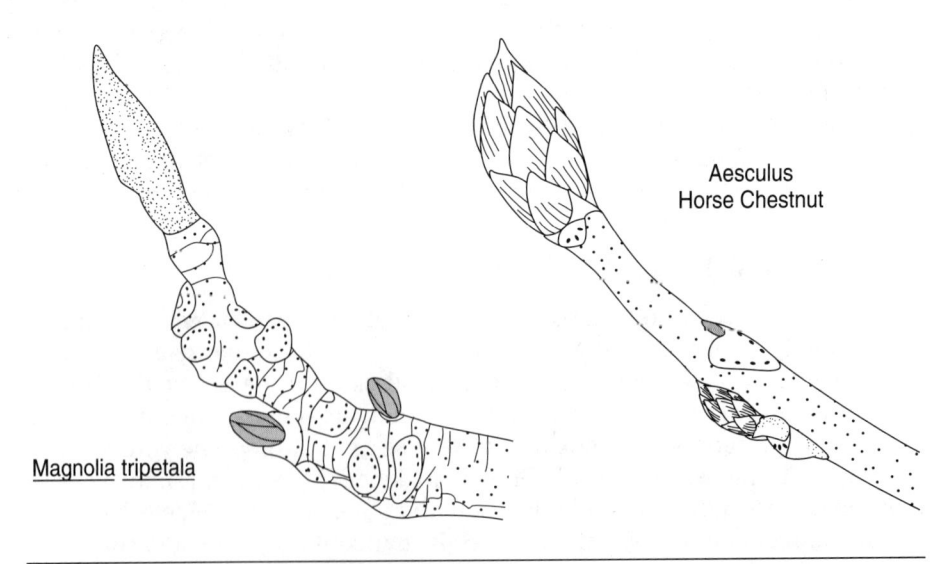

**Figure 32-1** Stem tips, showing terminal buds of *Aesculus*, the horse chestnut, and *Magnolia tripetala*. The abundant leaf scars of *Magnolia* seem to confirm that the leaves are crowded into an umbrella-like circle.

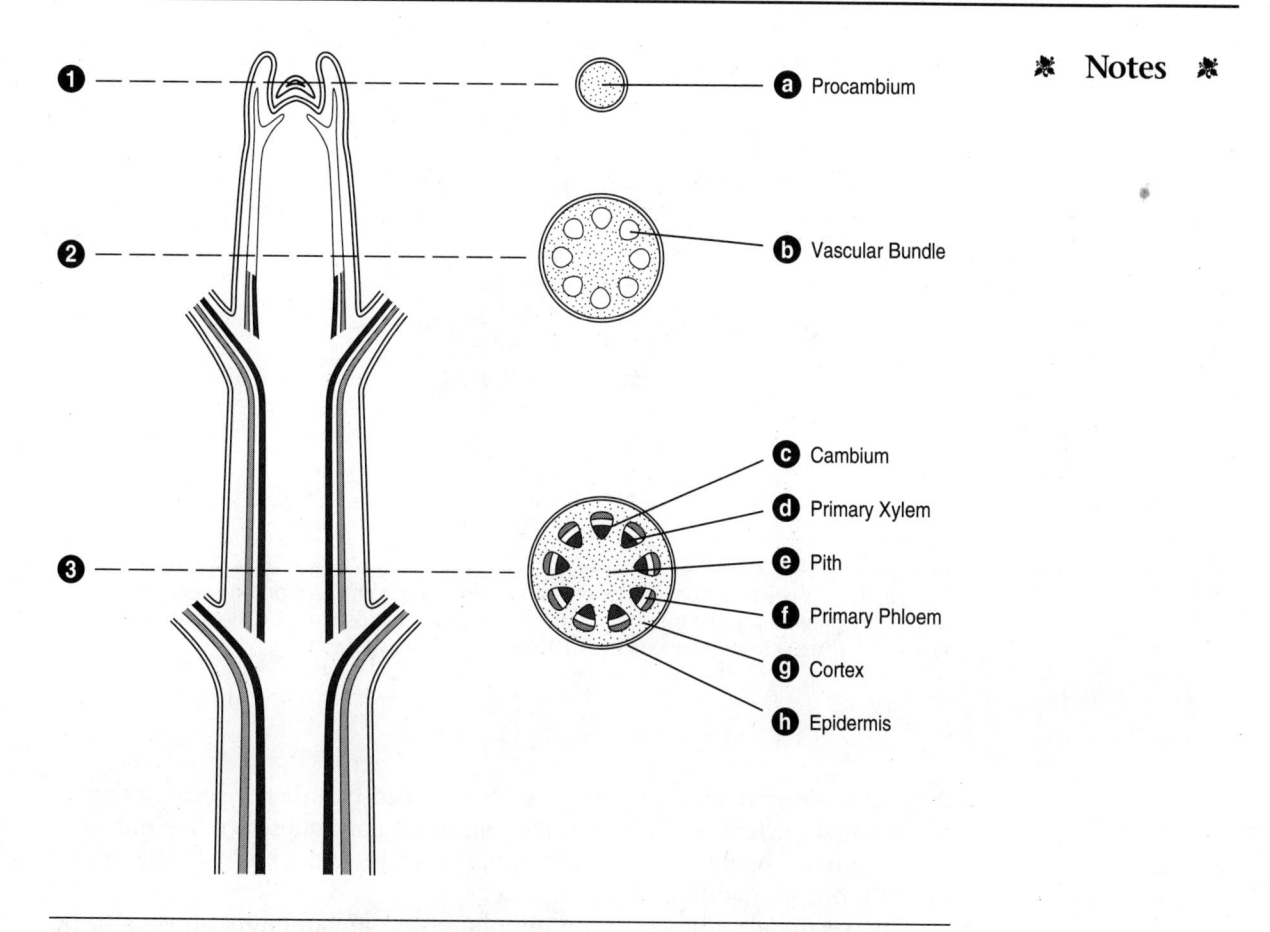

**Figure 32-2** At left, longitudinal section of a woody stem tip. At right, corresponding cross sections. (a) Procambium. (b) Vascular bundle. (c) Cambium. (d) Primary xylem. (e) Pith. (f) Primary phloem. (g) Cortex. (h) Epidermis.

In figure 32-3, a small amount of secondary growth has taken place. While the primary tissues are still detectable, the primary phloem, being pushed outward by the growth of secondary phloem, is becoming crushed and will soon be obscured. All tissues lying outside the cambium tend to be crushed, and the primary tissues in this area are soon entirely obliterated. Enough secondary growth has occurred that the tissues outside the cambium are clearly being destroyed. Primary tissues lying central to the cambium are not subject to crushing and, thus, remain in place through the life of the plant.

Figure 32-4 shows a portion of stem with three years of secondary growth. The primary tissues lying outside the cambium are now entirely obscured, as

### ❊ Notes ❊

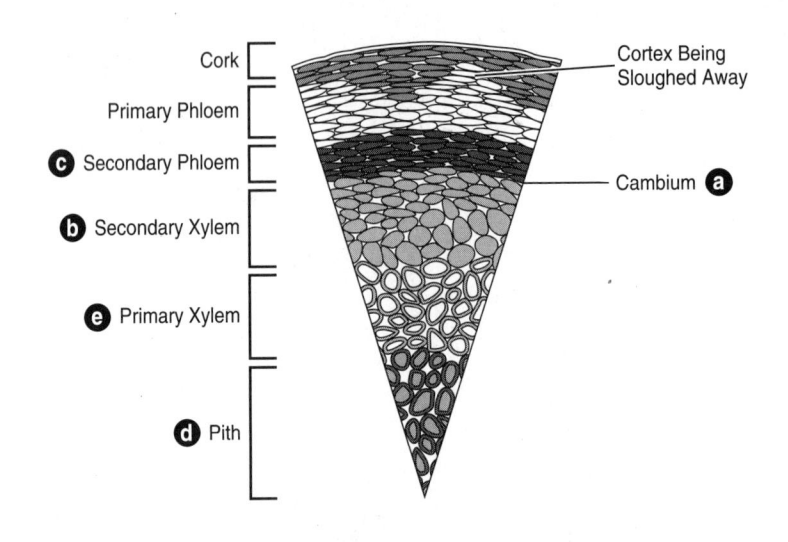

**Figure 32-3** Cross section of a woody stem exhibiting no more than one year of secondary growth. (a) The cambium, represented by a single line. (b) The secondary xylem. (c) The secondary phloem. (d) The pith. (e) Primary xylem.

they have either been sloughed away or replaced by other tissues. Note that as secondary xylem is deposited, the cambium continues to move outward. The cambium, being continuously enlarged, must make not only xylem and phloem, but also additional cambial cells.

The secondary phloem, lying peripheral to the cambium, appears to be compressed; and although just as much secondary phloem as secondary xylem is produced, the phloem occupies a comparatively small space (as shown in figure 32-3 of a stem having no more than one year of secondary growth).

Figure 32-5 shows a short segment of cambium (at top). When a cambial cell divides (as shown at bottom), one of the cells remains meristematic and the other becomes permanent. If the cell at (1) remains meristematic (and thus remains a part of the cambium), the cell at (2) will be added to the phloem. If the cell at (2) remains meristematic, however, the cell at (1) will become permanent and, thus, part of the xylem. These events alternate (that is, one division yielding a phloem cell, the next division yielding a xylem cell). The cambium also increases itself by producing additional cambial cells.

Secondary xylem is **wood**. As the wood grows (that is, as additional secondary xylem is produced), recognizable annual rings result. Four kinds of cells are produced in the xylem: vessel elements, tracheids (both of which are involved in conduction), fibers, and parenchyma. The wood formed in the spring tends to be abundant in vessel elements; that formed later in the year tends to be abundant in tracheids. In this manner, recognizable annual rings

Stems ♦ 287

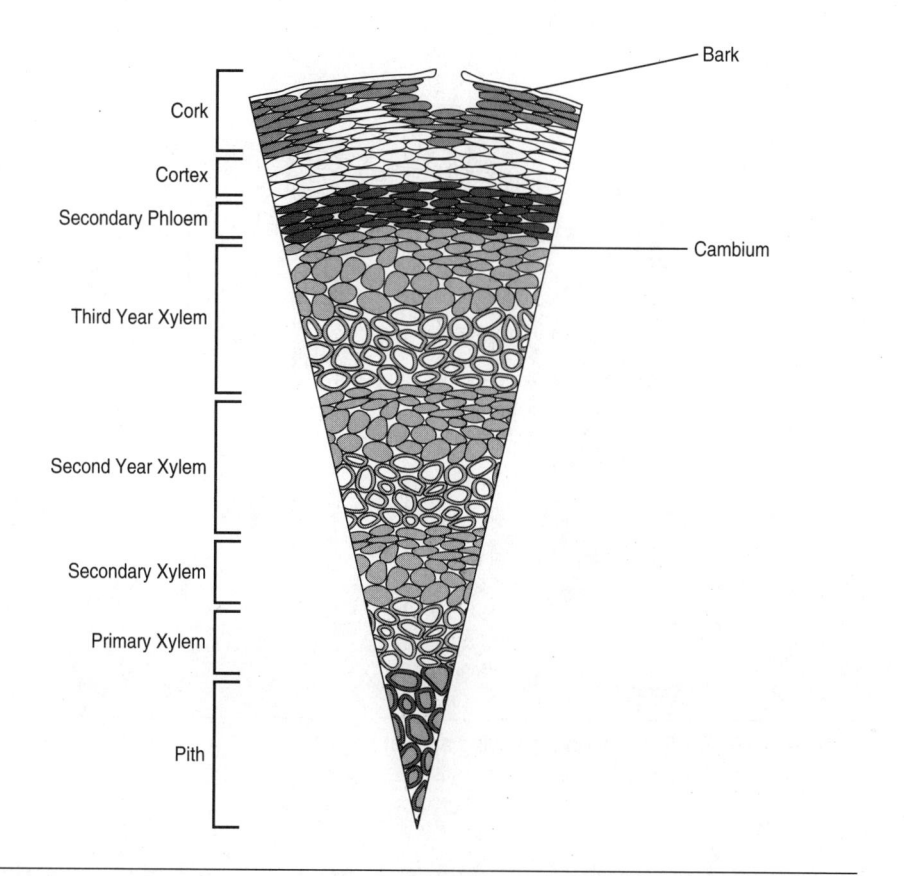

**Figure 32-4** Cross section of a woody stem exhibiting three years of secondary growth.

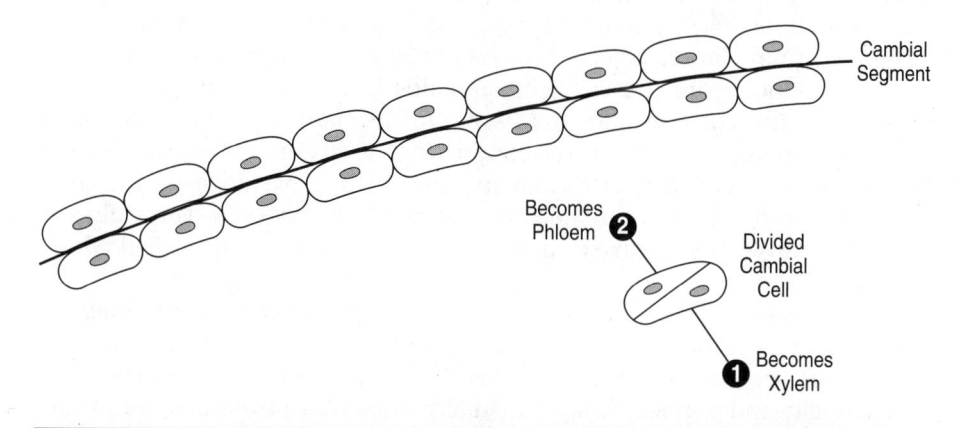

**Figure 32-5** At top, a short segment of cambium. At bottom, a cambial cell that has divided. One of the two resulting cells will become permanent; the other will remain meristematic. If cell (2) becomes permanent, it will become part of the phloem. If cell (1) becomes permanent, it will be added to the xylem.

## ✽ Notes ✽

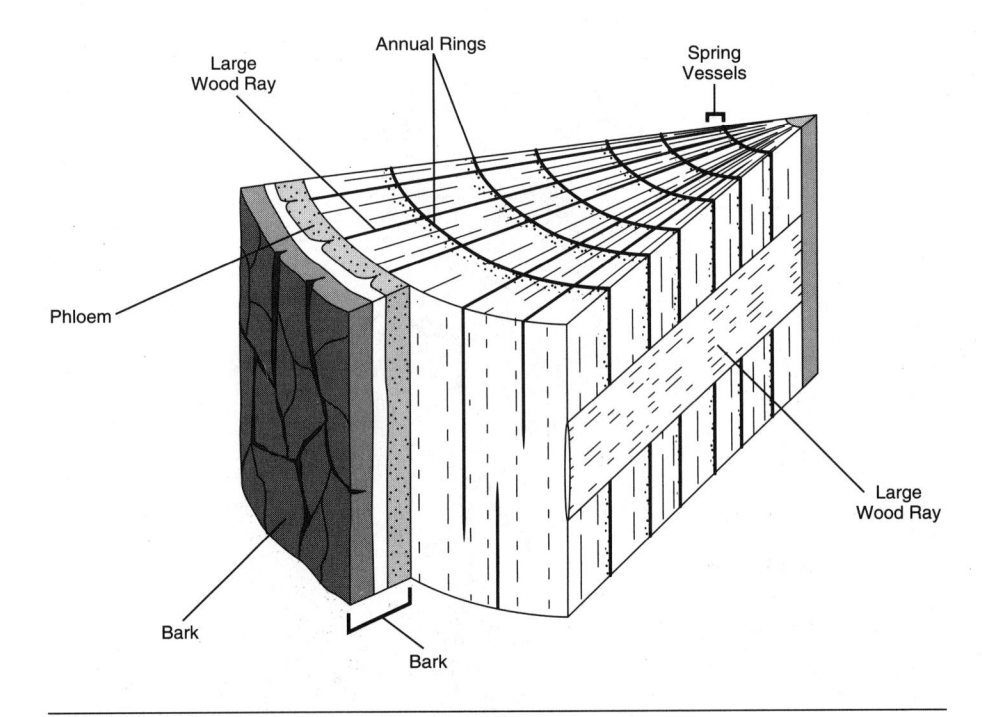

**Figure 32-6** Section of wood, showing annual rings.

are formed. A tree one hundred years old will have one hundred annual rings, the oldest of which will be the innermost and in contact with the primary xylem. The newest formed will, of course, be adjacent to the cambium.

The conducting cells of both xylem and phloem are short lived. Only those cells immediately adjacent to the cambium are alive, and only a zone a short distance from the cambium remains functional. The remainder of wood, regardless of tree diameter, is nonfunctional. The center portion of wood, which has lost its water-conducting ability, is called **heartwood**. Heartwood is often discolored, and its vessels often occluded by incoming seepage of protoplasm from neighboring parenchyma cells. The inflow of cytoplasm causes bladderlike formations in the vessel elements, called **tyloses** (figure 32-7).

A secondary cambium arises outside the phloem and in the remnant of cortex. Here, cells that had lost the ability to divide regain their meristematic ability. This secondary cambium, called the **phellogen**, produces the outer bark laterally and a greenish layer medially. The latter of these is the phelloderm. Phellogen is also called cork cambium.

Bundles of vascular tissue that pass out into leaves arise from the most recently formed xylem and phloem. Annual additions of xylem and phloem

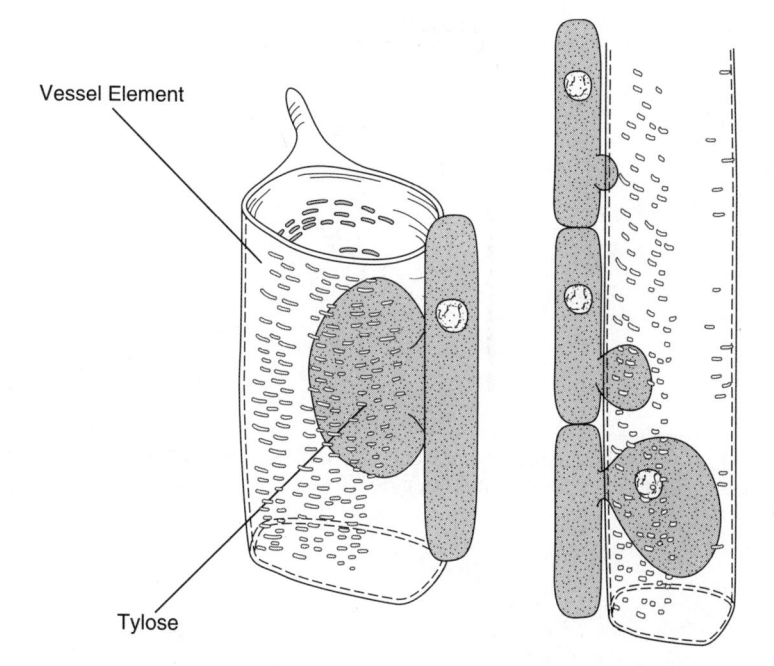

**Figure 32-7** Tyloses result from an inflow of cytoplasm from neighboring parenchyma cells to the vessel elements. Tyloses tend to stop the flow through vessels.

contribute to leaf bundles only once. The strands of vascular tissue that leave the recent zone of xylem and phloem to pass into a leaf are collectively called a **leaf trace**, and the vacant space immediately above the leaf trace and in the vascular cylinder is called a leaf gap (figure 32-8).

The most prominent elements of phloem are those that compose the sieve tubes. These elements are elongated cells arranged end to end and having perforated end walls. The end walls are called **sieve plates**. At the side of each sieve cell is a parenchymatous cell called the companion cell. There appears to be living cytoplasm in the sieve-tube element, although the mature cell does not have a nucleus.[1] Strands of cytoplasm run through the holes of the sieve plates to neighboring sieve-tube elements.

The sieve-tube elements, having cytoplasm, are living. In xylem, substances move through dead cells. The presence of cytoplasm is essential to

---

[1]The sieve tube elements could well be called sieve cells; but having previously defined a cell as a unit possessing a nucleus, it would be wise to avoid using the word *cell* here. Also, the term *sieve cell* is used in reference to the phloem of gymnosperms.

290 ◆ Chapter 32

## ❧ Notes ❧

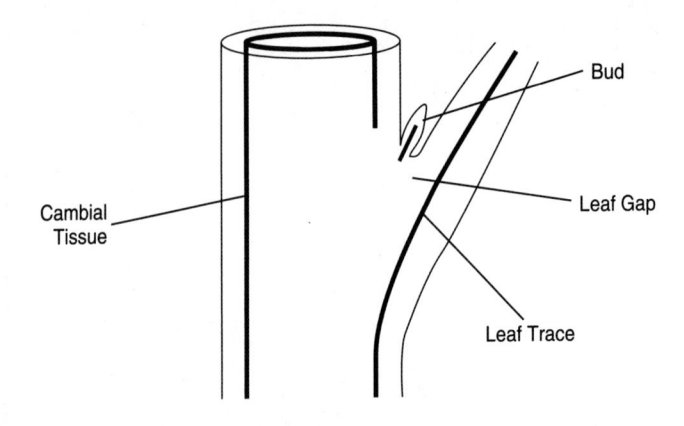

**Figure 32-8** The cambial tissue and the cells immediately adjacent to it. The leaf trace is the vascular tissue that leaves the main cylinder. The leaf gap is the space left above the leaf trace.

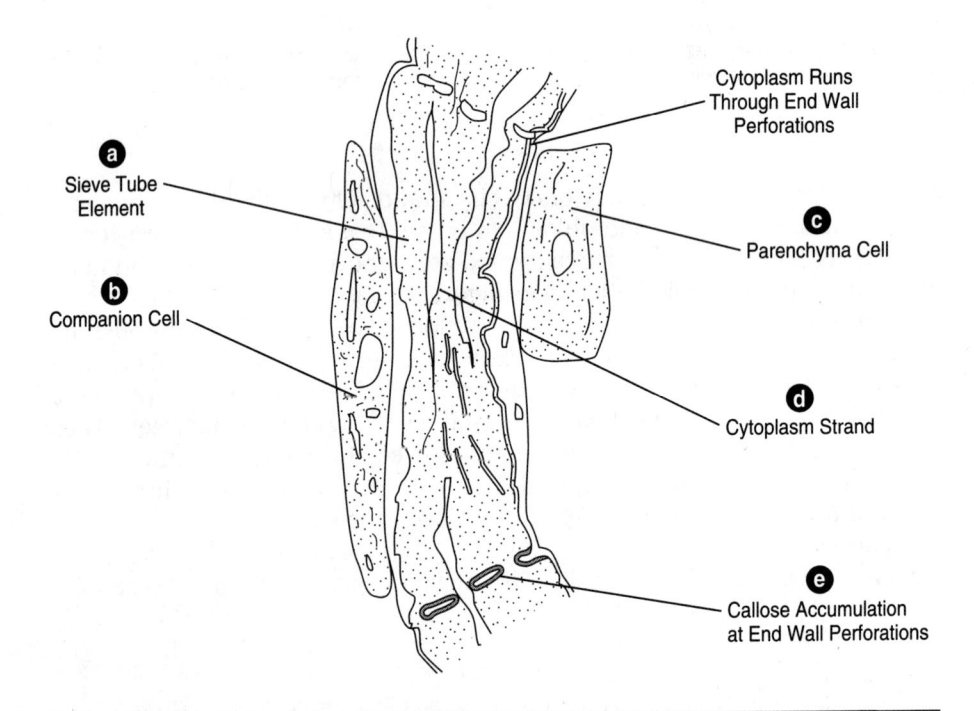

**Figure 32-9** (a) Sieve-tube element, (b) Companion cell, (c) Parenchyma cell, (d) Cytoplasm of sieve cell, (e) Callose accumulation.

the function of phloem cells. The flow of solutions through sieve tubes is hypothesized to result from pressure; the leaves develop a higher turgor pressure and push solutions through the phloem.

Phloem exudate contains 5 to 20 percent sugars, proteins, minerals, and hormones. Many substances can move through phloem only when combined with sugars. While the direction of fluid movement in xylem is primarily upward, fluids move both upward and downward through phloem.

Deposits of callose (a polysaccharide of glucose units) form at the sieve plates. In time, these deposits clog the holes in the sieve plates. Sieve-tube elements and their corresponding companion cells are daughters of a common cell. A sieve-tube mother cell undergoes division to produce the sieve-tube element and the companion cell that resides beside it.

In addition to the sieve tubes and companion cells, parenchyma and fibers are also present. The fibers may be impregnated with lignin.

The cells lying outside of the phellogen, the outer bark, become suberized and soon die. Suberin is a waxy material that makes the outer bark resistant to invasion of bacteria and mildew and retards the loss of water. Bark often possesses openings called **lenticels**, which are believed to allow an exchange of gases (although this has not been proven). A number of plant species, including woody plants, do not have lenticels.

**Notes**

**Figure 32-10** A lenticel, an eruption in the surface of the bark.

## Notes

## The Herbaceous Dicot Stem

The distinction between woody stems and herbaceous stems is not always absolute. In general, however, herbaceous stems do not produce enough secondary growth to make recognizable wood, and they die back to the ground at the end of the growing season.

An herbaceous dicot stem is very similar to the apical end of a woody dicot stem, where no secondary growth has taken place (figure 32-2b); the arrangement of tissues is the same. Figure 32-11 shows a cross section of an herbaceous dicot stem. Vascular bundles are arranged in a circle. This arrangement helps distinguish a central pith and an outer cortex, both of which are parenchyma. The cortex and pith are contiguous by the parenchyma that lies between vascular bundles. These areas are called **medullary rays**. The outermost layer is the epidermis. The vascular bundles have bundle caps lying outside the phloem. The cells of the cap are thick-walled sclerenchyma. Some of the vascular bundles pass out from the vascular cylinder to the leaves. These are leaf traces, and they may occur in threes, fives, occasionally more, and sometimes only one.

Herbaceous dicot stems are sometimes hollow. In some species, the hollow space is lined with an endodermis; in others, there is no endodermis, and the space is instead lined with parenchyma.

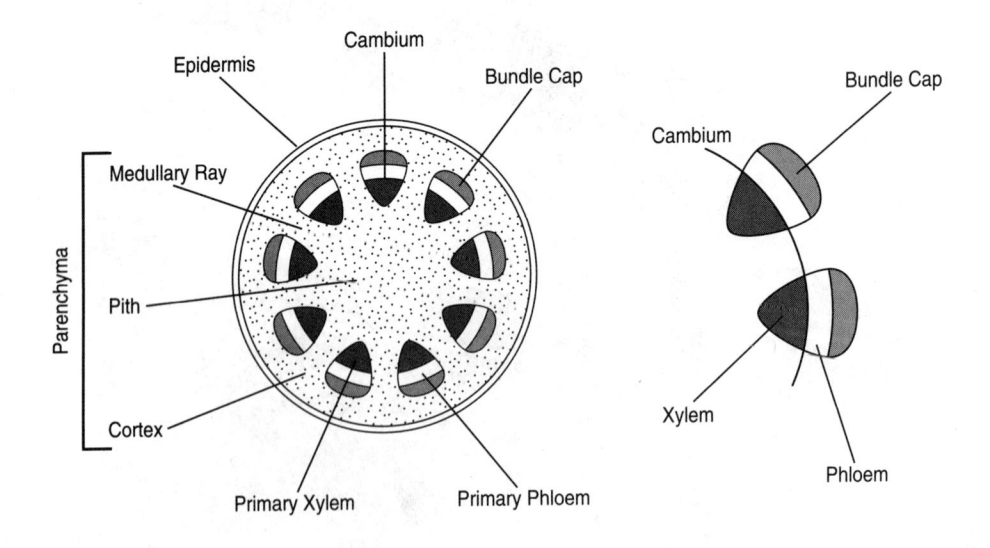

**Figure 32-11** A cross section of an herbaceous dicot stem and at the right two vascular bundles showing cambium running through and continuous between the bundles.

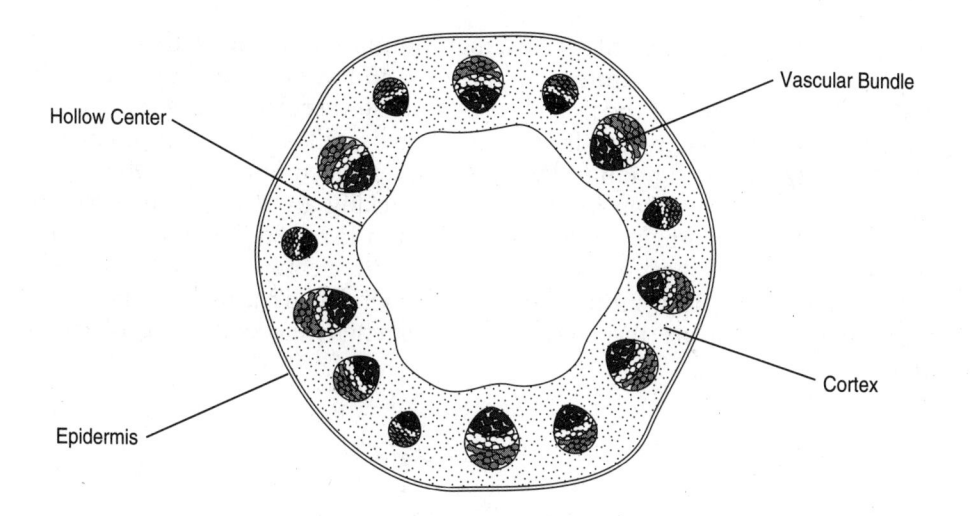

**Figure 32-12** Cross section of an herbaceous dicot stem having a hollow center.

## The Monocot Stem

Monocots are mostly annuals, meaning they live for only a single season. They are most readily recognized by their leaves; although there are exceptions, most monocot leaves exhibit **parallel venation**. A cross section of a monocot stem (figure 32-13) shows vascular bundles scattered in parenchyma. There may be a layer of sclerenchyma beneath the epidermis.

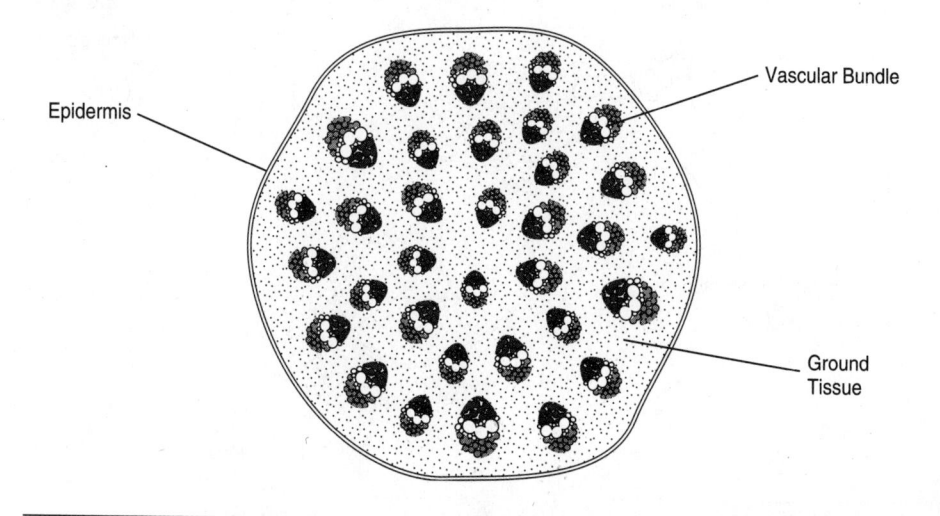

**Figure 32-13** Cross section of a monocot stem.

※ **Notes** ※

Although this arrangement of vascular bundles represents the simplest organization of stem structures, it is believed to have evolved comparatively recently. Examination of the xylem portion of the vascular bundle reveals two (or perhaps three) large vessels surrounded by small, thick-walled tracheids, see figure 32-14. In addition to the vessels, there is commonly another open space sometimes mistaken for a vessel but lacking a cell wall. This space is produced by a fracture; it is not a cell and is not involved in conduction. The phloem portion of the bundle shows sieve-tube elements with companion cells beside them. The entire bundle is surrounded by a **bundle sheath** of sclerenchyma. Outside the sheath is the parenchyma, consisting of large, thin-walled cells.

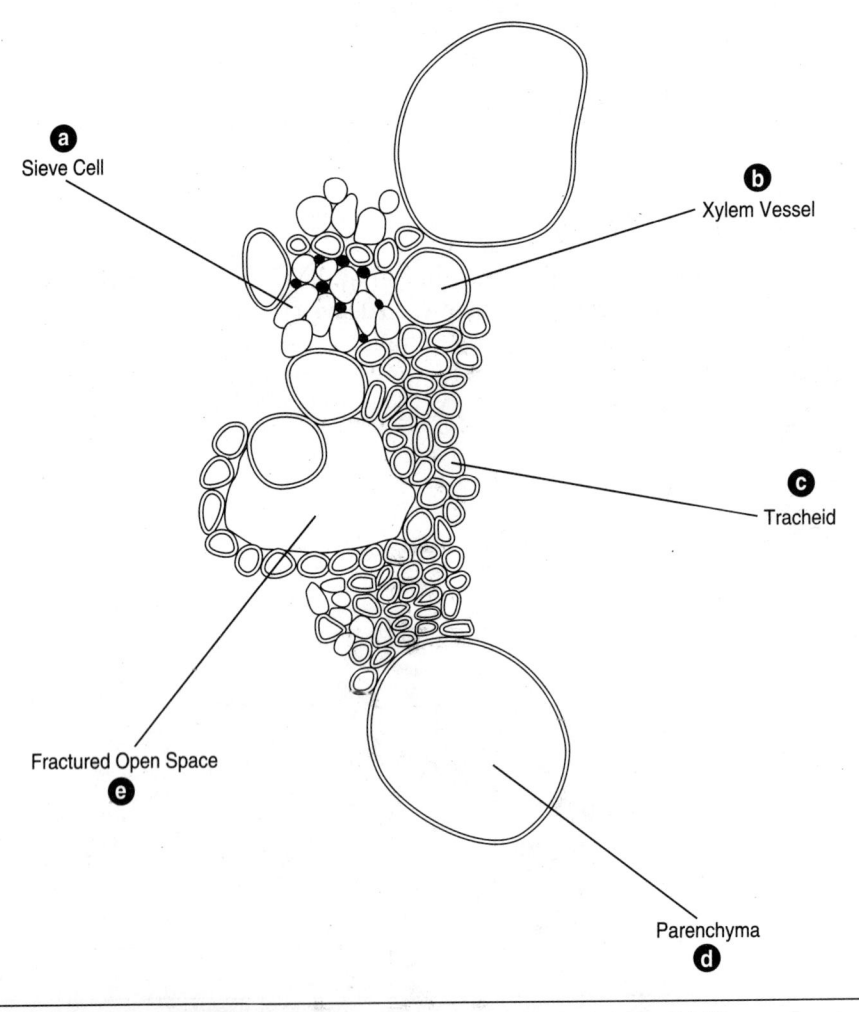

**Figure 32-14** Much enlarged portion of a monocot vascular bundle. (a) Sieve cell. (b) Xylem vessel. (c) Tracheid. (d) Parenchyma. (e) Fractured open space.

## Modified Stems

Several examples of modified stems are shown in figure 32-15. **Stolons** are horizontal stems that grow aboveground and develop new plantlets at the tips wherever the stems touch the ground. New shoots and roots are formed at **nodes**. Strawberry plants are an example of stolons. **Rhizomes** are horizontal stems that grow underground. They also produce new shoots and roots at the nodes. Lawn grasses and blueberry plants possess rhizomes. **Bulbs**, **corms**, and **tubers** are short, underground stems that take part in vegetative propagation. An example of a bulb is an onion. A bulb is a small mound of stem bearing overlapping fleshy leaves that store reserves of starch and sugar. The leaves are called bulb scales. Small bulbs may develop in the axils at the base of the bulb. A corm takes the form of a squatty, swollen stem. It lacks scale-like leaves. New, small corms originate from buds at the nodes. These small corms can be separated and planted. Examples of corms are gladiolus and crocus. Tubers are the expanded tips of rhizomes. The Irish potato (*Solanum tuberosum*) is the best known example. One can plant whole tubers or cut sections bearing one or more nodes. It is recommended that cut sections be allowed to dry for several days prior to planting in order to discourage decay.

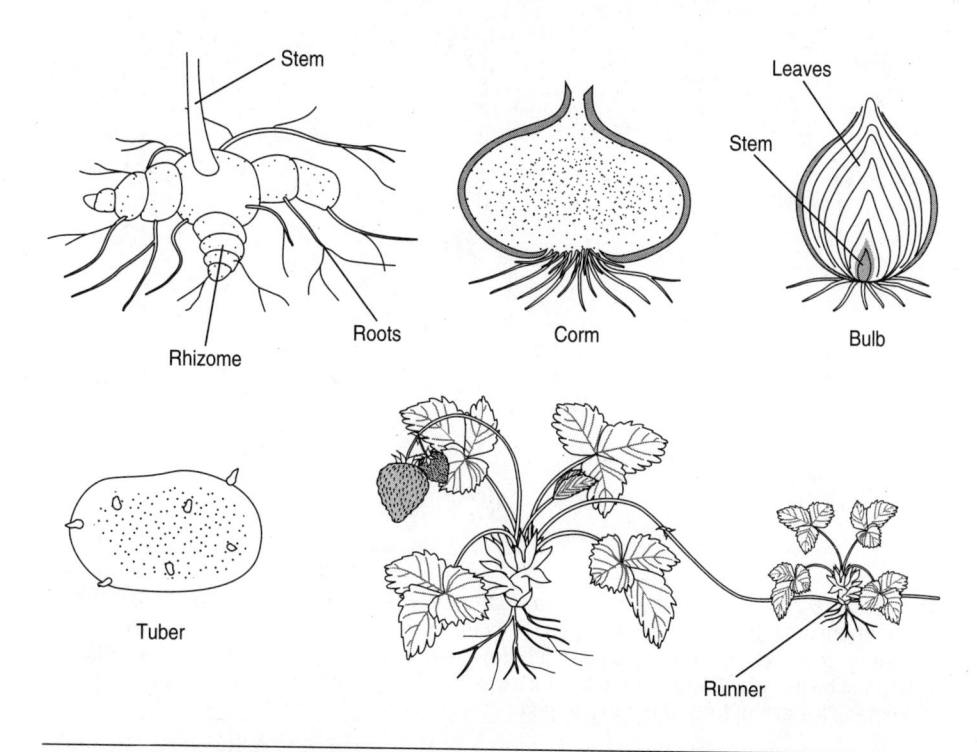

**Figure 32-15** Modified stems.

296 ◆ Chapter 32

## ❋ Notes ❋

A thorn is a modification of a stem. Thorns are present on honey locusts and hawthorns. Many kinds of plants produce climbing stems, which crawl over rocks or other stems. They are called vines or, if woody, lianas. Many vines and lianas have twining stems. The twining stems of some

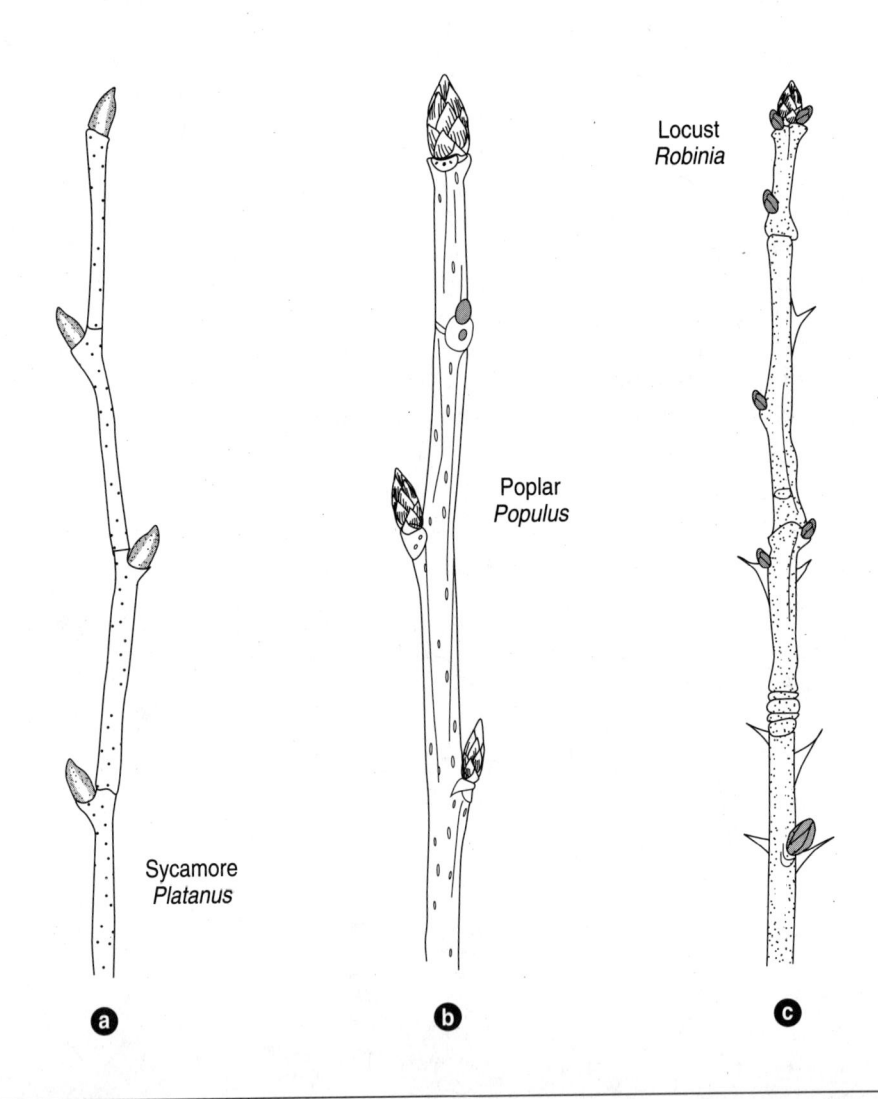

**Figure 32-16** (a) Stem of the sycamore, *Platanus*. The buds are enclosed in the dilated bases of petioles. (b) Stem of the poplar, *Populus*. The scaly buds shown here are frequently covered with a resinous varnish. (c) Stem of the locust tree, *Robinia*. It exhibits spinous stipules, which are paired appendages occurring at the bases of leaves. Bud characteristics are of frequent value in taxonomic work.

species turn clockwise, some counterclockwise, and some in both directions. Stem tips tend to turn in a spiral because of growth inequalities. This phenomenon is called **nutation**. When the young stem touches some surface, twining increases. The cells on the side of the stem that touches an object shorten, and the cells on the opposite side grow longer. The end result is a curving of the stem tip. Many vines have tendrils, which tend to grasp any touched surface. Grape vines have tendrils opposite each leaf. The tendrils of Boston ivy have adhesive discs, which enable it to adhere to stone walls. Poison ivy produces adventitious roots, which function in the same way.

A plant called "butcher's broom" produces flattened, leaflike stems called **cladophylls**. In the center of each cladophyll is a node bearing a scale-like leaf (figure 32-17). The feathery appearance of asparagus is caused by cladophylls.

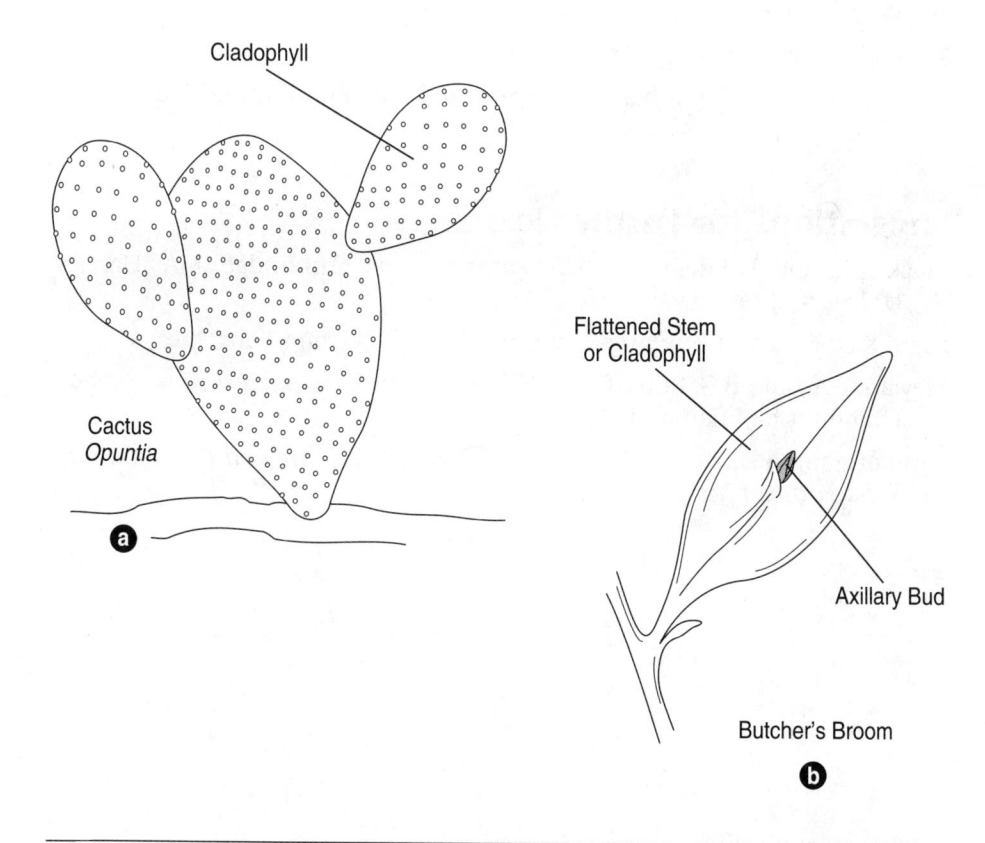

**Figure 32-17** (a) The cactus *Opuntia*, bearing cladophylls. (b) Butcher's broom. The flattened, leaflike structure is a modified stem called a cladophyll. Cladophylls are stems that assume the properties and functions of leaves.

❋ Notes ❋

## Questions for Review

1. Where are meristems located, and how do they function?
2. What are the functions of xylem and phloem?
3. Define the terms *node* and *internode*.
4. How can the age of a twig be determined?
5. How does a cross-sectional view of a monocot differ from that of a dicot?
6. From what tissues do white potatoes and sweet potatoes arise?
7. Characterize a corm, a bulb, and a tuber.
8. Define the terms *lenticel* and *leaf scar*.
9. What tissue is responsible for the length-wise growth of a stem? What tissue is responsible for diameter increase in a stem?
10. What causes the formation of annual rings?
11. Explain the meaning of the terms *hardwood* and *softwood*.
12. Does the removal of bark interfere with the ascent of water? Explain why or why not.

## Suggestions for Further Reading

Black, M., and J. Edelman. 1970. *Plant Growth*. Cambridge, MA: Harvard University Press.

Esau, K. 1965. *Plant Anatomy*. New York: John Wiley and Sons, Inc.

Meylan, B.A., and B.G. Butterfield. 1972. *Three Dimensional Structure of Wood*. London: Chapman and Hale.

Zimmerman, M.H., and C.L. Brown. 1971. *Trees: Structure and Function*. New York: Springer-Verlag.

# 33

# Roots

❋ Notes ❋

Roots are commonly understood to anchor the plant and take up water from the soil. The latter portion of this description leads to an equally common misunderstanding: that roots *seek* water. It is easy to devise a demonstration that dispels this idea. First, place a layer of moist soil in a pot and top this layer with a layer of dry soil followed by another layer of moist soil. Next, plant a seed in the uppermost layer of moist soil. When the seed germinates, the root will grow downward until it makes contact with the dry soil, where it will stop. It cannot, of course, grow through the dry soil in order to reach the moist soil below. The principle is easy to follow: "roots grow where they can grow." It might be called the *principle of the self-evident*.

## Contributors to Root Growth

Like stems, roots grow in length only at the tips. Although they grow where moisture occurs, moisture is not the only requirement for their growth. Sugar produced in the leaves by photosynthesis is also necessary, as are hormones, which are also produced in the leaves. Hormones must be transported from the leaves to the roots in order to influence root growth. Temperature; the presence of minerals and fungi; and acidity or alkalinity are other important factors in both root and plant growth. In addition to moisture, air in the soil is also significant. Interestingly, the roots of a pear tree can spread to nine or more times the diameter of the aboveground portion of the tree yet not reach much depth.

### ❧ Notes ❧

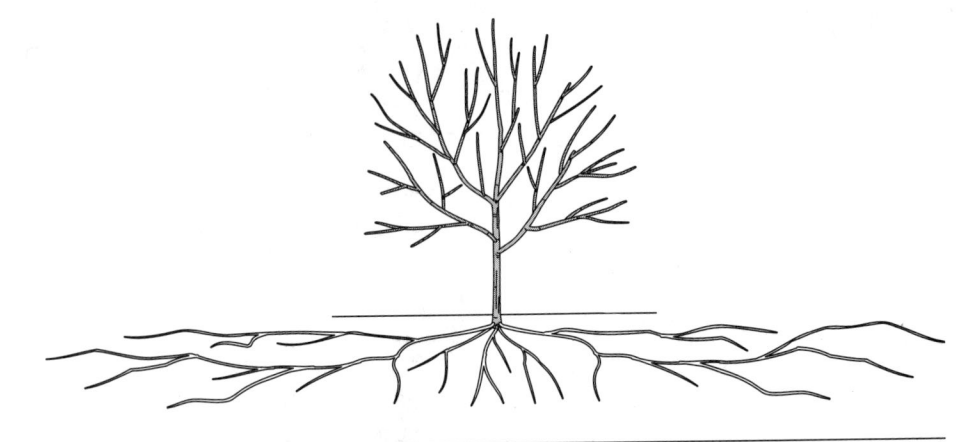

**Figure 33-1** The root system may be ten times the diameter of the aboveground portion of a tree.

## Root Hairs

Figure 33-2 shows a longitudinal section of the growing tip of a root. Meristematic activity, which increases the length of the root, occurs only at the tip. When the cells here divide, they produce both new root cells and root cap cells. The root cap cells are sloughed off as the root grows through the soil. The epidermal cells produce **root hairs** a short distance above the tip. Root hairs are part of epidermal cells.

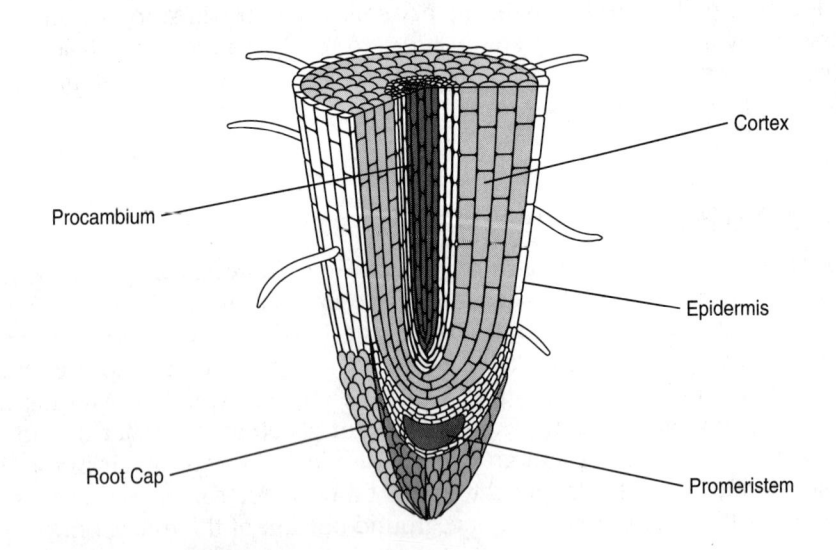

**Figure 33-2** Longitudinal section of a root tip.

In figure 33-3, note the location of the nucleus. This seems to indicate the nucleus is positioned at a point of greatest activity. The only portion of a root able to take in water is the root hair; and the total surface area of root hairs available to perform this function is much greater than the total surface area of the leaves. Because root hairs are short lived (living perhaps a day), root hair production must continue in order for the plant to be maintained. It is estimated that one hundred million root hairs are produced each day in a rye grass plant.

## Structure of a Root

Figure 33-4 shows a cross section of a root in a region where only primary growth has occurred. All tissues here derive from apical meristem. The innermost, star-shaped tissue is the primary xylem. Between the points of the primary xylem are areas of primary phloem. Immediately outside of these primary vascular tissues is a zone of parenchyma cells, called the **pericycle**. This is bounded by a layer of suberized cells one cell in thickness and called the **endodermis**. Peripheral to the endodermis is another zone of parenchyma cells known as the cortex. The cortex is bounded on the outside by the epidermis.

Figure 33-5 shows there to be a region in the root where one may trace a circle that will pass alternately through primary xylem and primary phloem. This alternating arrangement of the primary xylem and primary phloem is one of the principal differences between roots and stems.

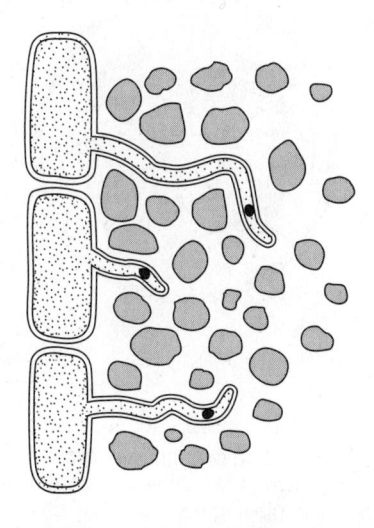

**Figure 33-3** Root hairs pushing through soil. Root hairs are a part of epidermal cells.

## ❧ Notes ❧

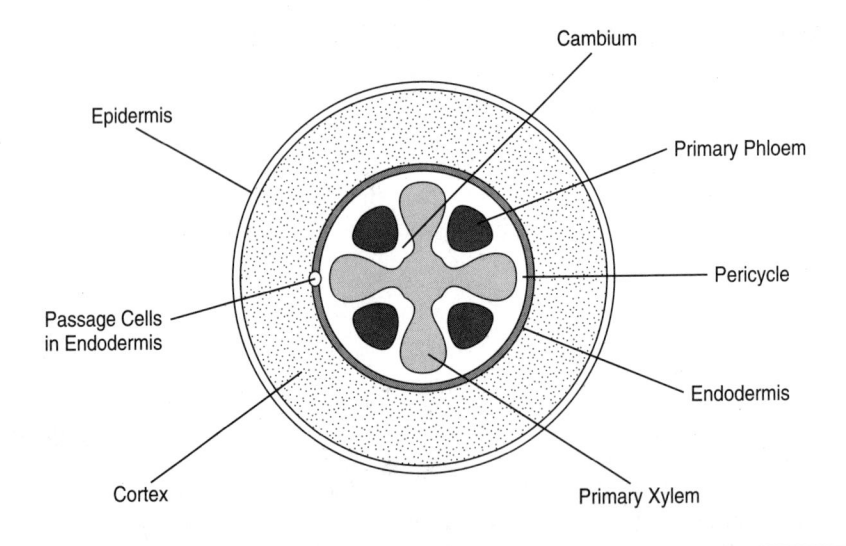

**Figure 33-4** Cross section of a root exhibiting no secondary growth. The position of the cambium lies between the primary tissues. (b) Passage cells lie in the endodermis.

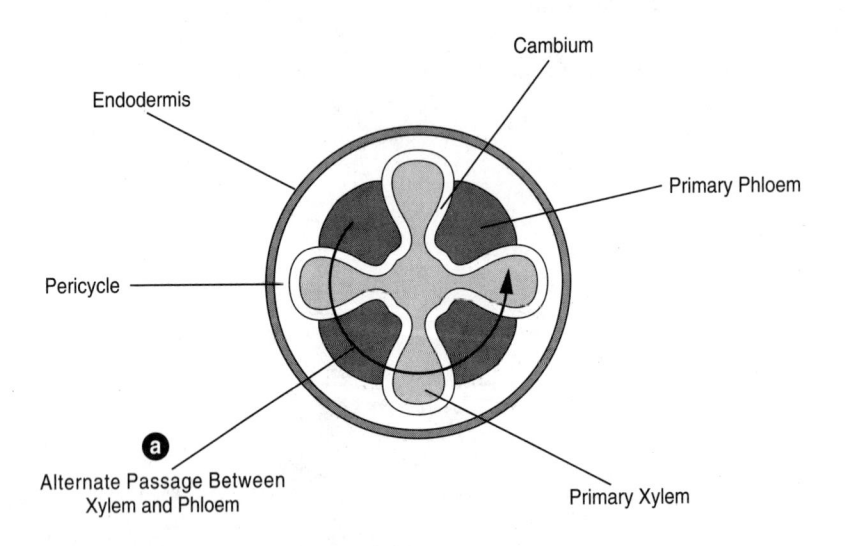

**Figure 33-5** Portion of a root exhibiting only primary growth and lying within the endodermis. A circle traced as shown at (a) passes alternately through primary xylem and primary phloem. Only a root exhibiting no secondary growth possesses this configuration.

## Casparian Strip

Close examination of the endodermis reveals closely packed cells having thickened walls and lacking intercellular spaces. There is also a strip of suberin called the **Casparian strip** (see figure 33-6). This strip lies on radial and transverse walls. Suberin is a waxy material that inhibits the passage of water. The positioning of the Casparian strip appears to block the lateral movement of water, both in and out, through the cell wall. Because the secondary wall does not exert selectivity, and the Casparian strip appears to occlude the passage of water through the walls, it is theorized that the protoplasm, or, more particularly, the cell membranes, exercise control over lateral water flow by selective permeability. Although it has been hypothesized that this may be in some way related to root pressure, the fact that not all of the endodermal cells have such a strip seems to undermine this hypothesis. Just beyond the points of the primary xylem, cells of the endodermis are not impregnated with suberin and water can move freely through the endodermis. Such cells can occur singly or in small clusters at other points around the endodermis. These cells are called **passage cells** because they allow the free passage of water and solutes to and from the inner region.

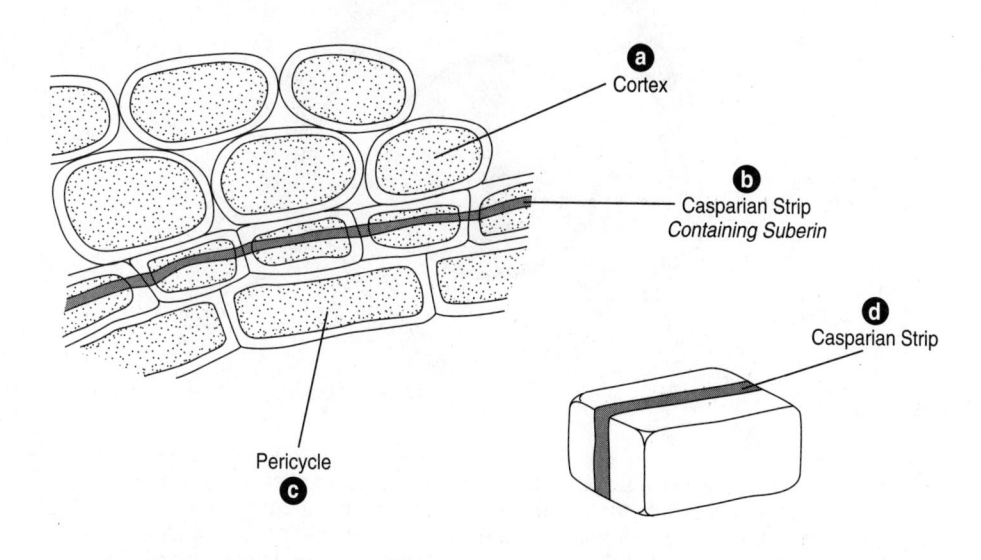

**Figure 33-6** The endodermis, a single layer of cells, has a band of suberin called the Casparian strip, (b), which lies along the radial and transverse walls. The Casparian strip is postulated to influence the inward and outward flow of water from the cortex, (a), to the pericycle, (c), and reverse.

## Notes

## Root Growth

When the root cambium produces secondary xylem and secondary phloem, the secondary xylem, formed medial to the cambium, lies in contact with the primary xylem. The secondary phloem, deposited peripheral to the cambium, pushes all tissues external to it outward, until the small patches of primary phloem are soon pushed quite away from their original positions between the points of primary xylem (see figure 33-7). Another cambial layer, the cork cambium, develops in the pericycle cells; this layer of cells produces the outer bark. The primary tissues lying outside of the cambium, having no way of increasing, fracture and slough away.

As a root ages, it becomes more like a stem. The original primary xylem remaining at its center serves as a clue to the root's identity, however. Another difference between roots and stems relates to the origins of branch roots and branch stems. Branch roots have their origin in the pericycle; thus, their origin is endogenous. Branch stems, on the other hand, derive from superficial tissues; their origin is therefore exogenous.

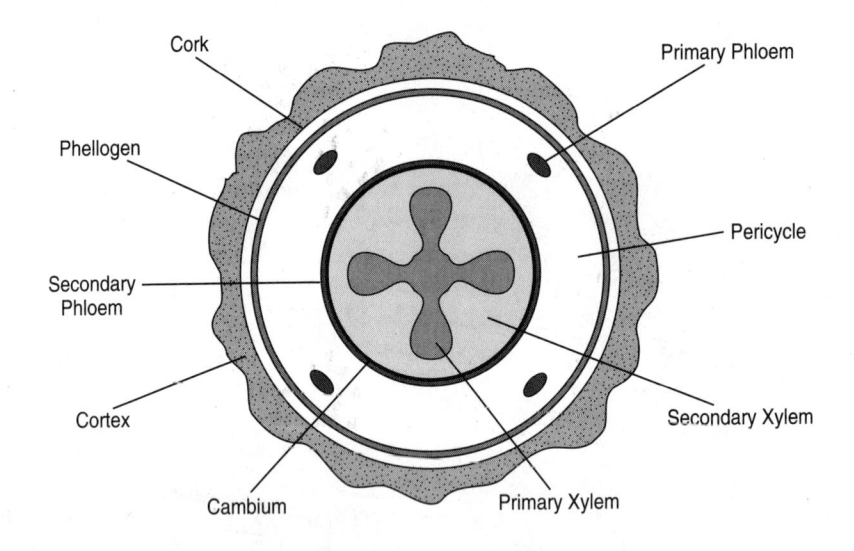

**Figure 33-7** Cross section of a root exhibiting a small amount of secondary growth. In depositing secondary xylem, the cambium moves outward to form a circle. This pushes the primary phloem outward, where its remnants are represented as four small islands. The primary xylem is unaffected. The secondary phloem is thin and compressed. The pericycle is still intact. A new cambium, the phellogen, takes form and produces a small amount of cork. The cortex and epidermis begin to slough away.

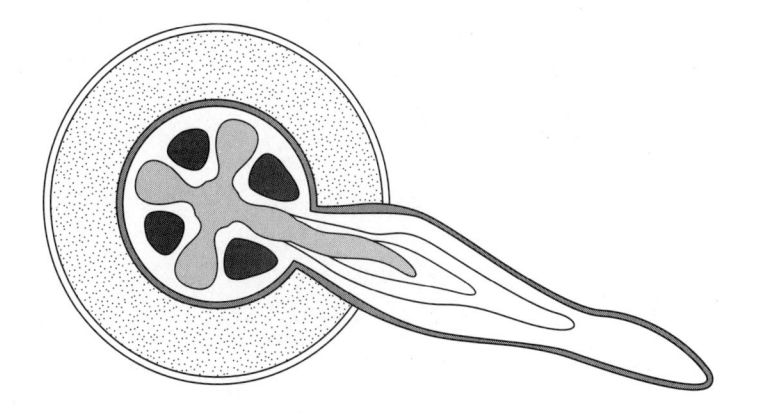

**Figure 33-8** Branch roots have their origin in the pericycle.

## Questions for Review

1. Distinguish between a very small root and a root hair.
2. Distinguish between a parasitic root and a mycorrhiza.
3. What is the most significant difference when comparing a cross section of a root and that of a stem?
4. What is the function of the root cap, and from what tissue does it arise?
5. How are endodermal cells distinguished?
6. From what tissue do branch roots arise?
7. Distinguish between a primary root, a secondary root, and an adventitious root.
8. Name several plants that store large amounts of food reserve in their roots.
9. Distinguish between primary tissue and secondary tissue.
10. What is cork cambium, and what is another name for it?

## Suggestions for Further Reading

Cronquist, A. 1971. *Introductory Botany*. New York: Harper and Row Publishers.

Epstein, E. 1973. Roots. *Scientific American* 228:48–58.

Esau, K. 1976. *Anatomy of Seed Plants*. New York: John Wiley and Sons, Inc.

# 34

# Leaves

**P**art of a definition of a leaf would likely be that it is broad, flat, and green. This could be expanded to mention that a leaf consists of a blade and a **petiole**, that it sometimes has **stipules** at the base, and that it has a bud in the axil between the petiole and the stem (unless the leaf is sessile, that is, lacking a petiole).

❋ Notes ❋

## Simple versus Compound Leaves

Having a bud in the axil is an important part of the definition of a leaf. Leaves are either simple or compound; a **simple leaf** consists of one unit, and a **compound leaf** is composed of separate **leaflets**. There are no buds in the axils of leaflets but found only at the base of a leaf. This feature can be used to distinguish a leaf from a leaflet.

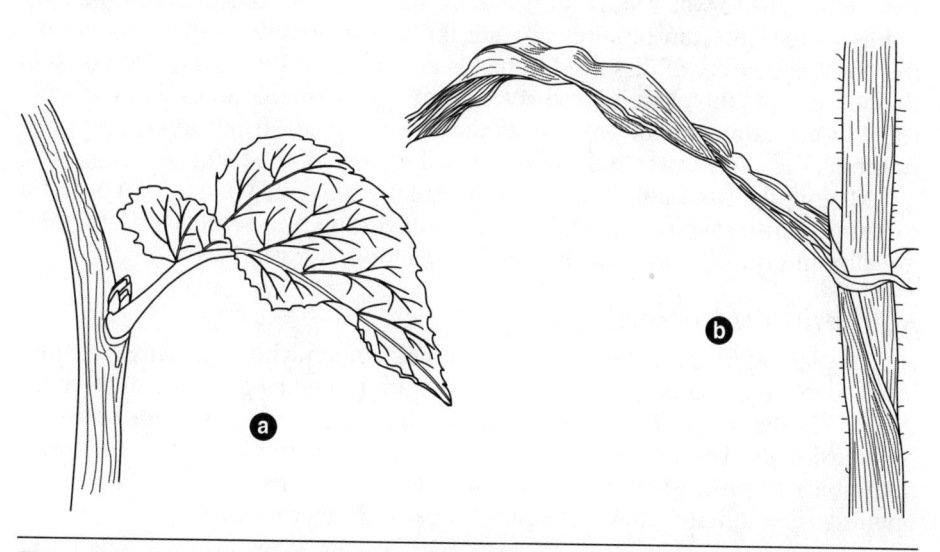

**Figure 34-1** (a) Leaves petiolate. (b) Leaves sessile.

## ✿ Notes ✿

**Figure 34-2** Two compound leaves: (a) palmately compound and (b) pinnately compound.

# Transpiration

Leaves are involved in three distinct plant functions: photosynthesis, respiration, and **transpiration**. Photosynthesis and respiration are discussed in chapters 9 and 10, respectively. Transpiration is the loss of water by evaporation. While transpiration occurs primarily through leaves, some water is also lost by young stems. Is there any value to water loss? Transpiration has a cooling effect, but the benefits of this are unclear. Transpiration also aids the upflow of water from the roots, and, thereby, the delivery of dissolved minerals and water to the cells of the plant. In this sense, then, the plant is well served by transpiration. Yet, transpiration, especially in the extreme, can prove hazardous to the plant. If not replaced by soil moisture, excessive water loss can result in the death of the plant.

Magnitude of water loss is often striking. A single plant of corn, for example, can transpire fifty gallons of water during one season. A date palm growing in an oasis can transpire 35,000 gallons in a year![1]

## *Stomates and Guard Cells*

Higher plants fortunately have evolved a mechanism whereby excessive water loss is deterred. The major portion of transpiration takes place through openings in the leaves, called stomates (figure 34-3). The stomates are bounded by **guard cells**. The guard cells are able to respond to changing conditions, thereby closing the stomate when water loss threatens to be excessive and opening the stomate when transpiration is no longer a threat.

---

[1] Gibbs, R.D. 1950. *Botany, An Evolutionary Approach*. Philadelphia: The Blakiston Co.

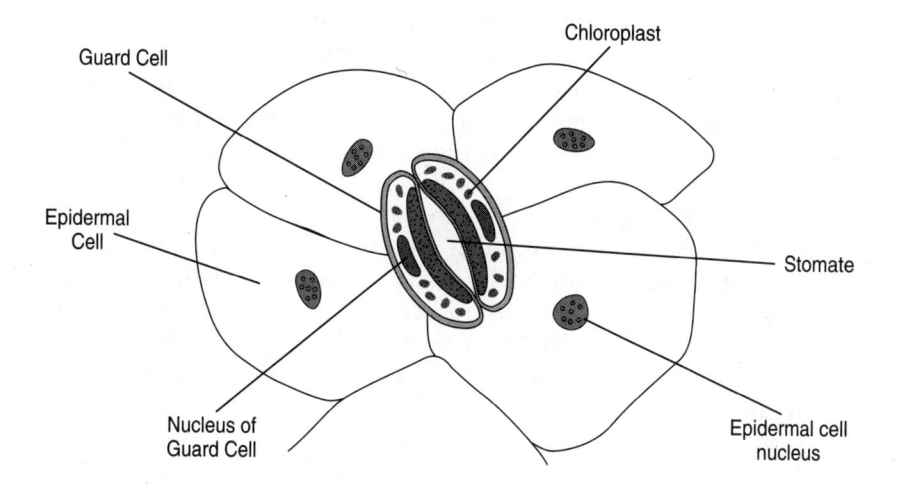

**Figure 34-3** A stomate in leaf epidermis.

Although there are numerous exceptions, stomates are usually concentrated on the lower surfaces of leaves. The water lily, having its leaves floating on the surface of a pond, has stomates entirely on the upper surface. Stomates are usually very numerous, there being hundreds per square millimeter of leaf surface, and millions on an entire leaf. Table 34-1[2] shows the number of stomates per square millimeter of leaf surface for several species.

| Species | Upper Surface | Lower Surface |
|---|---|---|
| *Abies balsamea* (balsam fir) | 0 | 228 |
| *Lilium bulbifera* (lily) | 0 | 62 |
| *Morus alba* (white mulberry) | 0 | 480 |
| *Syringa vulgaris* (lilac) | 0 | 330 |
| *Nymphaea alba* (water lily) | 460 | 0 |
| *Pinus strobus* (white pine) | 142 | 0 |
| *Helianthus annuus* (sunflower) | 175 | 325 |
| *Lycopersicon esculentum* (tomato) | 12 | 130 |
| *Phaseolus vulgaris* (bean) | 40 | 281 |
| *Triticum sativum* (wheat) | 33 | 14 |

**Table 34-1** Stomates Per Square Millimeter of Leaf Area

---

[2]Source unknown.

### Notes

The secondary wall thickenings of the guard cells are distributed in such a manner that when the cells lose turgor (that is, lose water), the guard cells change shape and close the stomates. When turgor is increased they again change shape to open the stomates. Although one might presume that the threat of water loss governs the closing of the stomates, this process may actually be governed by photosynthesis. The guard cells have chloroplasts, whereas other epidermal cells do not. When photosynthesis occurs in the guard cells, the concentration of carbon dioxide diminishes. Because carbon dioxide when dissolved in water creates an acidic condition, a decrease in carbon dioxide likewise decreases acidity. The solution thereby becomes more alkaline, a condition favorable to the breakdown of starch into sugars, and, thus, an increase of particles in the cytoplasm. This condition encourages the inflow of water to the guard cells, thus increasing their turgor, and causes them to change shape and open the stomates. While this explanation may not be entirely complete or correct, it is far from satisfactory to instead suggest that a stomate opens and closes simply because it needs to.

## Leaves and Transplanting

The nursery worker is concerned with water loss when transplanting—and with good reason. When a plant is taken up from the soil, a significant portion of the root system is lost. The capacity of the plant to take up water is thus much reduced. To combat this, the nursery worker must reduce the amount of water lost by transpiration. For this reason, transplanting is generally done in autumn or early spring, when leaves are not on the plant. This amounts to a compromise, however, because root restoration is most readily accomplished when active photosynthesis is taking place, which, of course, occurs when leaves are intact. An alternative, then, is to transplant when leaves are present and to remove some but not all of the leaves. If the plant is small enough, cutting away half of each leaf is sometimes done.

## Guttation

Figure 34-4 shows a strawberry leaf with droplets of water at the borders. These water drops represent neither transpiration nor condensation from the atmosphere. Rather, this water results from **guttation**, a process whereby water is exuded (forced out under pressure) from the leaf through special glands called **hydathodes**. Analysis of the water reveals that it contains sugars.

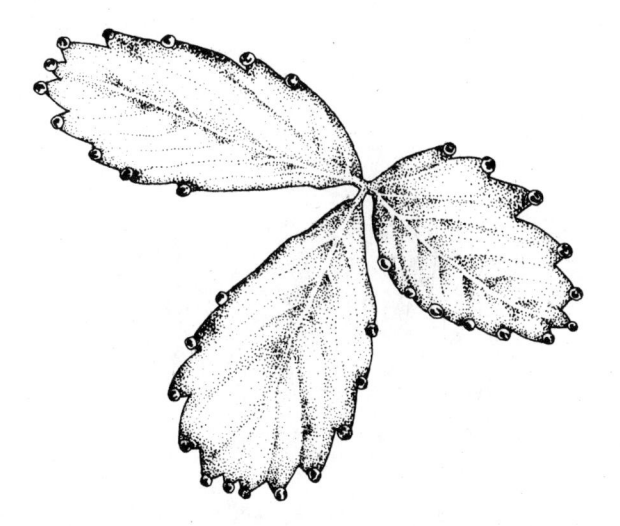

**Figure 34-4** Guttation in a strawberry leaf.

## Structure of a Leaf

In figure 34-5, a stomate appears on both the upper and lower surfaces of the leaf. A thin, waxy cuticle is also on the upper surface. This cuticle tends to retard water loss through epidermal cells. The figure also shows an upper epidermis and a lower epidermis. The lower epidermis has a hair called a trichome. The only epidermal cells bearing chloroplasts are the guard cells. Between the two epidermises are **mesophyll** cells abundant in chloroplasts. The mesophyll cells are of two kinds. In the upper portion of the leaf, the mesophyll cells are elongated and vertically arranged, and are collectively called the **palisade parenchyma**; and in the lower portion of the leaf, the mesophyll cells are characterized by conspicuous intercellular spaces, and are collectively called the **spongy parenchyma**, or **spongy mesophyll**. The intercellular spaces communicate with the substomatal spaces to allow gas interchange with the atmosphere. Carbon dioxide from the atmosphere enters the mesophyll cells through the intercellular spaces, while oxygen, a product of photosynthesis, is conveyed out of the leaf through the stomates.

## Leaves and Plant Classification and Identification

Plants can be classified into two groups according to whether the leaves are shed in the fall (**deciduous**) or retained through the winter (**evergreen**). While angiosperms are mostly deciduous and gymnosperms are mostly evergreen, there are some exceptions.

### ✽ Notes ✽

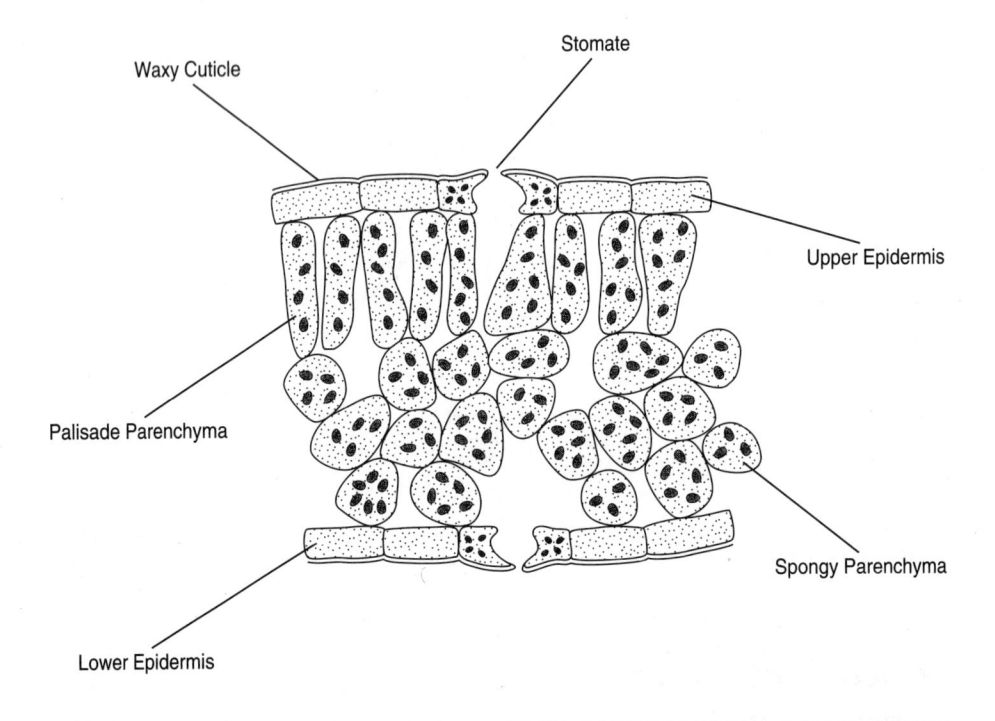

**Figure 34-5**  Cross section of a leaf. Stomates appear on both upper and lower surfaces.

What causes leaves to fall in autumn? While it might be supposed that diminishing temperature is the cause, the true cause is diminishing number of daylight hours. As the days grow shorter (and the number of hours of light thus decreases) a zone of corky cells, called the **abscission layer**, forms at the base of the petiole, figure 34-6. This layer becomes differentiated into a protective layer on the side toward the stem and a **separation layer**. At the latter region, the cells are weak, and the leaves break away from the tree when the wind blows.

While common leaf characters do not necessarily imply kinship among plants, leaf traits are nevertheless valuable in plant identification. Figure 34-7 shows variations in leaf shape; the names shown apply to the patterns.

Stipules may be present or absent (**estipulate**). The **venation** of leaves is important in identification. As stated earlier, monocots generally exhibit parallel venation, and dicots generally exhibit **net venation**, although there are exceptions. The vestiture (presence of hairs) is also valuable in identification, and there is great variety in vestiture. In fact, approximately forty different terms exist to describe leaf hairs. The margins of leaves are significant, as is leaf arrangement on the stem. The latter may be opposite (alternate) or whorled.

Leaves ◆ 313

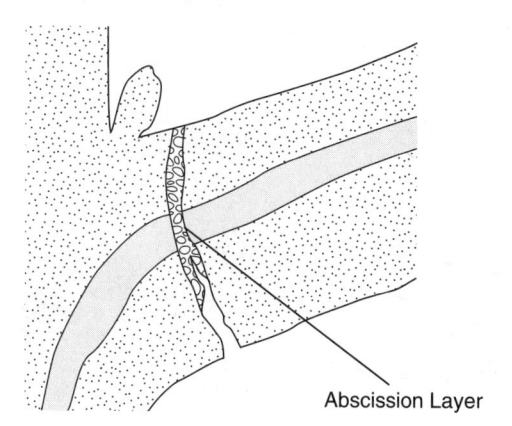

❀ Notes ❀

Figure 34-6  Abscission layer (corky cells that contribute to leaf fall) at the base of a petiole.

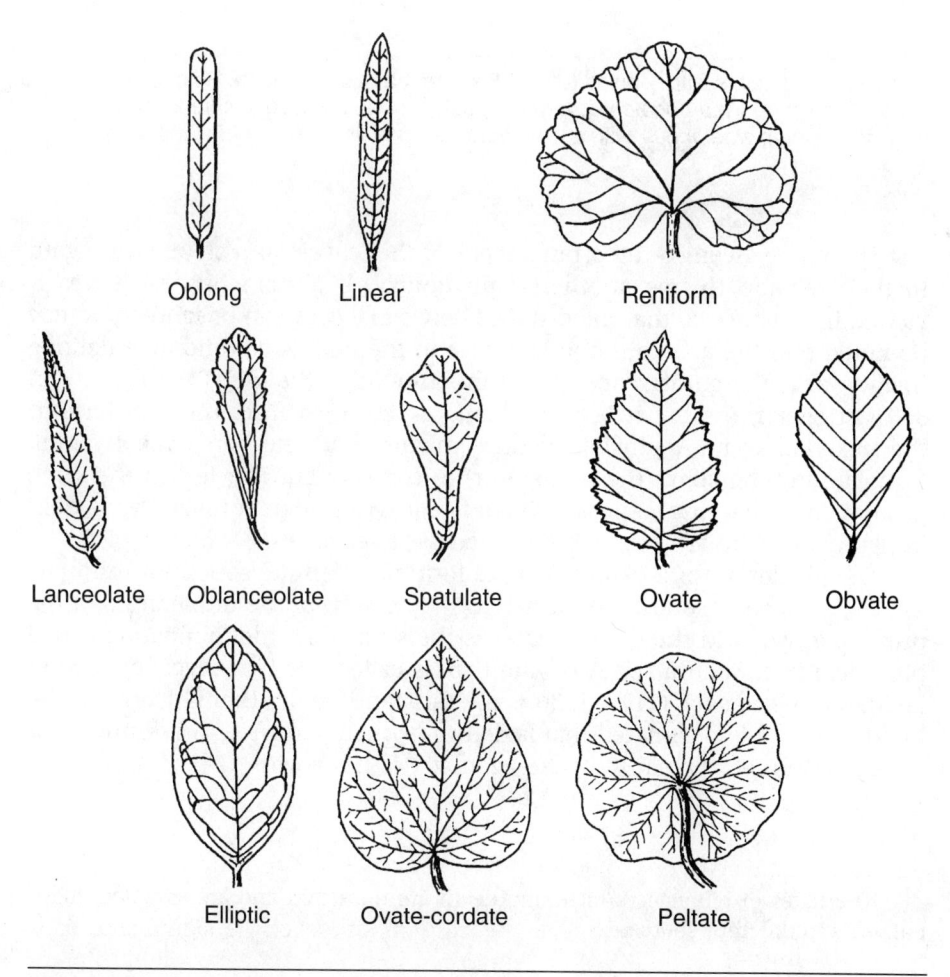

Oblong    Linear    Reniform

Lanceolate    Oblanceolate    Spatulate    Ovate    Obvate

Elliptic    Ovate-cordate    Peltate

Figure 34-7  Several leaves, showing variations in shape (of value in identification).

## ✺ Notes ✺

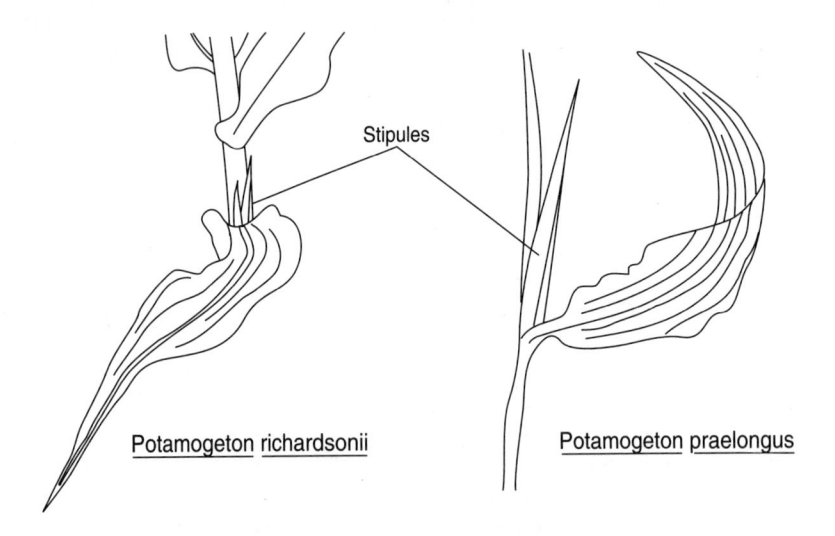

**Figure 34-8** Leaves with stipules (appendages at the base of a leaf). Both are from aquatic plants: (a) *Potamogeton richardsonii*, with stipules reduced to shreds, and (b) *Potamogeton praelongus*, with a long, persistent stipule.

The name given to the arrangement of the leaves on a stem with regard to their relation to one another is **phyllotaxy**. If a stem bearing leaves is viewed from above so that the distance between leaves can be ignored, it may be noted that the arc from one leaf base to the next is commonly a definite fraction of the circumference. At left in figure 34-9, the arc is ½, that is, ½ of 360°. At center, the arc is ⅓, there being a 120° turn from one leaf base to the next below it. A phyllotaxy of ⅖ means that one can trace two full turns around the stem (the numerator) to come to another leaf in the same position, and in doing so pass through five leaves (the denominator). At right in figure 34-9, the arc between two successive leaves is ⅖ × 360 = 144°.

The phyllotaxy of a plant takes the form of a definite series: for example, ½, ⅓, ⅖, ⅜. In each series, the numerator is the sum of the numerators of the preceding two, and the denominator is the sum of the denominators of the preceding two. Such a series of numbers is called the Fibonacci series after Leonardo Fibonacci (1170–1230)[3], an outstanding mathematician of the Middle Ages. It is postulated that leaf arrangement is related to hormone distribution, though the cause of the geometry is unknown.

---

[3]The dates of Fibonacci's birth and death are uncertain and are assigned different years in different sources.

1/2 Phyllotaxy                1/3 Phyllotaxy                2/5 Phyllotaxy

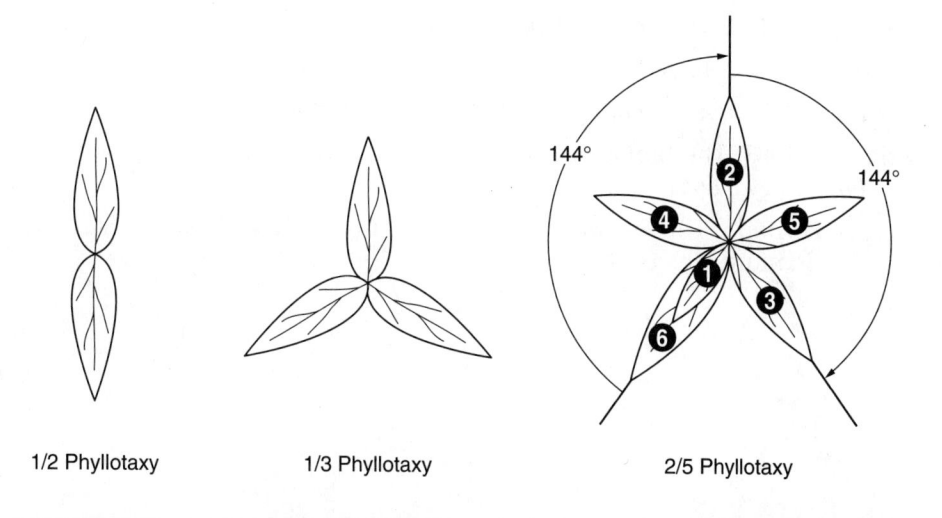

**Figure 34-9** Phyllotaxy: at left, ½ phyllotaxy; at center, ⅓ phyllotaxy; and at right, ⅖ phyllotaxy. Note that the leaves are numbered. If you trace from leaf 1 to leaf 2, it measures 144°; from leaf 2 to leaf 3 also measures 144°. If you continue to leaf 6, you will have traced two full passes around to reach the position of beginning at leaf 6, and you will have passed through five leaves. This is ⅖ phyllotaxy.

## Questions for Review

1. How do the lower epidermis and upper epidermis of a leaf differ?
2. What modifications of leaves enable a plant to live in arid conditions?
3. Describe several types of leaf arrangement.
4. Distinguish between pinnate and palmate venation.
5. Distinguish between simple and compound leaves.
6. Describe the structure of guard cells and relate what factors are involved in their opening and closing.
7. What is mesophyll, and of what kind of cells is it composed?
8. What causes the leaves of angiosperms to be shed in the autumn?
9. Define the term *transpiration*. Is it of any value to a plant?
10. Define the terms *hydrophyte* and *xerophyte*.
11. Define the term *phyllotaxy*.

❈ Notes ❈

## Suggestions for Further Reading

Darwin, C. 1896. *Insectivorous Plants.* New York: D. Appleton and Company.

Esau, K. 1977. *Anatomy of Seed Plants.* New York: John Wiley and Sons, Inc.

Zeiger, E. 1983. The biology of stomatal guard cells. *Annual Review of Plant Physiology* 34:441–475.

Zeigler, H. 1987. "The Evolution of Stomata." In *Stomatal Function.* Stanford, CA: Stanford University Press.

# 35

# Flowers

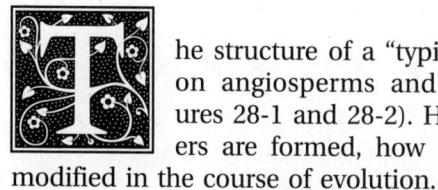

The structure of a "typical" flower was described in chapter 28 on angiosperms and should be reviewed (specifically figures 28-1 and 28-2). Here, consideration is given to how flowers are formed, how they vary, and how they have become modified in the course of evolution.

## How Flowers Are Formed

A flower is made of modified leaves. When a plant reaches a certain stage of maturity, something happens that causes the plant to stop making ordinary leaves and start producing the modified leaves that we recognize as flowers. Specifically, the **internodes** do not elongate. Thus, several whorls are crowded together and inserted on an expanded receptacle. As mentioned previously there are four whorls: the sepals (which compose the calyx), the petals (which compose the corolla), the stamens (also called **androecium**, meaning "male house"), and the **carpels** (also collectively called the pistil). At the base of the pistil is the ovary, which bears ovules. Another name for this structure is **gynoecium** (meaning "female house").

Although the flower is part of the sporophyte generation, it is a common practice to refer to the pistil as the female part of the flower because the female gametophyte develops in the ovules that reside in the ovary. In like manner, it is common to refer to the stamens as the male part of the flower because pollen grains develop into male gametophytes. Megaspores, which are produced in the ovule, represent the culmination of the sporophyte generation. One of the megaspores develops into the female gametophyte. Microspores are produced in anthers and grow into pollen grains, which, in turn, develop into male gametophytes.

❈ Notes ❈

## ✸ Notes ✸

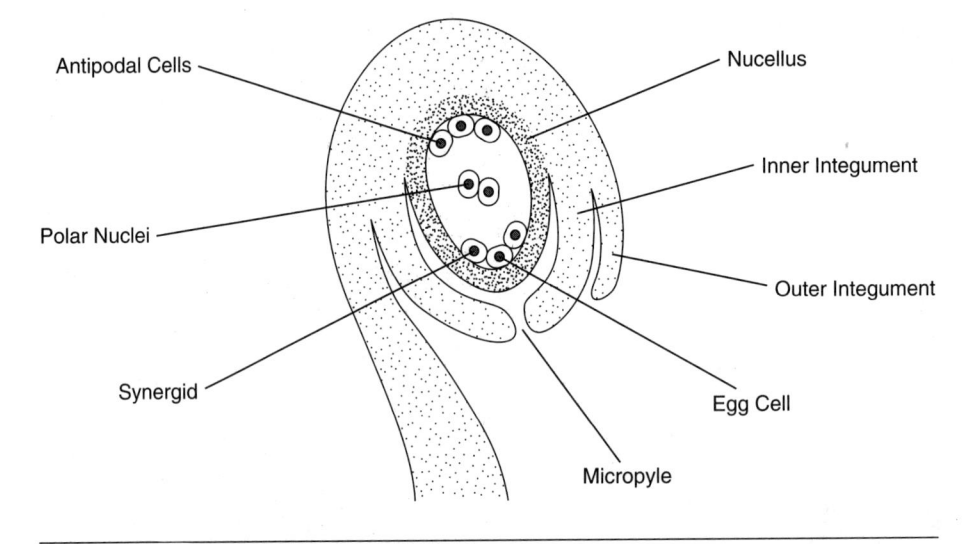

**Figure 35-1** The female gametophyte in an ovule.

The pistil, composed of one or more carpels (modified leaves), has three parts: the expanded ovary at the base (which bears ovules), the elongated style, and a sticky stigma at the tip of the style (which is receptive to pollen grains). A cross section of an ovary (figure 35-2b) reveals one to several chambers, called **locules**. The ovules reside in these chambers and are attached to the ovary wall. The place of attachment is the **placenta**. A cross section of an

Cross Section Through Anthers
**(a)**

Cross Section Through Ovary
**(b)**

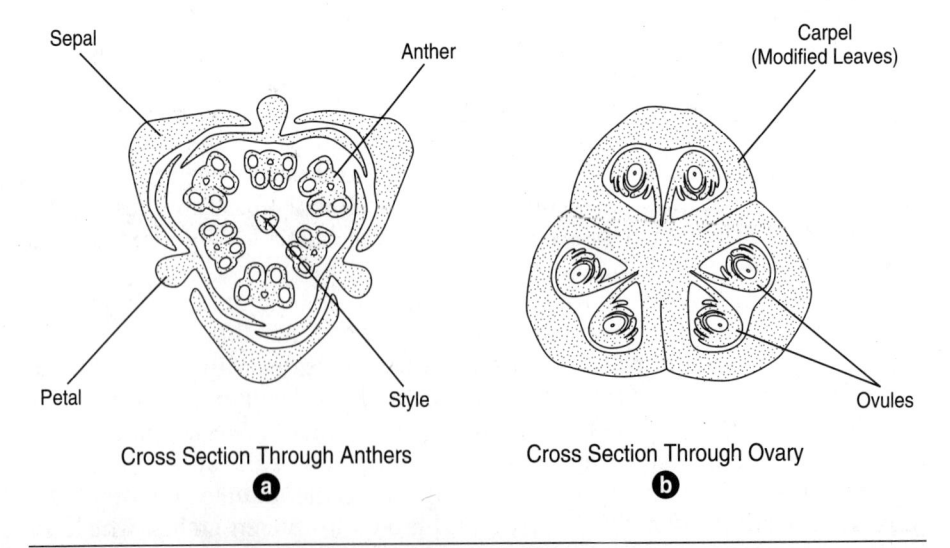

**Figure 35-2** (a) Cross section of a flower bud through the region of the anthers, which contains a sepal (part of the calyx), petal (part of the corolla), anther, and style. (b) Cross section of an ovary consisting of three carpels, each of which contains two ovules.

anther (figure 35-2a) reveals chambers that contain the microspore mother cells. These cells undergo meiosis to produce the microspores, which become pollen grains (figure 35-3).

## Variations in Flowers

There is great variation in flowers. Some flowers are so small that they can scarcely be seen without a lens. One example is duckweed, *Lemna minor*, an aquatic plant found floating on water. At the opposite extreme is a plant of the

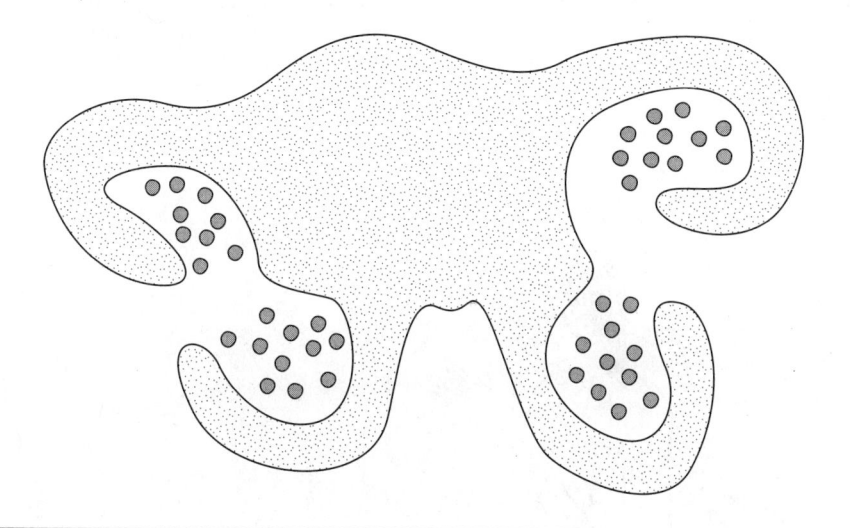

**Figure 35-3** Cross section of an anther, showing pollen grains.

**Figure 35-4** Two species of duckweed, minute flowering plants that float on water.

❋ Notes ❋

**❋ Notes ❋**

Malay peninsula. *Rafflesia* produces flowers measuring three to four feet in diameter. It is called the "stinking corpse lily" and attracts carrion-eating beetles. Another giant flower, called *Amorphophallus titanum*, is shown in figure 35-6.

Sometimes the calyx or corolla (known as **perianth** parts) is missing; in this case, the flower is called **incomplete**. If both stamens and pistil are present, the flower is said to be perfect. If one or the other of these is lacking, the flower is unisexual and said to be imperfect. There are **pistillate** (having a pistil but lacking stamens) flowers and **staminate** (having stamens but lacking a pistil) flowers. If both pistillate and staminate flowers are present on the same plant, the plant is said to be monoecious (that is "living in one house"). If pistillate and staminate flowers are on different plants, the plants are said to be dioecious ("living in two houses"). In monocots, the flower parts are

**Figure 35-5** *Rafflesia*, a leafless parasitic plant. The flower can measure one meter in diameter. It is malodorous and, thus, attracts carrion-eating beetles. (Illustration by Donna Mariano)

typically borne in threes or multiples of three. In dicots, the flower parts are generally borne in fours or fives (see figure 35-7).

If the arrangement of the flower parts, particularly the perianth (calyx and corolla), exhibit radial symmetry, the flowers are said to be **actinomorphic** (star shaped). If the flower parts exhibit bilateral symmetry, they are called **zygomorphic**. If the perianth is missing altogether, the flower is said to be **apetalous**, or naked.

A distinction needs to be made between a flower and an inflorescence. In some plants, the flowers are borne singly; in others, they are borne in clusters. A dandelion is not a flower but, rather, an aggregate of flowers on a single receptacle, that is, an inflorescence. Other examples of plants having inflorescences are goldenrod, daisy, and marigold. Figure 35-9 shows a hanging inflorescence called a *catkin*. An inflorescence of this type occurs on willows.

❉ **Notes** ❉

**Figure 35-6** *Amorphophallus titanum*, another gigantic flower. This plant grows in Sumatra. (Illustration by Donna Mariano)

## ❋ Notes ❋

*Sagittaria*
Aquatic Monocot
Flower Parts in 3's

*Silene*
Dicot Flower
Flower Parts in 5's

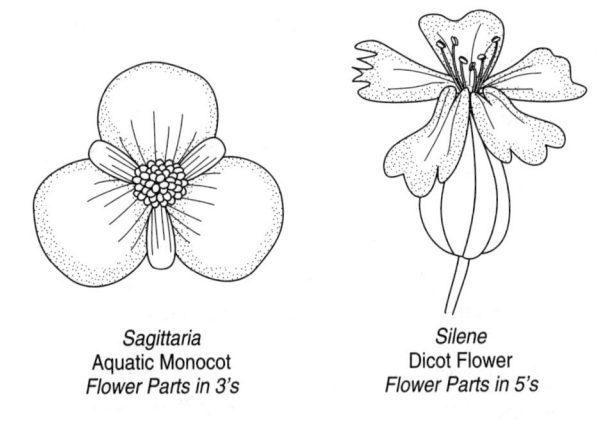

**Figure 35-7** *Sagittaria*, an aquatic monocot having flower parts in threes. The rhizome of *Sagittaria* is sometimes used as food, and the plant sometimes goes by the name "swamp potato." *Silene*, a dicot having flower parts in fives. It is sometimes called "catchfly" because, being sticky, it traps insects. It is not, however, insectivorous.

Actinomorphic Flower

Zygomorphic Flower

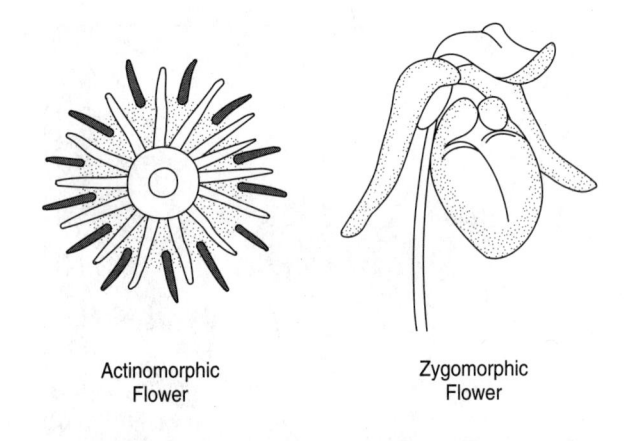

**Figure 35-8** An actinomorphic flower and a zygomorphic flower.

When the several whorls of flower parts are inserted on the receptacle beneath the ovary and are free from the ovary, the parts are said to be **hypogynous**, and the ovary is said to be superior (that is, above). This is characteristic of the so-called typical flower. When the perianth parts grow on the summit of the ovary, the perianth parts are said to be **epigynous**, and the ovary is said to be inferior (that is, below). When the perianth parts are around the ovary but not at its base, the parts are said to be **perigynous**. Figure 35-10 illustrates these various arrangements.

Flowers ◆ 323

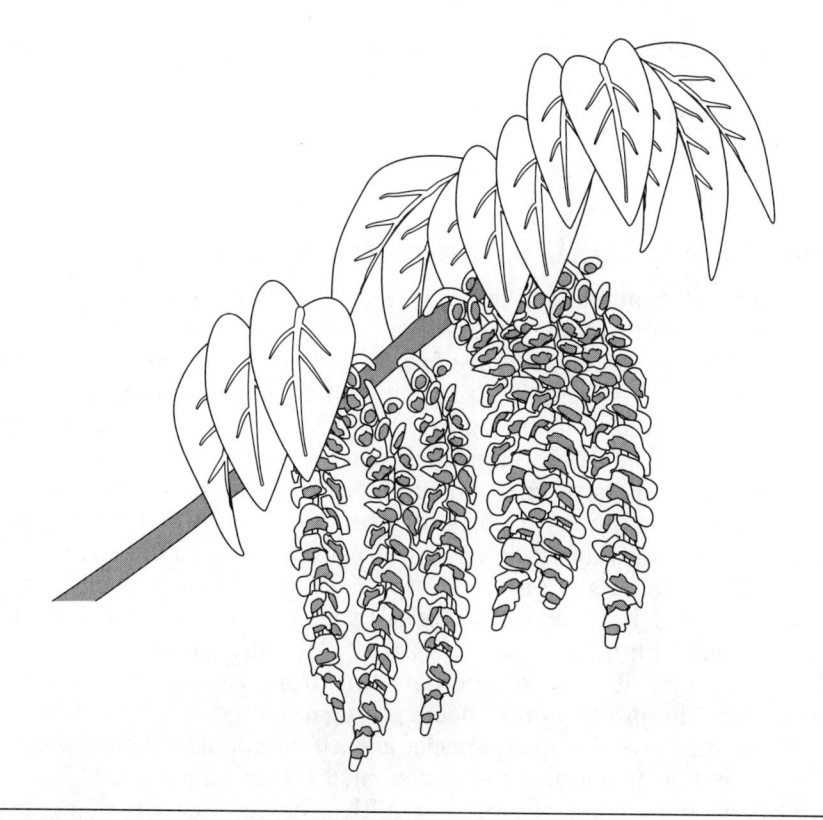

※ **Notes** ※

**Figure 35-9**  Catkins, a type of inflorescence.

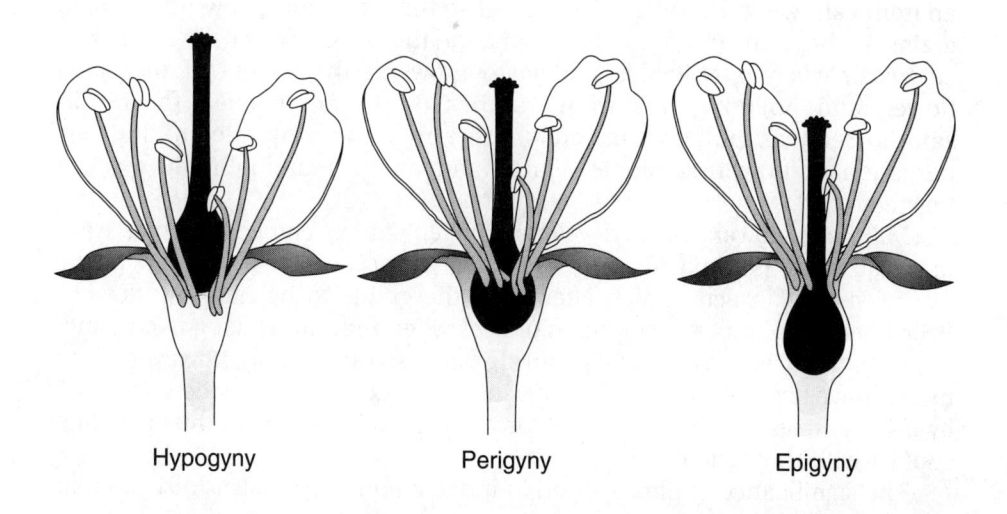

Hypogyny          Perigyny          Epigyny

**Figure 35-10**  The insertion of flower parts.

### ✤ Notes ✤

The characteristics of flowers and fruits are more stable than those of leaves and are thus more relied on in taxonomic studies (in the identification and classification of plants).

## Evolutionary Modifications in Flowers

Flowering plants made a "sudden" appearance during the late **Mesozoic** era, approximately one hundred million years ago, and then rapidly increased during the Eocene period. Primitive flowers had numerous and separate parts and a superior ovary. Magnolias and the tulip tree, *Liriodendron*, exhibit these characteristics and are thus recognized as primitive plants. In time, numerous modifications in flower parts took place. Flowers lacking the perianth depend on wind for transferring pollen. Flowers of the poplar tree are of this sort. When present, perianth parts may be free (that is, not united), as in primitive flowers, or they may be partly or wholly united in the calyx, corolla, or both. The morning glory is an example.

Many flowers have evolved mechanisms that invite visitations from animals, insects, birds, and mammals that play roles in the transferring of pollen from one flower to another and from one plant to another (**pollination**). The production of nectar, a sugary liquid, is one such mechanism. While the flowers of many species are self-compatible, the pollen of one flower being able to fertilize the ovules of the same flower, the flowers of many other species are **self-incompatible**, the fertilization of ovules in one flower being dependent on pollen being transferred from another flower. Self-incompatibility among flowering plants is of two kinds. One is anatomical, wherein differences of structure do not allow the pollen grains to be transferred to the stigma of the same flower. The other is chemical, wherein pollen is unable to grow on the stigma of the same flower, and pollen grains from another flower are needed for fertilization. Bees, as pollen transferers, play an important role in the perpetuation of flowering plants. Plants and animals undoubtedly evolved together.

Milkweed flowers have developed pollen grains embedded in a waxy material called **pollinia** (figure 35-11). The grains become entangled in the feet of visiting insects, which later carry the grains to neighboring flowers. Pollen becomes stuck to the bodies of bees when they are seeking nectar and, in this manner, is carried to neighboring flowers. Interestingly, however, some insects have bypassed this system of what might be called "intended purpose" by drilling holes at the base of the corolla to search for nectar, thus avoiding contact with the pollen.

The significance of photoperiod in inducing flower initials is discussed in chapter 19 on hormones.

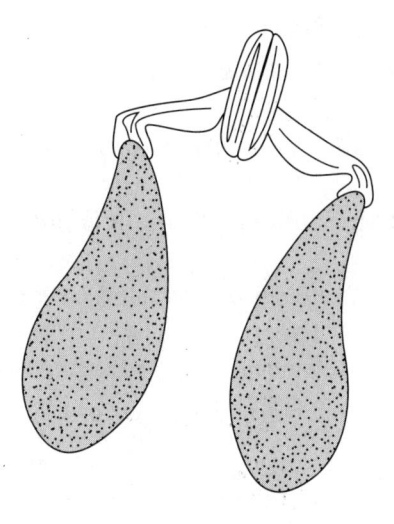

**Figure 35-11** Pollinia

## Questions for Review

1. The parts of a flower are derived from _____.

2. The outermost whorl of these parts is the _____, and the next inner whorl is the _____.

3. The pollen sacs are called _____.

4. The ovules reside in the _____.

5. Another name for the pistil is _____; likewise, the pollen sacs may also be called the _____.

6. The culmination of the sporophyte generation is to produce _____.

7. A flower is said to be perfect if it has _____ and _____.

8. If both staminate and pistillate flowers are present on the same plant, the plant is said to be _____.

9. Flower parts in monocots commonly occur in what number?

10. A name indicating a radially symmetrical flower is _____, and a name indicating a bilaterally symmetrical flower is _____.

❋ Notes ❋ **Suggestions for Further Reading**

Condon, G. 1977. *The Complete Book of Flower Preservation*. Englewood Cliffs, NJ: Prentice-Hall.

Edsall, M.S. 1984. *Roadside Plants and Flowers*. Madison, WI: University of Wisconsin Press.

Ferguson, M., and R.M. Saunders. 1976. *Wildflowers*. New York: D. Van Nostrand Co.

Hillman, W.S. 1962. *The Physiology of Flowering*. New York: Holt, Rinehart, and Winston, Inc.

Janick, J. 1979. *Horticultural Science*. San Francisco: W.H. Freeman Co.

Salisbury, F.B. 1963. *The Flowering Process*. Oxford, England: Pergamon Press.

# 36

# Fruits and Seeds

When the egg cell in the ovule is fertilized, changes take place in both the ovule and the ovary. The fertilized egg undergoes divisions to produce an embryo plantlet, and the ovule, which contains the plantlet, matures into a seed. The ovary, which houses the seeds, matures into a fruit. A fruit is therefore defined as a ripened ovary.

❈ Notes ❈

## Forms of Fruit

The fruits of different species of plants exhibit a variety of forms, and fruit form is a valuable trait in classifying plants. The wall of a ripened ovary is called the pericarp. The pericarp may consist of two or three layers: the exocarp, mesocarp, and endocarp. A **berry** has a fleshy pericarp with a thin skin. Examples of berries are grapes, tomatoes, peppers, and blueberries. A **drupe** is a single-seeded fruit having a stony endocarp (that is, a pit). Examples of drupes are cherries, peaches, and olives. A **legume** derives from a single carpel and splits along two lines. Examples of legumes are beans and peas. A **follicle** is similar to a legume but splits along one line. An example is milkweed. An **achene** is a small fruit having a single seed and a hard pericarp. Buckwheat and dandelion are two examples. The list of fruit characteristics goes on, and those who study taxonomic botany make much use of fruit characteristics in the course of classification, see figure 36-1.

## Seed Structure and Characteristics

The seed containing the embryo plant also contains endosperm, a food storage tissue. The whole is enclosed in one or two seed coats. Seeds range in size from dustlike, such as the seeds of orchids, to a length of one foot or more and a weight of as much as 40 pounds, such as the coconut.

## Notes

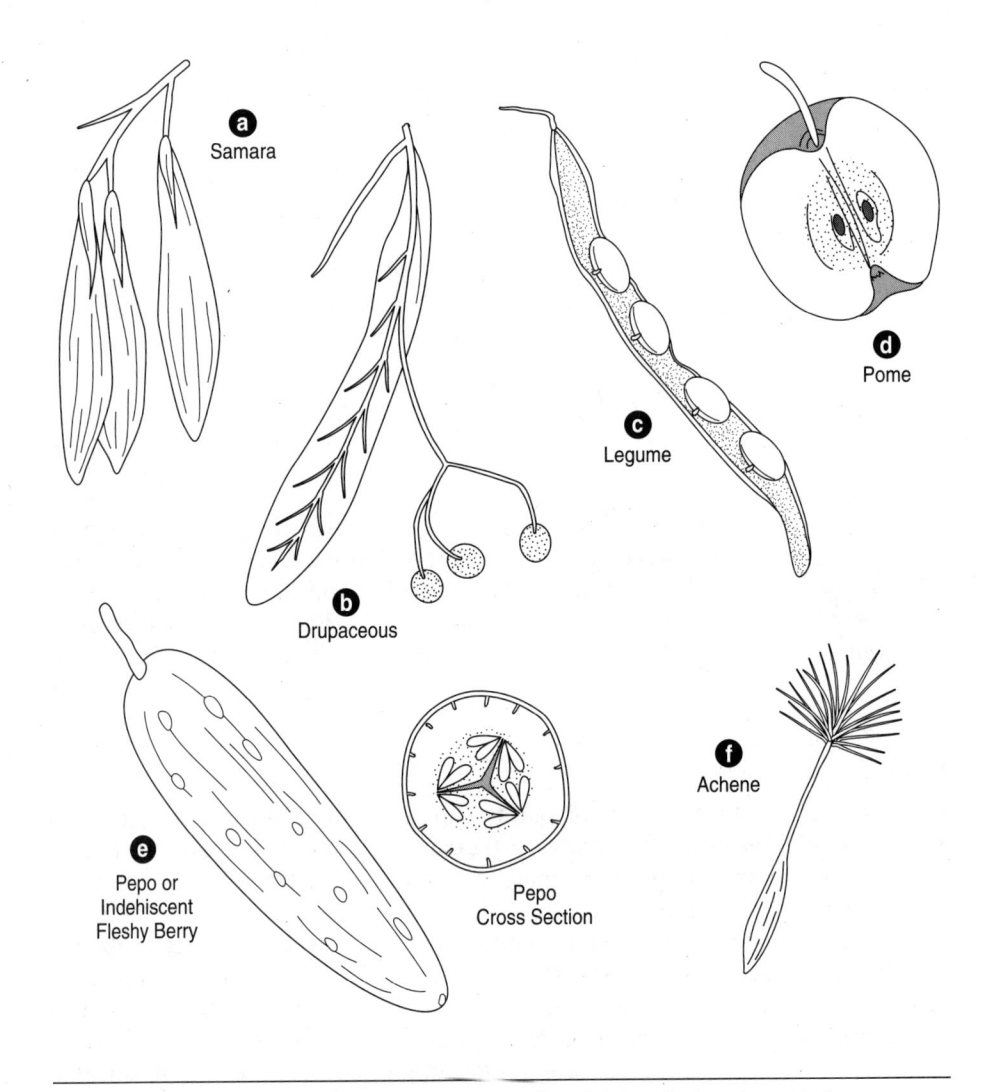

**Figure 36-1** Examples of fruits. (a) A samara, sometimes called a winged achene (e.g., ash). (b) A winged fruit described as drupaceous (e.g., bass-wood or linden). (c) A legume (e.g., beans and peas). (d) A pome (e.g., apples and pears). (e) A pepo, or indehiscent fleshy berry (e.g., cucumber). (f) An achene (e.g., dandelion).

The fertilized egg gives rise not only to an embryo plantlet, but also to some **extraembryonic** cells. In early cell divisions, an embryo portion is distinguishable from a basal cell and the suspensor. The basal cell is at the micropylar end of the ovule. The basal cell and the suspensor are not part of the embryo, although they are derived from the zygote.

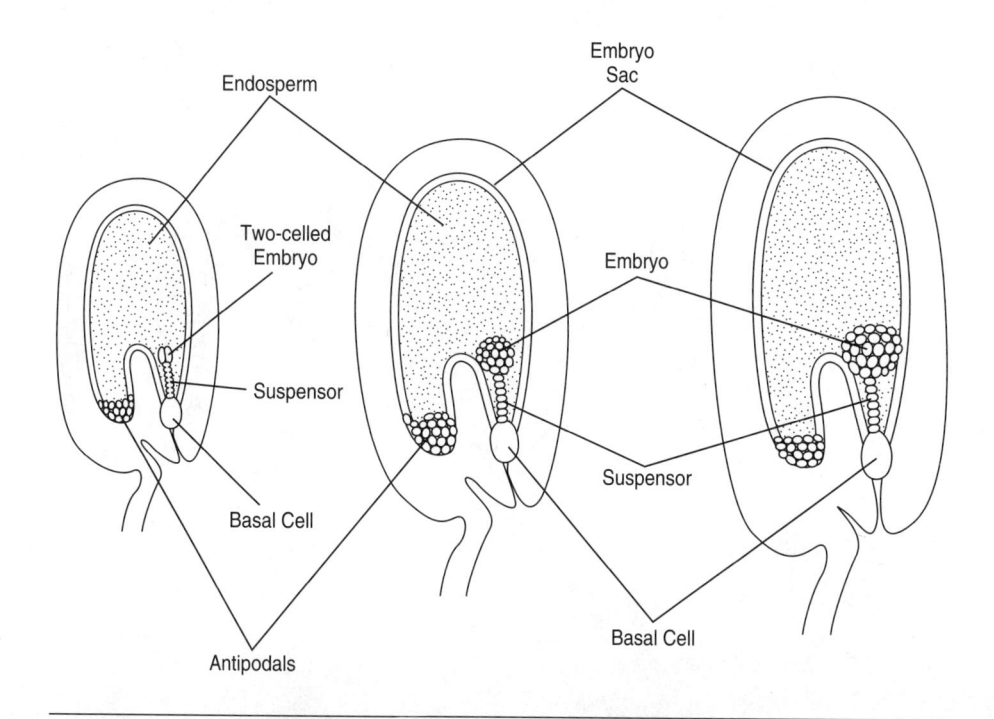

**Figure 36-2** Early embryonic development, showing extraembryonic cells.

There are two kinds of angiosperm seeds: monocotyledonous and dicotyledonous. The cotyledons are the first-formed leaves that are a part of the seed. They often contain a reserve of nutrients. An important character of angiosperm seeds is the endosperm. Recall that the endosperm, being a consequence of triple fusion, is 3N. Although all seeds contain endosperm before they are mature, the endosperm in many kinds of seeds is consumed by the growth of the embryo plantlet during the maturation process. Whereas monocots always have a residue of endosperm in the mature seed, only some species of dicots exhibit endosperm residue. Squash and castor bean, for example, retain endosperm, whereas the endosperm in lupine (a commonly studied plant) is entirely consumed. There is no significant value to the 3N condition of endosperm. It is simply regarded as an evolutionary event with no special value. The endosperm of gymnosperms is always haploid.

When a seed germinates, the portion extending above the cotyledons is the epicotyl; the portion extending below the cotyledons is the hypocotyl. One may therefore suppose that the epicotyl produces the stem, and the hypocotyl produces the root. This is partially true because the hypocotyl may produce a portion of the stem in addition to producing the root. The part of the hypocotyl that produces the root is called the radicle. The growing point of the embryo plant is called the plumule.

### ❋ Notes ❋

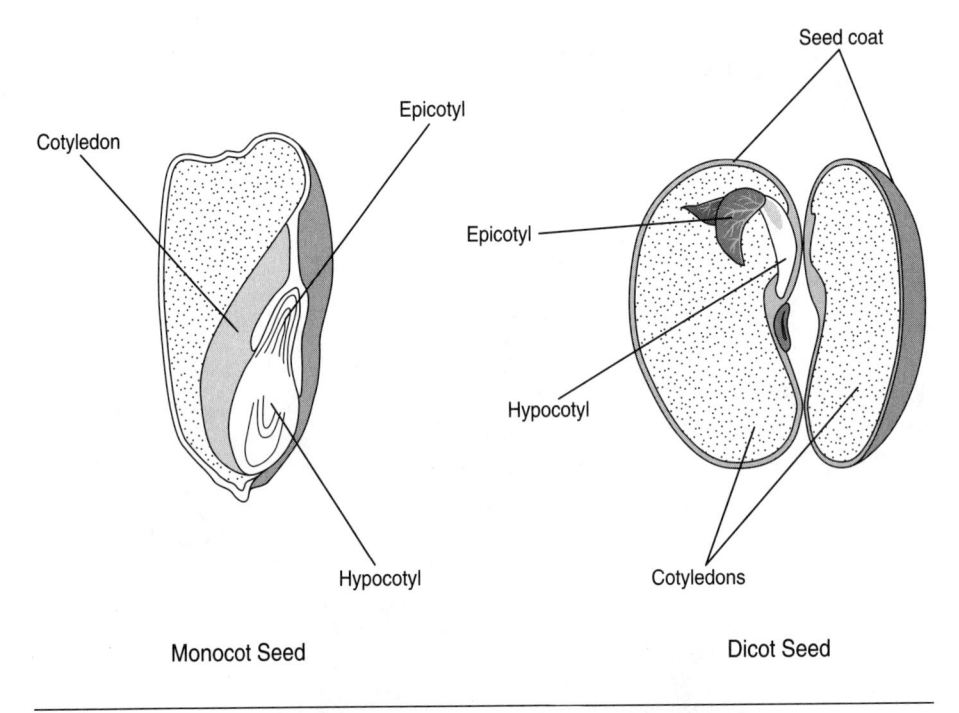

**Figure 36-3** (a) A monocot seed (e.g., corn). (b) A dicot seed, the cotyledons separated (e.g., beans).

As a seed grows toward maturity, the seed coats continue to grow, covering the outside but never completely enclosing the internal portion. A small hole called the micropyle remains. The micropyle allows the entrance of the pollen tube. Although embryo development continues to progress, it slows and in maturity is essentially suspended. In maturity, the water content is much reduced, and the seed may remain in this dry state and dormant for months or even years.

The student may wish to review the discussion of double fertilization in chapter 28 on angiosperms.

## Functions of Seeds

The seed is important in carrying plant life through periods of unfavorable conditions as well as in distributing plants from place to place. Certain seeds are adapted to distribution by wind; some of these have tufts of hairs (for example, poplar and milkweed), and some are winged (for example, maple and catalpa). Certain other seeds are adapted to distribution by water (for example, coconut), while still other seeds are carried by animals, in either the fur or the digestive tract. Seeds distributed by water require coats that are impervious to water in order to avoid premature germination. Such seeds eventually germinate when the seed coats are eroded away, as by wave action at the beach or by bacterial action. In some cases, such seeds germinate only after being in a fire.

The fruits of certain plants exhibit unique adaptations that aid in distribution. For example, some seeds are forcefully discharged from the fruit as a result of a buildup of internal pressure (touch-me-not, for instance). The seeds may be thrown a distance of several feet (as is the case with Impatiens). In such cases, a mere touch will cause a sudden outburst. The seeds of pansies and violets are shot from their pods one at a time. Mistletoe produces seeds that are sticky and adhere to the beaks of birds. The birds later scrape off the seeds on the bark of a tree elsewhere.

## Variations in Seed Composition

There is great variation in seed composition. In some seeds, the cotyledons have a high oil content (for example, in soy beans and peanuts). In others, the cotyledons store starch. Many bean-family plants have seeds with a high protein content. Some seeds contain sugar (onion seeds, for instance). The seeds of sweet corn contain sugar that changes to starch in maturity. Seeds are important sources of food for people and other animals. Rice, wheat, oats, barley, corn, beans, peas, and nuts are all important food sources.

## Seed Longevity

There is also great variation in seed longevity among plants. The seeds of elms, magnolia, and soft maple remain alive for only several days; the seeds of willow and poplar survive for several weeks; and the seeds of a number of species, especially weeds, remain viable for many years. In 1879, Professor W.J. Beal of Michigan State College (now Michigan State University) initiated a long-range experiment whereby seeds of twenty common plant species were mixed with sand and put in storage. Every five years, seeds of each species were taken up and tested for viability. Because the viability proved more than expected, the time period was extended to ten years so as not to run out of stored seeds. The results of this experiment are shown in table 36-1.

Seeds of certain bean-family plants have been removed from herbarium specimens known to be more than one hundred years old and have been successfully germinated. An incredible example of seed longevity was exhibited by *Nelumbo nucifera* seeds recovered from an ancient bog in Manchuria, which were germinated and grown to full flowering. Radio carbon dating of such seeds confirmed them to be no less than one thousand years old. The life expectancy of seeds is much affected by storage conditions. Keeping seeds dry during storage is vital to longevity.

❋ Notes ❋

### Notes

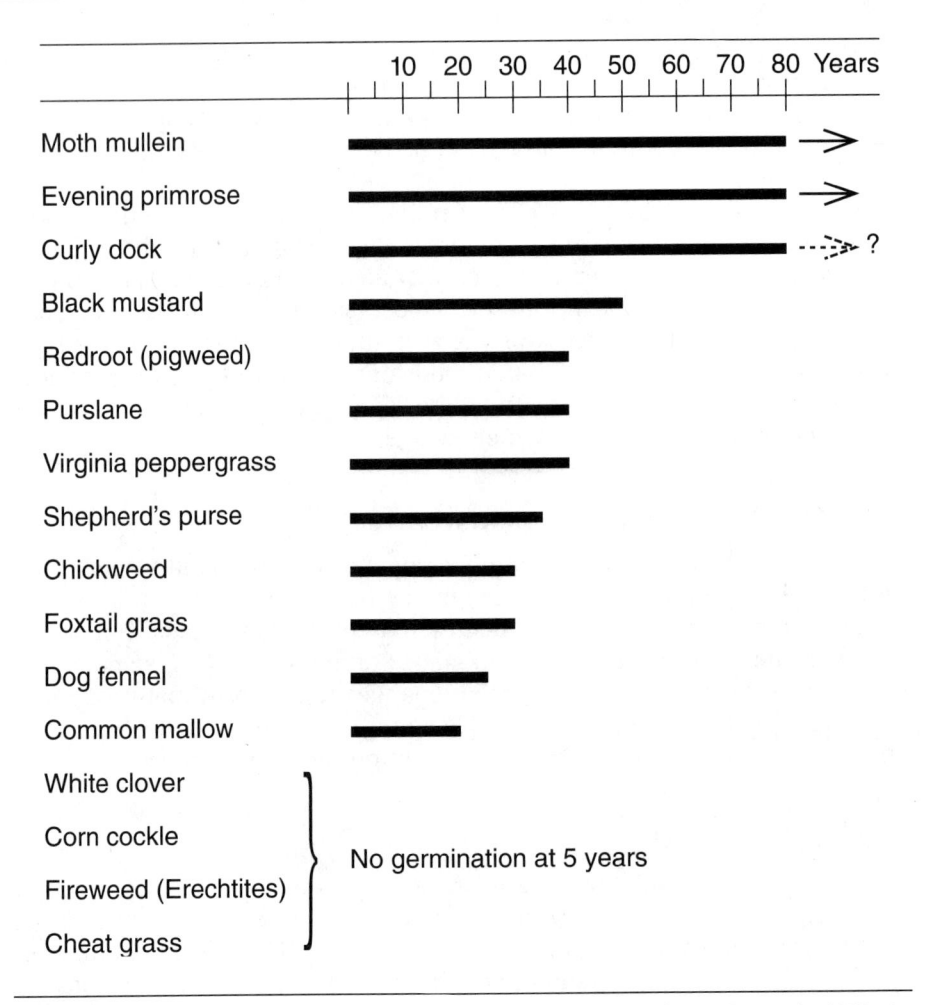

| Species | Years |
|---|---|
| Moth mullein | ~80+ → |
| Evening primrose | ~80+ → |
| Curly dock | ~80 --?→ |
| Black mustard | ~50 |
| Redroot (pigweed) | ~40 |
| Purslane | ~40 |
| Virginia peppergrass | ~40 |
| Shepherd's purse | ~35 |
| Chickweed | ~30 |
| Foxtail grass | ~30 |
| Dog fennel | ~25 |
| Common mallow | ~15 |
| White clover | No germination at 5 years |
| Corn cockle | No germination at 5 years |
| Fireweed (Erechtites) | No germination at 5 years |
| Cheat grass | No germination at 5 years |

**Table 36-1**  The Longevity of Seeds Based on W.J. Beal's Experiment Begun in 1879 at Michigan State College.

## Seed Germination

Some kinds of seeds require special treatment to initiate germination. Those having impervious seed coats may be soaked in concentrated sulfuric acid for several minutes and then thoroughly washed with water in an effort to make the coats permeable. Another approach is to dip such seeds in scalding water. Timing of such treatments is critical.

Light is important in germination. Some kinds of seeds will not germinate when buried in soil because they require exposure to light after having absorbed moisture. They therefore must be brought to the surface.

Germination of light-sensitive seeds is promoted by red light but inhibited by far-red light.

While some seeds are able to germinate as soon as they are mature, many seeds first require a conditioning treatment, most commonly, a period of low temperature. This makes sense because otherwise, seeds would germinate in the fall and the resulting delicate seedlings be killed by winter conditions. Supplying low temperature, a process known as **cold stratification**, is done in conjunction with moisture. In nature, this means overwintering; otherwise seeds can be put in sphagnum moss with moisture and stored in the refrigerator. Honeysuckle requires 30 to 60 days of such treatment, pears 60 to 90 days, hemlock 60 to 120 days. During this period of low temperature, a number of biochemical and physical changes occur before the seeds gain the capacity to germinate. Another name for cold stratification is **vernalization**. This name is also used in reference to high-temperature treatment, though less commonly so. Vernalization is practiced on cereal grains more often than on other types of seeds. It is especially significant for winter wheat, given that the treatment speeds the life cycle and thus allows the seeds to mature in northerly climates. The turkey red variety of winter wheat can be first soaked in water to raise the moisture content to 60 percent and then treated at 33 to 37°F for nine weeks or more. Grain can thus be grown to maturity in approximately thirty days. Such grains can be grown in Siberia. In fact, another name for the process honors a Russian scientist: **Jarovisation**.

Germinating seeds also require oxygen. The soil in which seeds sprout should be well aerated and porous. Careful attention must be given to watering.

Germinating seeds are often damaged by fungi residing in the soil. It is sometimes necessary, therefore, to either sterilize soil in an autoclave or treat soil with a disinfectant such as bichloride of mercury or formaldehyde. In addition, seeds can be coated with zinc oxide or mercury or copper compounds to minimize loss from fungi and bacteria.

## Reproduction

Both sexual and asexual methods of reproduction occur among seeds. **Amphimixis** is the name given to sexual processes of reproduction (normal seed development), whereas apomixis is the term given to asexual reproduction. Dandelion produces seed without fertilization. Although dandelion flowers produce pollen, the pollen is not effective in fertilization. The seeds develop without this process; they develop apomictically. **Sporophytic budding** is another form of apomixis. The sporophytic cells, which surround the embryo, sometimes grow to new sporophytes within the seed. This tissue is called the nucellus, and the production of an embryo from such cells is called **nucellar embryony** (see figure 36-4). This process may yield two embryos in one seed: one apomictic and the other by fertilization of the egg.

❋ **Notes** ❋

## ❊ Notes ❊

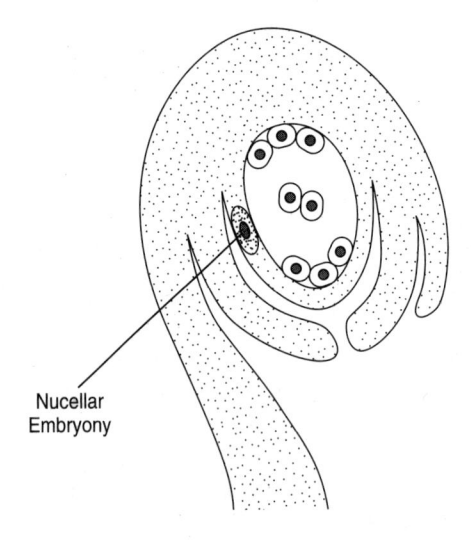

**Figure 36-4** Nucellar embryony, showing a small section of nucellar tissue (sporophytic tissue) that may produce an embryo plantlet.

There is no alternation of generations because the sporophytic cells give rise to the new sporophytes. In **apogamy**, another form of apomixis, cells of the gametophyte other than the egg cell produce a new sporophyte. A new plantlet arises from either synergids or antipodal cells. Polyploidy (having more than two sets of chromosomes) is often observed in apomictic species. The condition appears to result from abnormal chromosome behavior during meiosis.

Parthenogenesis is the name given to the formation of a new sporophyte from an egg cell without the egg cell being fertilized by a sperm nucleus. One might expect the sporophyte resulting from parthenogenesis to be haploid, and this is sometimes the case. More often, however, a doubling of chromosome number occurs in the first cell cleavages, and a diploid sporophyte results. It should be noted that in such a case, all of the chromosomes are of maternal origin.

Parthenogenetic development of an egg cell can be caused by a proximity stimulus. For example, a pollen tube may grow down through the style and approach the egg but not penetrate it. The mere closeness of the pollen tube in such a case may cause an unfertilized egg cell to develop into a new sporophyte. In other cases, foreign pollen incapable of fertilization may grow a pollen tube through the style. Again, a new plantlet may develop because of the proximity of the pollen tube. Insects, chemical stimuli, or wounding can yield similar results, as can low- or high-temperature treatments, pinching, pricking, x-rays, and radium. Parthenogenesis can be stimulated in the brown alga *Fucus* by treatment with fatty acids. All this may lead one to ask, "Is sperm necessary?"

## Questions for Review

1. What are the requirements for seed germination?
2. Define the terms *plumule, hypocotyl, radicle*.
3. What is cold stratification, and what does it accomplish?
4. Define the following terms: *fruit, drupe, achene, endosperm,* and *micropyle*.
5. Name two plants whose fruits are dispersed by wind.
6. How can seeds be protected from the ill effects of fungi and bacteria?
7. Define the terms *amphimixis, apomixis, polyploidy,* and *parthenogenesis*.

## Suggestions for Further Reading

Eames, A.J., and L.H. McDaniels. 1947. *An Introduction to Plant Anatomy*. New York: McGraw-Hill Book Company, Inc.

Esau, K. 1953. *Plant Anatomy*. New York: John Wiley and Sons, Inc.

Kozlowski, T.T. 1972. *Seed Biology*. Vols. I, II, and III. New York: Academic Press.

Maheshwari, P. 1950. *An Introduction to the Embryology of Angiosperms*. New York: McGraw-Hill Book Company, Inc.

Ridley, H.N. 1930. *The Dispersal of Plants Throughout the World*. Ashford, Kent, England: L. Reeve and Co.

Stebbins, G.L., Jr. 1941. Apomixis in angiosperms. *Botanical Review* 7:507–542.

# 37

# Other Methods of Propagation

Plants can be propagated by means other than seeds. In fact, some cultivated varieties do not even produce seeds. Bananas, pineapple, and sugar cane are always propagated by vegetative means. While potatoes produce seeds, seeds are not used in furnishing the marketplace with potatoes. If seeds were used, the product would in fact be undesirable.

Some species of horticultural plants have separate sexes when only one sex is desirable. The female *Ginkgo* tree, for example, produces "fruits"[1] that are a nuisance because they are malodorous. Female poplar trees give forth clouds of seeds that look like snow and collect in drifts on the ground. In such cases, asexual means of reproduction are used to obtain only the desired sex.

## Division

Division is one method of vegetative propagation. An aggregate of roots is cut into pieces and the pieces planted as separate parts. Rhubarb, sweet potatoes, and dahlias are propagated this way. In the case of dahlias, care must be given to ensure that a bud is present on each root piece. Many species of plants have so many buds on the roots that special attention is not required. White potatoes are derived not of roots but of underground stems; they are always propagated by dividing the tubers and leaving a bud on each tuber piece. Scales that grow at the base of bulbs such as lily bulbs can also be divided. Because many root pieces can produce adventitious stems, root cuttings are commonly used by nurseries for propagation purposes. Cherries, plums, and figs are all produced this way.

❋ **Notes** ❋

---

[1]The word is placed in quotation marks because *Ginkgo*, being a gymnosperm and thus naked seeded, does not produce fruits.

### Notes

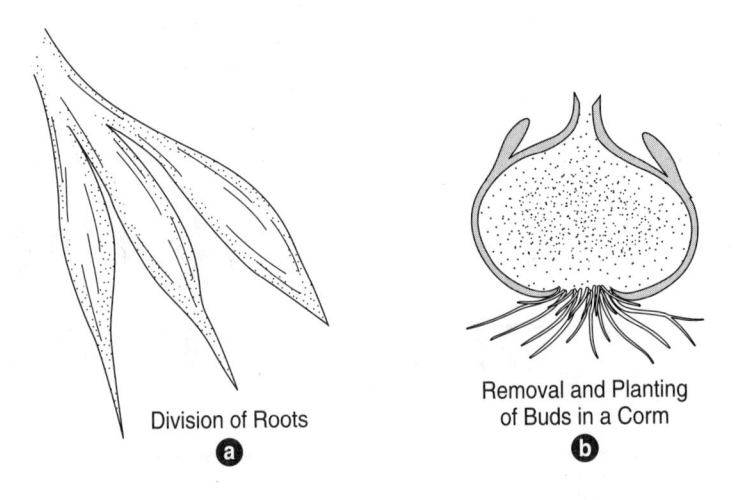

**Figure 37-1** (a) Propagation may be done by dividing roots. (b) A corm. Propagation is achieved by the removal and planting of buds.

## Layering

**Layering** is another vegetative method of propagation. This technique is successful because a majority of plants can be induced to produce roots on their stems. Such roots are called adventitious roots, and many plants do this naturally. Blackberry canes and forsythia branches may hang down so that their tips come in contact with the ground and take root there. Success can be best ensured by placing a stone on the portion of the stem that is in contact with the ground, thereby preventing the tips from swinging back and forth with the wind. Several types of layering are represented in figure 37-2.

## Cuttings

Because stems are able to produce adventitious roots, **cuttings** of stem pieces are commonly planted. This is in fact the preferred method of vegetative propagation. Both herbaceous and woody cuttings are successfully used. The new roots grow on the cut surface of the stem piece. A root-growth promoting substance such as indole acetic acid, naphthalene acetic acid, or a phenoxyacetic acid compound is commonly used. The cut surface of the stem piece is dipped in a powder containing such a substance just prior to planting. The act of cutting may itself produce a wound hormone, which plays a role in producing adventitious roots. Sand and vermiculite are the preferred media for the rooting of cuttings. Very small cuttings are sometimes placed on agar in a petri dish in order to circumvent the problems caused by fungi and bacteria. Leaves or pieces of leaves may be successfully used as cuttings in propagating a certain few species of plants. Begonia, Sansevaria, Kalanchoe, and African violets are all propagated in this way. Rhizomes are

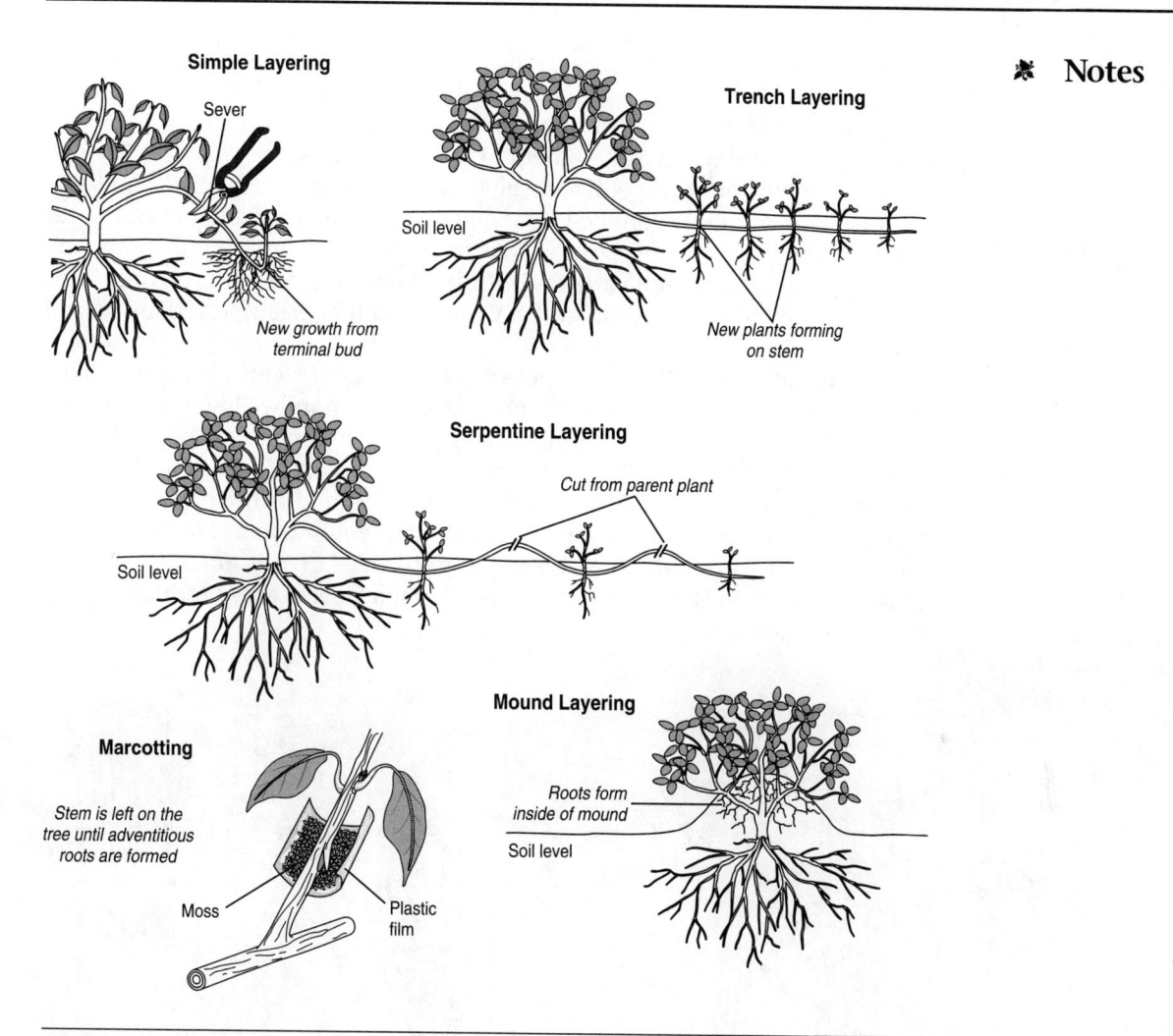

**Figure 37-2** Some methods of vegetative propagation.

planted to increase irises, and strawberries are increased by the use of stolons (stems that grow on the surface).

# Grafting

Among horticultural plants, there are numerous times when a desired shoot system does not develop well on its own root. In such situations, great improvement can be achieved by grafting the desired shoot to a root of higher quality, a root system that is more disease resistant. European varieties of grapes are regularly grafted to American-type root stock.

### ❦ Notes ❦

A graft is made by fusing a **scion** (twig) of one variety to the base of a stem bearing the desired root system of another variety. When the two pieces are brought together, careful attention must be given to aligning the cambiums so that the xylem and phloem of the scion and stock connect. The two parts are then bound together to allow fusion. When the healed cambium produces a continuous connection of new xylem, water conduction can occur between the parts. It is also possible to graft on higher, lateral branches, thereby producing several varieties of fruit on one tree. Sometimes, only buds are grafted to the stock (a **bud graft**). Various techniques employed are shown in figure 37-3.

Grafts between different species of the same genus can usually be done, as can grafts between different genera of the same family. Tissue compatibility is necessary. Lilacs can be grafted to privet; almonds can be grafted to peach stock; peaches can be grafted to cherry stock; and pecans and English walnuts can be grafted to black walnut. Potatoes and tomatoes are related and can thus be grafted. They are in the same family, Solanaceae. If a tomato scion is grafted to a tobacco rootstock, the leaves of the tomato will have nicotine, demonstrating that the nicotine in tobacco leaves is produced in the roots of the tobacco plant.

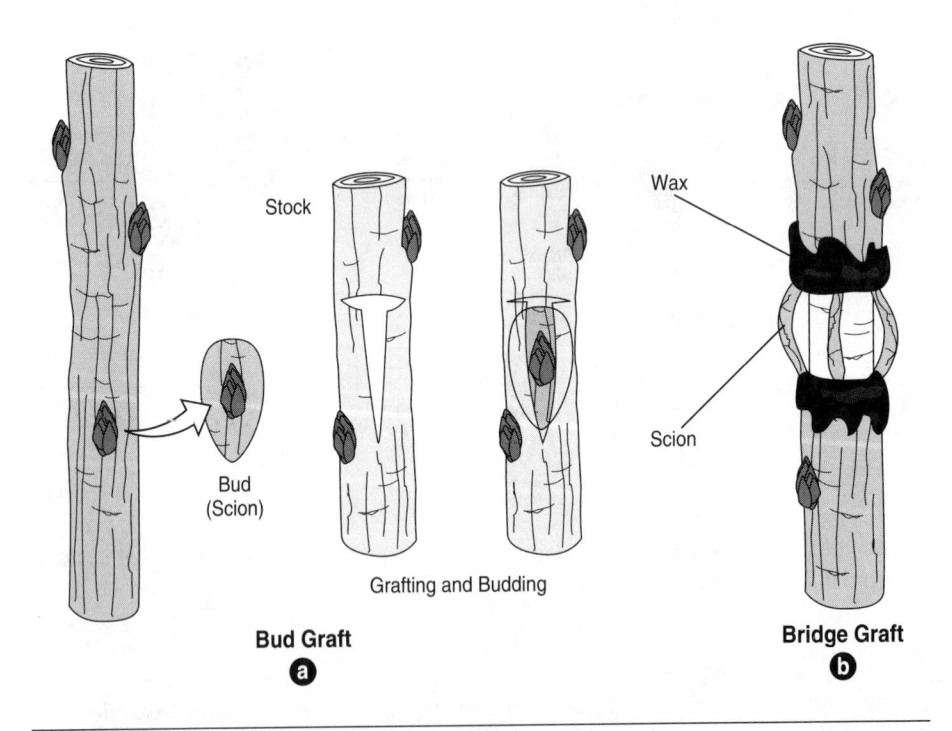

**Figure 37-3** (a) Bud grafting: a bud excised from a stem is placed under bark having been opened by either a T- or a door-shaped cut. (b) A bridge graft can be used to repair an injured stem.

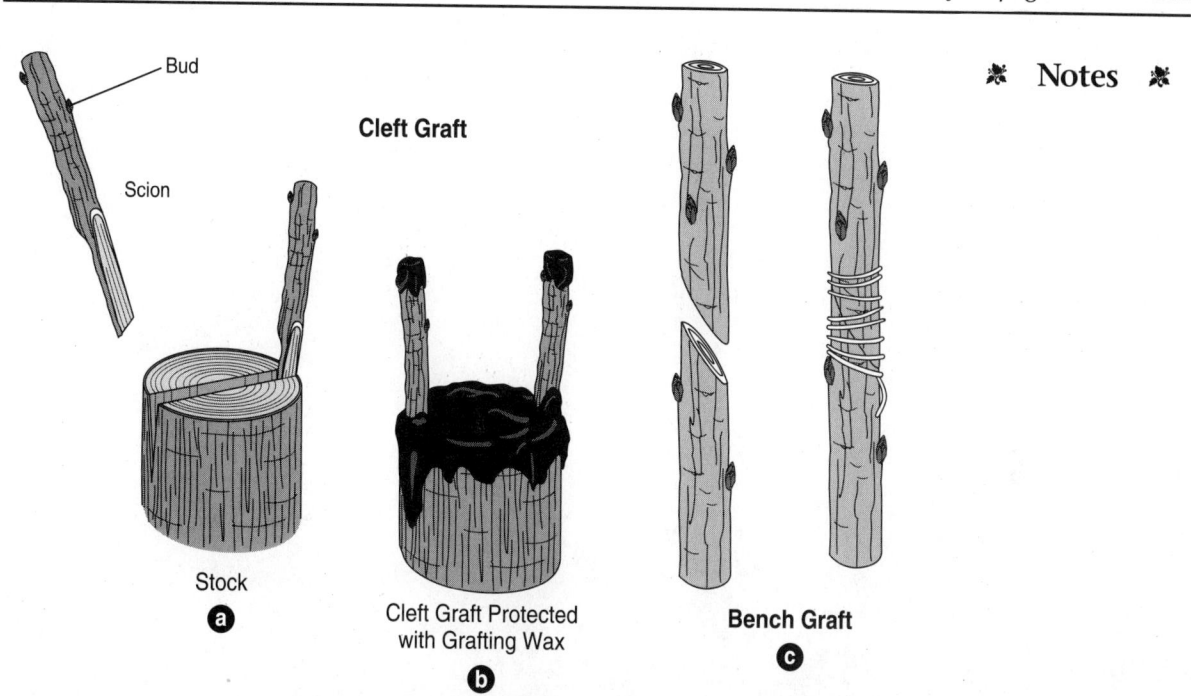

**Figure 37-4** Grafting of a scion on a stock. (a) A cleft graft (used in top-working a tree). (b) The cleft graft protected with grafting wax. (c) A bench graft secured by a waxed string.

## Questions for Review

1. What can be done following transplanting to reduce the danger of water loss in a plant?
2. Describe several methods whereby plants can be reproduced without seeds.
3. What is vegetative reproduction? Name a few plants that are regularly propagated in this manner.
4. Define the terms *adventitious, layering,* and *marcotting.*

## Suggestions for Further Reading

Haldane, J.B.S. 1955. Some alternatives to sex. *New Biology,* vol. 19.

Hartmann, H.T., and D.E. Kester. 1975. *Plant Propagation: Principles and Practices.* Englewood Cliffs, NJ: Prentice-Hall.

# 38

# Evolution

※ Notes ※

For most of historical time, the dominant view was that all creatures, plant and animal, were the product of separate, individual creations. Aristotle (384–322 B.C.) believed that both lowly and highly organized organisms arose spontaneously from mud, and later others wrote recipes for the generation of flies, bees, and mice from nonliving precursors. When Aristotle cataloged and gave names to species, he believed he was cataloging creation. Hundreds of years later, Carolus Linnaeus (1707–1778), originator of our modern system of classification, also held the view that every species was the result of individual acts of creation. Only later in his career did Linnaeus begin to consider the possibility of evolutionary change.

## Early Changes in Thought

It was not until the late eighteenth and early nineteenth centuries that views began to change. Buffon (Comte de Georges Louis LeClerc Buffon [1707–1788]) suggested that the environment had a molding influence on organisms, and Lamarck (Chevalier de Jean Baptiste Pierre Antoine De Monet Lamarck [1744–1829]), following the same trend of thought, proposed the theory of the inheritance of acquired characteristics. According to Lamarck's theory, organisms respond to their needs by achieving changes, which they then pass on to their progeny.

Progress toward a concept of evolution (that is, of changes occurring and accumulating over the long span of time) did not come without setbacks. Baron Georges Leopold Cretien Frederic Dagobert Cuvier (1769–1832) stood fast against such speculations. Cuvier was a man of superior intellect. His great interest in natural history began in childhood. In 1795 he became assistant professor of comparative anatomy at the Museum National d'Histoire Naturelle. In 1802 he became professor at the Jardin des Plantes. Later, he was given an appointment at the Institut Nationale. Although Cuvier discovered

## Notes

numerous fossils, he interpreted these as evidence of cataclysms followed by new creations rather than of evolution.

Etienne Geoffroy Saint-Hilaire (1772–1844) believed that each species was created for its own special purpose; that nature had only one plan of construction; and that the plan of creation was permanent. He interpreted new species and higher categories as occasional appearances of monsters.

### Charles Darwin

It was left to Charles Darwin (1809–1882) to establish the first sound theory of organic evolution. With the publication of *On the Origin of Species* in 1859, a revolution in thought spread through the scientific community, first in England and then in France and Germany. In little time, Darwin's theory created shock waves that reverberated around the world. Darwin purported that species came into being by gradual changes and descent from other species. The doctrine of evolution declares that organisms are related to each other through common descent and that the differences that exist between species have arisen through hereditary alteration of pre-existing forms.

Darwin's work created such upheaval largely because it applies not only to lower life-forms, but also to human beings. His ideas thus conflicted with biblical teachings, and created trouble for religious fundamentalists. It is not the goal of this text to negotiate this conflict; rather, this text does not intend to present the argument that evolution takes place. The theory has virtually universal acceptance in the scientific community, and attention here should be focused on how evolution works.

### The Tenets of Darwinian Theory

Darwinian theory is based on several tenets:

1. Species tend to produce more individuals than can be accommodated.

2. Variations occur among offspring. Some variations yield a greater capacity to adapt to the environment and, thus, to survive. Such variations are significant when they are inheritable.

3. Certain forms are thus better able to survive than are others. This is referred to as natural selection, or survival of the fittest. Variations that are inheritable make evolution inevitable.

### Other Theories of Evolution

Various theories of evolution were formulated during the latter part of the nineteenth century. Karl Wilhelm Nageli (1817–1891), who devoted much of his life to the microscopical study of plants, proposed the presence of an inner-directed force that guides the course of evolution. He coined the word

**Figure 38-1** Charles Darwin (1809–1882) laid the foundation for our concepts of evolution. (Illustration by Donna Mariano)

*idioplasm* for the material of heredity. Theodor Eimer was an advocate of so-called straight line evolution and a phenomenon called **orthogenesis**, wherein evolution proceeds in an undeviating manner toward a particular outcome that has nothing to do with environment. Hugo DeVries (1848–1935) contributed significantly to our understanding of evolution by his theory of mutations (those sudden, unexpected changes in organisms that cannot be accounted for by the laws of heredity and that can be used to explain the abrupt appearances of new species). DeVries' greatest contribution to science was likely his use of scientific method, rather than the older method of observation and inference, in the study of evolution. He also was one of those who rediscovered Mendel's work in 1900. These theories emerged before any

### Notes

detailed knowledge of the laws of heredity. These laws once uncovered by Gregor Mendel had to wait yet another thirty-four years before finding a widespread audience and acceptance in the scientific community.

The work of Gregor Mendel made possible the application of proven science to what had previously been educated speculation. The labors of William Bateson (1861–1926), who extended our understanding of the principles of segregation and gene interaction, and Thomas Hunt Morgan (1866–1945), who demonstrated that genes are located on the chromosomes, while technically falling under the realm of genetics, significantly reinforced the Darwinian concepts of evolution.

But *how* does evolution occur? Charles Darwin's theory of the tendency toward overpopulation; competition; gradual, cumulative changes; and the survival of the fittest is a highly favored one. Yet much can be added to facilitate our understanding of this process. Evolution is not caused by urge, seeking, or need. How organisms are able to accommodate to environment and how they relate to each other is important. Many questions are yet unanswered. Does evolution progress at a steady rate, or do evolutionary changes occur in bursts? Is evolution triggered by challenge? An understanding of genetics is helpful in understanding the presumed mechanism of evolution. (For those interested in review, Gregor Mendel's work is discussed in chapter 5, whereas DNA and its role in genetics are discussed in chapter 6.)

## The First Organisms

The first organisms to appear on earth were the blue-green algae, the cyanophytes. If life had only one beginning, then every other living thing (and all those that are extinct) derived from this single beginning. Thus, life would be monophyletic in origin. Some are highly skeptical of such a theory, however, and prefer to postulate that life arose a number of times, perhaps many times. In this case, life would be polyphyletic in origin. For the time being, the answer remains in the realm of philosophical speculation.

Early plant forms are known only through the fossil record, and, thus, much is uncertain. Not only is the fossil record incomplete, but the kinds of plants that were subject to becoming fossils grew in low elevations; in places of abundant moisture; and where they could be carried downstream and deposited at the bottoms of lakes. Plants that grew in higher, dryer places rarely became fossilized. It is with such plants, however, that evolution took place most rapidly. And it is at higher elevations that angiosperms had their origins. Unfortunately we have thus far been denied the chance to study their beginnings in detail.

Another handicap of the fossil record is that plants composed entirely of soft tissues decay; their chances of fossilization are thereby also largely denied. Fossil algae and fossil thallophytes are, thus, less widely known, although there does exist some evidence of them.

## Prokaryotic Life

Examine the upper line of the geological chart in table 38-1. The span from the left edge to the right edge represents the vast stretch of time from the formation of the Earth's crust to present day—a span of approximately 4.5 billion years. Just more than seven-eighths of that time comprises the Precambrian era—from 4.5 billion years ago to somewhat fewer than 600 million years ago. Within this era and approximately 3.2 billion years ago, prokaryotic organisms (bacteria and cyanobacteria) came into being. For approximately five-sixths of the time since then, bacteria and the blue-green algae were the only organisms on Earth. The oxygen in the atmosphere was progressively increased by the photosynthetic activity of Cyanophyta (the blue-greens).

## Eukaryotic Life

The lower line of the geological chart in table 38-1 covers the time represented by one-eighth of the upper line—the time beginning with the Paleozoic era (600 million years ago). Near the beginning (that is, the left end) of this "ribbon," an event of singular significance took place: the origin of eukaryotes. That event was followed by the origin of sex. From that time forward, a great number and a great diversity of forms developed.

## The Emergence of Seed Plants

Seed plants appeared toward the end of the Devonian period. It is striking to learn that gymnosperms appeared long before angiosperms. Evidence of gymnosperms is seen in coal deposits and other fossils of the Carboniferous period, whereas angiosperms made their appearance in the **Cretaceous** period. Two hundred million years of time lie between the early gymnosperms and the appearance of angiosperms. Charles Darwin referred to the appearance of these plants as an "abominable mystery." There is wide agreement that angiosperms descended from a primitive gymnosperm, perhaps a shrub. Certain gymnosperms of Paleozoic time display several angiosperm traits. Part of the mystery relates to the fact that angiosperms had their origin in the dryer environment of the upland, where fossil formation is not favored and, yet, where more rapid evolution occurs. Angiosperms became dominant and the gymnosperms regressed around the time that the Rocky Mountains were forming. Today there are approximately 250,000 species of angiosperms and 700 species of gymnosperms. One may infer that gymnosperms are on their way toward extinction. If this is so, it will not be for a long time!

The magnolia is regarded as a primitive angiosperm. The single characteristic that distinguishes angiosperms from other groups is, of course, the flower. The flowers of the magnolia show primitive characters: numerous carpels, numerous stamens, numerous and separate perianth parts, spiral arrangement, and a **superior ovary**. The trend in evolution has been toward

❋ Notes ❋

## ✻ Notes ✻

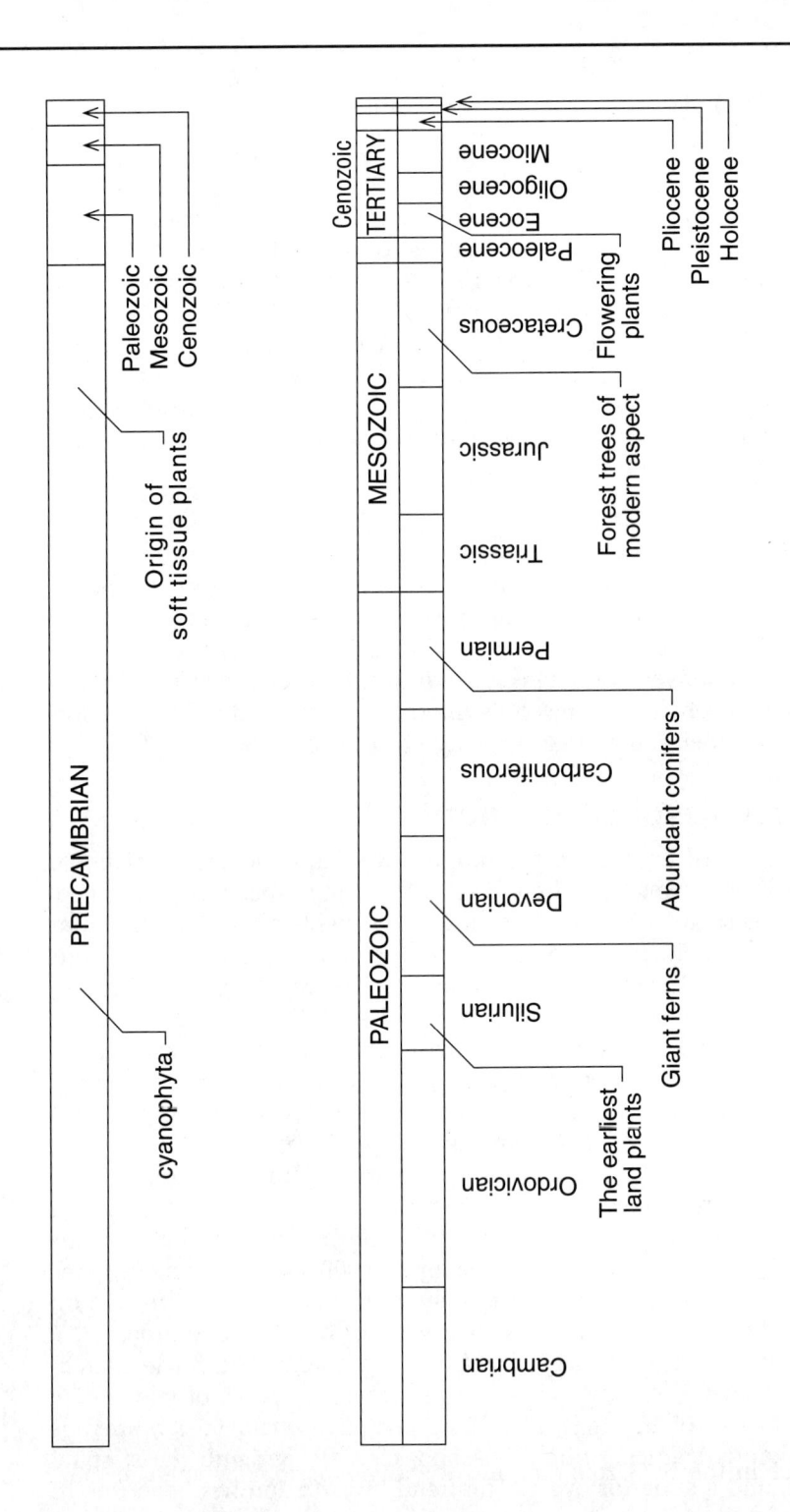

**Table 38-1** The Geologic Time Chart

fewer parts of definite number, bilateral symmetry, and an **inferior ovary**. Wind pollination is primitive, and insect pollination is more recent.

The earliest land-dwelling plants appeared in the Silurian period, 420 million years ago. Giant ferns are found in strata of 350 million years ago (from the Devonian period). Coniferous trees began to increase during the Permian period, reached their zenith in the Cretaceous period (100 million years ago), and have since declined. Flowering plants appear in the fossil record comparatively recently: in the Eocene period of the Cenozoic era, or somewhat earlier. Subtropical forests grew in Alaska during the Eocene period.

## Grasses

Grasses appeared in the middle of the Mesozoic era. Woody plants appeared earlier than did herbaceous plants. Woody is primitive, and herbaceous derived from woody ancestors. While most plants have their meristematic cells at the tips, grasses have their dividing cells near the nodes. This is a splendid development because it makes possible continued growth after grazing or cutting. We owe our very lives to this singular characteristic of grasses. There are approximately 5,000 species of grass. Blue grass, bent grass, timothy, fescue, crab grass, and quack grass are all grasses as are the grain cereals, wheat, rice, oats, barley, corn, sugar cane, sorghum, millet, and bamboo. There are vastly more species of extinct forms of grass than there are of living forms. It is postulated that nine-tenths of all species that have ever lived are extinct. Extinction, then, is more common than is survival—for both plant and animal forms.

## Human Life

At the far right of the geological chart in table 38-1 are two narrow zones, each less than one millimeter wide; these represent the Pleistocene and Holocene periods. Human origins trace to the Pleistocene; Stone-Age cultures and learning to control fire trace to the Holocene.

## Life over Time

Table 38-2 shows life-forms in relation to geological time in a manner somewhat different from that employed in table 38-1. The bacteria, algae, and fungi are shown to be stabilized, remaining essentially constant in numbers of forms across time. The club mosses were much more abundant in the Devonian and Carboniferous periods, and became much diminished as the Mesozoic era approached. The same is shown for the horsetails. The ginkgoes were more abundant during the first half of the Mesozoic era, and today are represented by a single form, *Ginkgo biloba*. Coniferous trees had their time of dominance in the Mesozoic era, and the angiosperms appeared in more recent times, at the beginning of the Cenozoic era.

## ❋ Notes ❋

| ERA | PERIOD | MILLIONS OF YEARS AGO | MAJOR DEVELOPMENTS |
|---|---|---|---|
| CENOZOIC | Quaternary | 0 | |
| | Tertiary | 63 | short life cycle (annuals) |
| MESOZOIC | Cretaceous | 135 | special pollinating mechanisms less woodiness |
| | Jurassic | 181 | double fertilization seeds in ovaries flowering habit |
| | Triassic | 230 | |
| PALEOZOIC | Permian | 280 | seed-bearing cones; perennial habit; woodiness: treelike evergreen habit |
| | Pennsylvanian | 310 | seed-bearing ferns sporophyte dominant |
| | Mississippian | 345 | |
| | Devonian | 405 | development of stems roots and leaves |
| | Silurian | 425 | |
| | Ordovician | 500 | growth on land; eukaryotes fungi and algae; evolve sexual reproduction |
| | Cambrian | | blue-green algae chemosynthetic bacteria |
| | Precambrian | 600 | prokaryotes |

**Table 38-2** Geological Time Chart Showing the Rise and Decline of Various Plant Forms

## Questions for Review

1. Approximately when was it suggested that organisms might vary and undergo changes, which led, in time, to the idea of evolution?

2. Among the several tenets to the concept of evolution is survival of the fittest; what are the others?

3. Hugo DeVries contributed to our understanding of the concept of evolution through his theory of _____, which may explain the abrupt appearances of new species.

4. Who demonstrated that genes are located on the chromosomes?

5. If life had only one beginning, every species is derived from _____.

6. Life seems to have had its beginning in the _____ period; land-dwelling plants are believed to have appeared approximately _____ years ago.

7. Which of the two seed-plant groups came into being first: gymnosperms or angiosperms?

8. Are there currently more or fewer species of angiosperms than of gymnosperms?

9. Which is the more primitive condition: woody or herbaceous?

## Suggestions for Further Reading

Darwin, C. 1962. *The Voyage of the Beagle.* Garden City, NY: Doubleday and Company, Inc.

Grant, Verne. 1963. *The Origin of Adaptations.* New York: Columbia University Press.

Moody, P.A. 1962. *Introduction to Evolution.* New York: Harper and Row.

Stebbins, G.L. 1950. *Variation and Evolution in Plants.* New York: Columbia University Press.

# 39

# Ecology

In recent years, **ecology** has become a household word as widespread concern has developed regarding destruction of the environment, despoliation of tropical forests, pollution, and extinction. In the midst of such alarm, the meaning of the term *ecology* has become blurred and confused with the term *environment*. In fact, "preserve the ecology" is an often heard phrase. We should begin, then, by understanding the term *ecology*. Ecology is the science that focuses on the relationship between organisms and the environment. Plant ecology encompasses plant and animal interactions, plant and plant interrelationships, and the responses of plants to other aspects of the environment. Plants, animals, and human beings all affect one another.

## Plant Ecology

There are various facets of plant ecology. Some divide the field into two types: **autecology** (focusing on individual plants and how the environment affects them) and **synecology** (focusing on plant communities, how the environment influences them, and how plant communities influence the environment). Some others divide the field into several parts: **floristics** (focusing on the distribution of plants and considering plants from a numerical point of view), plant geography (focusing on the distribution of plant communities), and plant sociology (the same as synecology). All of these require an awareness of constant change and interactions.

## Adaptation

It is frequently declared that plants are engaged in competition. Plants neither "struggle," "seek," nor "try"; thus, if the word *competition* is to be used, it must be understood to be entirely passive. At the beginning of chapter 33 on roots,

✺ **Notes** ✺

### Notes

it was stated that roots grow where they can grow. The same can be said of plants. Plants grow where they grow because of various adaptations, which came about through natural selection. Plants growing in arid regions, for example, tend to be grayish in color, thick leaved, thorny, and have sunken stomates. These adaptations allow them to survive in locations lacking in water. The giant Sequoia trees of coastal California occupy a relatively limited area where they have survived in their ecological niche for millions of years. In the long range, they may be on their way to extinction. Some would say that they are holding on long after their time. The ubiquitous dandelion, on the other hand, does not have such limitations. It tolerates a wide range of temperature and moisture conditions; grows well in clay and sandy soils, and in light and shade; and often seems to dominate the landscape.

Plants grow in a variety of environments, some of which are harsh: hot springs, salt flats, icebergs. The commonplace environments are marine, fresh water, and terrestrial.

## Environment

When we speak of environment, we are referring to more than just location. Environment also includes factors such as light, temperature, minerals, pH, and for land plants, the availability of water. For instance, lichens can grow in a great many places: on bare rock; in the arctic tundra at high altitudes and high latitudes (where nearly 500 species are found); in bogs and forests; on rotting logs; and in rural areas on the trunks of trees. Yet they seem not to grow on the trunks of trees in the city, suggesting an aversion to the pollution often found in cities.

Certain species of blue-green algae thrive in hot springs, where the temperature may reach 85°C. Algae live in deep water where no other plants grow. They thrive where only short-wavelength, blue light can reach and out of the reach of longer wavelengths of light. Their capacity to grow in such an environment may relate to their phycocyanin pigment. Some species of algae grow in shade, some in open sun. Many species of algae are so cosmopolitan, they seem to know no boundaries.

## Climate

While several species of sphagnum moss grow in bogs, each one occupies its own ecological niche. One species grows at the bottom of wet hollows, another species at the sides of hummocks, and still another at the drier portions on the tops of hummocks. Because they each have their own differing requirements and separate places, no one of the species tends to crowd another. Although they exist in close proximity, they do not grow in the same climate. Again, plants grow where they can grow.

The only climate to which a plant can respond is a climate that touches it. A plant does not respond, for example, to the temperature one inch away or to water and minerals two millimeters away from its root tips. The climate to which a plant responds, then, is the microclimate, the climate that touches the plant. The previously cited examples of sphagnum moss speak to this.

## The Global-Warming Controversy

A discussion of ecology and climate naturally leads one to think about global warming. It is purported by some that the carbon dioxide in the atmosphere is increasing to an alarming level, resulting in a so-called "greenhouse effect," or interference with radiational cooling leading to an increase in average temperature. Proponents of this contention claim that should the carbon dioxide level continue to rise, dire consequences will result: the polar ice caps will melt, the level of the sea will rise, and the borders of continents will be altered.

If you were to go to the library to read about global warming (a worthy endeavor), you would undoubtedly find a number of articles on the subject. For instance, you might find an article regarding the effect of carbon dioxide levels on a certain species of butterfly or on deep ocean water. But your attention would more likely be more drawn to those articles proclaiming doom: "Global warming will turn the nation's midsection into a desert"; "Coastal cities will be submerged under the sea." It is important to note, however, that on your visit to the library you would just as readily find an equal number of articles on global cooling. For example, Dr. Fred Singer, Professor of Environmental Science at the University of Virginia, has written on this subject. Among his observations are the following:

> A potentially dangerous cooling trend has been under way most of the Earth's history. There was five or ten times more carbon dioxide in the atmosphere a hundred million years ago, and average temperatures were 9 to 18 degrees warmer. The climate optimum occurred about 1100 A.D. Greenland was then green. Labrador was warm enough to grow grapes. There has been a 9 degree rise in temperature since 1880, which probably indicates a "return to normal" following the "Little Ice Age" of 1600 to 1800.
>
> It is not clear that man is responsible for any warming trend. Human activity accounts for only 5 percent of all carbon dioxide.
>
> Global measurements indicate any rise in temperature occurs mainly at night, mainly in the winter, mainly in the northern hemisphere. If the pattern continues, it will mean a longer growing season, fewer frosts, and no increase in drought.
>
> Instead of melting, the polar ice caps appear to be increasing.

At one time, certain "climatologists" were predicting a thirty-foot rise in sea level (from melting ice), an estimate that has since been revised to twelve inches. Interestingly, it may well be that temperature changes cause changes

❋ **Notes** ❋

in carbon dioxide levels, rather than the reverse; rising temperatures cause the land and sea to release carbon dioxide. At the least, the subject of global warming is certainly worthy of more thought and inquiry.

## Ecological Interrelationships

Some plants grow in symbiotic associations with other organisms. The lichen is a classic example. The nodules of bacteria that grow on the roots of bean-family plants are another example as is the growth of fungal hyphae on tree roots (mycorrhiza). The reliance of many species of flowering plants on visitations from insects can also be considered symbiotic. The range of the yucca plant is limited by the range of the Pronuba moth, which is involved in the plant's pollination and seed formation. The development and range of flowering plants is intimately related to the insects associated with them.

Some plants are parasitic, such as Dodder (*Cuscuta gronovii*) and Indian Pipe (*Monotropa uniflora*). Indian Pipe obtains its nutrients from the roots of other plants through a mycorrhizal association linking the two. Such plants have entirely lost the capacity to form chlorophyll, and, thus, the capacity to produce their own nourishment. These plants survive through their parasitic dependency on hosts.

Some plants are **epiphytes**, meaning they grow on other plants. Although green in color, such plants partially depend on the plants on which they grow. Mistletoe (*Phoradendron* sp.) is such a plant. Dr. Daniel Janzen of the University of Michigan describes an epiphyte that has developed a reliance on ants. Ants deposit parts of dead insects in cavities of the swollen stems, where the dead insects provide some nutritional benefit for the plant. The ants also harvest the seeds and plant them in the stems of young plants.

**Insectivorous** plants may derive benefits from ingesting insects that they "capture." An insectivorous plant that never captures an insect, however, may yet fare quite well. It may be that insects merely provide some nitrogen for such plants. Some of the so-called insectivorous plants possess mechanisms that capture insects in the way that flypaper captures flies; the sundew, *Drosera rotundifolia*, is an example. Others capture insects by drowning them. One example is the pitcher plant (*Sarracenia flava*). *Nepenthes* (another form of pitcher plant), while having the capacity to trap and digest insects, furnishes the home for certain other species of insects said to be able to live in the liquid at the base of the "pitcher" and nowhere else. Some insectivorous plants, such as the Venus flytrap (*Dionaea muscipula*), capture their "prey" by clamping the victim in a "vise" (the leaves). An aquatic form called the bladderwort (*Utricularia vulgaris*) sucks its victims through a trap door, which then quickly shuts behind them. This form also secretes proteolytic enzymes, which digest the insects. Thus, two traits are necessary in order for a plant to be considered insectivorous: having a type of trapping mechanism and producing protein-digesting enzymes. Plants that can entrap insects but that do not produce protein-digesting enzymes are not considered to be insectivorous. Likewise, plants that produce such enzymes but do not capture insects are also not considered to be insectivorous.

## Natural Recycling

Recycling is another facet of ecology. Plant material built up by photosynthetic activity is broken down by the reverse processes of respiration and decomposition (which is respiration of microorganisms). Plant materials that came from the ground and atmosphere are thereby recycled and returned to the environment. The cycle is one of composition and decomposition. The decomposers are mainly fungi and bacteria. Each year 80 billion tons of carbon dioxide enter into the cycle by land plants, and 600 billion tons of carbon dioxide are processed in the oceans by mostly one-celled algae. Thus, nine-tenths of the world's photosynthesis occurs in the seas.

## Plant Succession

Ecologists speak of **plant succession**, meaning an orderly, progressive series of changes in the type of plant community, one community type gradually replacing another until a stabilized condition called **climax** vegetation is achieved. Each community in the series alters the environment, thereby creating conditions favoring the next type of growth. Animals play a significant part in this process.

Some take the position that succession occurs because some species are more efficient than others, the survival of the fittest. Whether a plant thrives or fails is more likely related to its microenvironment than to its genetic fitness, however. A seed may grow successfully in a certain spot but fail should it fall one inch to the side of that spot.

Plants having widely dispersed seeds may be "opportunists" that can grow well on denuded (stripped barren) soil. When a section of land is denuded, the first plants to reappear are those considered to be the fittest of all: weeds, grasses, and other herbaceous plants. In the eastern part of the United States, poplar and sumac would likely follow; gooseberries would likely come next in the western part of the country. Although the fittest may come first, they will in time be crowded out by other plants able to germinate and begin growth in the environment produced by the weeds, grasses, poplars, and sumac. The final, stabilized vegetation, called the climax forest, is determined by environment, climate, and soil. Maple and basswood is one kind of climax forest; oak and hickory is another. The soils deposited by the ice of the last glacier that came from the northeast support an oak-hickory climax. The soils deposited by the ice of the last glacier that came from the northwest support a basswood-maple climax.

### ❋ Notes ❋

The northern peninsula of Michigan is covered with coniferous forest. If these trees were to be cut down or destroyed by fire, the same species would not grow again. Rather, poplar trees would likely come first and would eventually be followed by other hardwood, deciduous trees. The climax forest of northern Michigan is said to be neither pine, spruce, nor hemlock but, rather, maple, oak, or hickory. So why are coniferous trees growing there now? The argument is that they have held on from the time that they were the climax vegetation after the retreat of the last glacier; but that if some event were to take them away, they would not return.

## Questions for Review

1. What is the meaning of the term *climax vegetation*?
2. Define the word *ecology*.
3. What seems to be an environmental requirement for lichens?
4. Name one or two plants that are parasites on other plants.
5. What traits do some plants exhibit that may cause them to be called *insectivorous*?
6. What is the meaning of the term *plant succession*?

## Suggestions for Further Reading

Gabriel, M.L., and S. Foge. 1959. *Great Experiments in Biology*. Engelwood Cliffs, NJ: Prentice-Hall.

Gleason, Henry, and Arthur Cronquist. 1964. *The Natural Geography of Plants*. New York: Columbia University Press.

Odum, Eugene. 1959. *Fundamentals of Ecology*. Philadelphia: W.B. Saunders Company.

Treshow, M. 1970. *Environment and Plant Response*. New York: McGraw-Hill Book Co.

# 40

# Plants and Human Welfare

All life on Earth runs on sunlight by the use of light energy in the process of photosynthesis. The food we eat is either plant or animal, and the energy derived from this food is easily traced back through the food chain to green plants. While there are nearly 500,000 species of known plants, all of the food on which people rely comes from twenty species, and 80 percent of our food comes from six species. These are all grasses: wheat, rice, corn, barley, rye, and oats. The grass family feeds the world.

All civilizations have depended on the cultivation of grasses. The peoples of ancient Babylonia, Egypt, Rome, and Greece grew wheat, barley, rye, and oats. In China and Japan, rice is the staple food. The pre-Columbian peoples of America—the Incas, Mayas, and Aztecs—looked to corn for their daily bread.

## Feeding an Increasing Population

Problems of population increase must be considered whenever discussing food resources. The rate of population increase is far greater than that of food production. This tendency portends alarming possibilities. For a long stretch of time, the population of the world remained relatively stable. The population began to increase significantly around the year 1650, doubling in 200 years. Another doubling of the population occurred in the 80 years following, and still another in the 45 years following. While one may be inclined to use this knowledge to project another doubling of the population in 20 years, followed by one in 10 years, and followed by one in 5 years, such projection borders on the absurd. Nevertheless, it is quite proper to be concerned about population increase and our capacity to increase food production to meet demands. Today, more than one billion people are not assured sufficient food, and perhaps as many as twenty million people die of

❈ Notes ❈

※ **Notes** ※

starvation each year. Some help comes in the form of plant breeding. Better plants are continually being produced by genetic methods. Hybrid corn is a notable example of success in this realm. Unfortunately, however, it is not enough.

## Other Human Uses for Plants

Food production is not the only way whereby plants contribute to human welfare. Coal, oil, and gasoline all, either directly or indirectly, trace their beginnings to plants of long ago. Thus, much of our machinery actually runs on plant products. This book is a product of plants, as are spices, tea, stimulants, sedatives, pain killers, oils, waxes, gums, alcohol, linen, brooms, and Christmas trees. Plants have been used at varying times and in varying cultures to create visions, depose demons, incur blessings, improve fertility, cast spells, cure toothache, and ensure falling in love.

## Cultivated Plants

Alphonse Louis Pierre Pyrame de Candolle (1806–1893) spent many years studying the origins of cultivated plants. The majority of our knowledge concerning cultivated plants comes from the in-depth studies of de Candolle, whose father, Augustine, was also a highly regarded botanist who contributed much to the formulation of a natural system of classification. Alphonse's son Anne (sic) collaborated with his father in botanical studies.

Several plants that have been cultivated for more than 4,000 years include almonds, apples, apricots, bananas, cabbage, cucumber, eggplant, figs, grapes, lentils, peaches, pears, soybeans, tea, and wheat. Although the grass family is said to feed the world, only one of these foods is in the grass family: wheat. Apples, peaches, and pears belong to the rose family. Thus, while the grass family feeds the world, the rose family furnishes the dessert!

Alphonse de Candolle also cataloged several Old World species that have been cultivated for 2,000 years, were known to the Greek botanist Theophrastus (371/370 280/287 B.C.), and have been found at ancient, lake-dweller sites. These include alfalfa, asparagus, beets, carrots, cherries, cotton, grapefruit, lemons, limes, lettuce, oats, plums, sugar cane, and yams. Again, only one of these is a grass.

Old World plants that have been cultivated for fewer than 2,000 years are interpreted in drawings found at Pompeii and are mentioned by Dioscorides.[1] They include buckwheat, coffee, currents, gooseberries, horseradish, spinach, parsley, strawberries, and oranges.

Some plants long cultivated in the New World and before the time of Columbus include tobacco, **maize**, sweet potatoes, peanuts, pumpkins,

---

[1]Pedanius Dioscorides (c. A.D. 50) detailed the properties of approximately 600 medicinal plants.

squash, and vanilla. Several New World forms that came into cultivation after the time of Columbus are black cherries, black walnuts, blueberries, cinchona (the source of quinine), pecans, and rubber.

Much of our knowledge regarding genetics, metabolism, and morphology comes from the study of plants. People have learned how to produce seedless fruits; prevent the premature falling of fruit; greatly increase yield by breeding; create new species; cause plants to flower according to a specified schedule; develop selective herbicides; and produce drugs and vitamins—all through plant research.

## Viruses, Bacteria, and Fungi

Including a section on viruses, bacteria, and fungi in a chapter on plants and human welfare is contrary to a five-kingdom system, wherein none of these are considered plants. But because many of these significantly affect human welfare, they are included here for the want of a better place.

Viruses are known chiefly for their capacity to produce disease. A number of human ailments are caused by viruses; polio, influenza, chicken pox, smallpox, measles, mumps, yellow fever, warts, and the common cold are among these. Much evidence points to viruses as also causing some cancers.

A number of viral diseases affect both wild and domestic animals: the foot and mouth disease of cattle; rabies; distemper; hog cholera; psittacosis (parrot fever); and Newcastle disease (which infects fowl).

Nearly one thousand kinds of plant diseases are caused by viruses. Mosaic diseases tend to inhibit chlorophyll production in a definite pattern, often along the veins. There are mosaic diseases that affect beans, cucumbers, peas, peaches, sugar beets, sugar cane, tobacco, turnips, rice, potatoes, tomatoes, and wheat. Rice **necrosis** stunts rice plants. Aster yellows, potato paracrinkle, potato leaf role, sugar beet curly top, and tomato bushy stunt are all mosaic diseases. Such diseases are generally transmitted from one plant to another by leaf hoppers and aphids. **Nematodes** also sometimes transmit viral diseases from plant to plant. Viruses do no harm to these insect vectors. Viral diseases can also be transmitted mechanically, by injury, or grafting.

Although viruses have a bad reputation because of the diseases they cause, viruses sometimes have desired effects. For example, the rice necrosis mosaic virus, which stunts rice plants, inexplicably improves growth of jute plants (used in making burlap sacks and rope). And the highly esteemed Rembrandt tulip derives its distinctive markings from a virus.

Bacteria play a significant role in human events. Being decomposers, they are involved in the recycling of organic materials. They are able to convert atmospheric nitrogen into compounds usable by plants. Bacteria are used in the production of cheese, yogurt, sauerkraut, and pickles. They also participate in making vinegar and acetic acid. Bacteria also deserve praise for their role in making useful antibiotics such as bacitracin, tyrothrycin, subtilin, and polymixin.

### Notes

At the same time that bacteria are essential to our existence, they can also present hazards to human welfare. Tuberculosis, cholera, anthrax, diphtheria, gonorrhea, and tetanus are all caused by bacteria. Food poisoning is caused by the bacterial organisms *Salmonella*, *Staphylococcus*, and *Clostridium*. *Rickettsiae*, minute bacteria that dwell in the bodies of arthropods, cause two serious diseases: Rocky Mountain Spotted Fever and Tsutsugamuchi Fever.

Many plant diseases can be traced to bacteria, including fire blight, soft rot, wilts, and crown gall. They cause the death of young apple and pear trees and destroy vegetables in storage. Citrus canker is caused by the bacterium *Xanthomonas*.

Fungi, like viruses and bacteria, are both friend and foe to humans. They are just as necessary for the continued existence of human life as are the green plants. They are decomposers that unlock carbon dioxide and restore it to the atmosphere, where it can again be used in photosynthesis. If fungi were selective in what they decompose, they might be less scorned. But just as they decompose old leaves, dead trees, and garbage, they also decompose cloth, paint, leather, waxes, jet fuel, wire insulation, cartons, photographic film, and timber. They also are enemies of food producers, growing on bread, fruits, vegetables, and meat. *Ustilago zeae*, in particular, has taken a great toll on corn crops.

As recounted in chapter 18, the tragic Irish Potato Famine, which was caused by the fungus *Phytophthora infestans*, took millions of human lives. Ergotism, caused by the fungus *Claviceps purpurea*, was a scourge of the Middle Ages.

While there are 20,000 species of pathogenic fungi, there are also many valuable fungi forms. Perhaps as many as 5,000 species are economically important. Yeasts play a role in the manufacture of bread and the production of alcohol. The mycorrhiza associated with plant roots are necessary for the welfare of certain trees. And some fungi are esteemed as food.

Cyclosporine, a so-called "wonder drug," is isolated from a soil-dwelling fungus. This drug suppresses undesirable immune-system responses to organ transplants. While other drugs do this as well, they have the undesirable side effect of also killing bone marrow cells, thereby leading to leukemia. Cyclosporine is thus of enormous benefit in organ transplants.

## Questions for Review

1. Name the six grass species that serve as the main food sources for all people.
2. Discuss briefly some methods whereby food production can be increased.
3. Name two species of plants that have been cultivated for more than 4,000 years.
4. Name four species of plants that are of American origin.
5. Alphonse de Candolle made significant contributions to botany; name one of these.

## Suggestions for Further Reading

Anderson, E. 1969. *Plants, Man and Life.* Berkeley, CA: University of California Press.

Heiser, C.B. 1973. *Seed to Cultivation: The Story of Man's Food.* San Francisco: W.H. Freeman Co.

Luria, S.E. 1953. *General Virology.* New York: John Wiley and Sons, Inc.

Richardson, W.N., and T. Stubs. 1978. *Plants, Agriculture, and Human Society.* Reading, MA: Benjamin Cummings Publishing Co.

❃ Notes ❃

# Glossary

## A

**abscisic acid.** A stress hormone that inhibits the action of auxins and gibberellins.
**abscission.** The falling away of leaves, flowers, or fruits, usually following the formation of an abscission layer constituted of cork cells.
**absorption spectrum.** Graph of absorption of varying wavelengths of light by a pigment.
**abstriction.** In ascomycetes, a form of conidiospore formation wherein spores are cut off from the ends of hyphae.
**acellular.** Not being constituted of cells. In a mass of cytoplasm the nuclei are not separated from each other by membranes.
**achene.** A one-seeded, dry, indehiscent fruit.
**acrasin.** A hormone produced in a myxomycete. Influences the coalescence of swarm spores.
**acrasinase.** An enzyme involved in the breakdown of the hormone acrasin.
**actinomorphic.** Radially symmetrical. Star-shaped. Said of flowers.
**actinomycetes.** A group of filamentous bacteria that reproduce by fission or conidia. Some cause diseases of plants and some produce antibiotics (e.g., *Streptomyces*).
**actinomycosis.** An infection caused by an actinomycete.
**action spectrum.** A measure of the rate of photosynthesis at varying wavelengths of light.
**active transport.** The movement of substances across a cell membrane. Requires the expenditure of energy.
**ADP.** Adenosine diphosphate.
**adventitious.** Characterized by a root arising from stem or a stem arising from a root.
**aeciospores.** Binucleated spores of a rust fungus and formed by a fusion of cells without the fusion of nuclei.
**aecium.** In rust fungi, a cup-like structure that produces aeciospores.
**aerobic.** Requiring oxygen for respiration.
**akinete.** An algal spore produced by a change in a vegetative cell.
**algin.** A product derived from brown algae and used commercially (e.g., the manufacture of ice cream).
**aliphatic.** An organic compound of open-chain structure (e.g., paraffin, fatty acids).
**amino.** Relating to the group $-NH_2$.

**amino acid.** An organic acid in which hydrogen has been replaced by an amino group. It exhibits both acidic and basic properties, and is a building block in the manufacture of protein.
**amoeboid.** Relating to changing shape, moving, or eating, or ingesting by means of pseudopodia (temporary cytoplasmic protrusions).
**amoeboid movement.** Movement caused by flowing protoplasm accompanied by changing cell shape.
**amphimixis.** The union of germ cells in sexual reproduction.
**anaerobic.** Respiration accomplished in the absence of oxygen.
**anaphase.** That phase in mitosis when chromatids separate and move toward opposite poles. In meiosis, the paired chromosomes move apart.
**androecium.** A collective term for the stamens.
**aneuploidy.** Having a chromosome number that is not an exact multiple of the haploid number, N.
**angiosperm.** The group of vascular plants having seeds enclosed in an ovary.
**anisogamous.** The fusion of unlike gametes, the difference usually being one of size.
**annual.** A plant that completes its life cycle in one growing season.
**annual ring.** A growth layer that is seen in a cross section of a woody stem and that reveals the growth for a single year. The term applies primarily to wood grown in temperate zones.
**annulus.** In ferns, the specialized ring of cells around the sporangium.
**anther.** That part of the stamen consisting of pollen sacs that bear the pollen.
**antheridiophore.** The stalk on which an antheridium is borne.
**antheridium.** A male gametangium (a spore-bearing organ of plants other than seed plants).
**anthocyanin.** Water-soluble pigments. Often found in central vacuoles.
**antibiotic.** A substance that prevents or retards the growth of organisms. Often produced by microorganisms and prevents the growth of other microorganisms.
**anticodon.** The three nucleotide bases in m-RNA (the three bases in DNA representing a codon).
**antipodal cells.** The three cells of a mature embryo sac. Located at the end opposite the micropyle.
**apetalous.** Lacking petals.
**aphid.** A small insect that sucks the juices of plants, causes wilting, may cause gall formation, and may serve as a vector for virus diseases of plants. Secretes a sweet liquid attractive to insects.
**apical meristem.** In vascular plants, the dividing tissue located at the tip of a root or stem.
**apogamy.** The development of a sporophyte from the gametophyte without the union of gametes. The embryo arises from cells other than the egg.
**apomixis.** Production of seed without the union gametes.

**apothecium.** In ascomycetes, a cup- or saucer-shaped structure bearing asci.
**archegoniophore.** The stalk on which an archegonium is supported.
**archegonium.** A female gametangium. A structure in which the egg is protected by a jacket of cells.
**ascogonium.** The female, gamete-bearing structure of ascomycetes.
**ascospore.** In ascomycetes, the spore produced within an ascus.
**ascus** (plural, **asci**). In ascomycetes, the spore-bearing structure in which the ascospores are borne.
**asexual.** A form of reproduction involving neither the union of gametes nor meiosis.
**aspergillum.** A structure used in dispersing holy water. The spore-bearing structure of *Aspergillus* is the same shape, hence the name.
**ATP.** Adenosine triphosphate.
**attachment site.** In the t-RNA molecule, the site where an amino acid is attached so that it may be carried to the ribosome.
**autecology.** A branch of ecology focusing on individual plants and their effects on the environment.
**autotrophic.** Self-nourishing. Pertains to plants that manufacture their own nourishment.
**auxin.** A plant growth-regulating substance that allows cell elongation.
**avirulent.** Lacking the capacity to cause disease.

# B

**bacillus.** A rod-shaped bacterium.
**back cross.** Also called a **test cross**. The crossing of an individual with a homozygous recessive.
**bacteria.** Microscopically sized organisms of the class Schizomycetes.
**bacteriochlorophyll.** Pigment that occurs in photosynthetic bacteria and is related but not identical to the chlorophyll of higher plants.
**bacteriophage.** Also called **phage**. A virus that attacks bacteria.
**basal body.** A cytoplasmic organelle that organizes cilia or flagella. Identical in structure to a centriole.
**basal cell.** In an ovule, an extra-embryonic cell lying next to the micropyle.
**basidia.** In basidiomycetes, spore-bearing structures within which nuclei fuse and meiosis then occurs, producing haploid basidiospores.
**basidiospore.** In basidiomycetes, a type of meiospore borne on the basidium.
**bast.** Fibers obtained chiefly from phloem but sometimes from pericycle or cortex. Used in the manufacture of rope or cordage.
**berry.** A pulpy fruit with immersed seeds.
**beta-carotene.** A carotene molecule that forms two molecules of vitamin A upon cleavage.
**bioassay.** A test for the presence or the concentration of a substance as indicated by the substance's effect on a living organism.

**biochemical.** Chemical reactions occurring in living organisms.
**bioluminescent.** Light that is given off by a living organism and results from oxidative changes.
**bivalent.** A pair of synapsed chromosomes.
**blastomycosis.** Disease caused by a blastomycete, a pathogenic, yeastlike fungus. Such fungi are usually classified among the moniliales.
**blue-green algae.** Algae of the division Cyanophyta.
**body cell.** Any cell of an organism other than the cells of reproductive tissues.
**bordered pit.** In a tracheid or other xylem cell, a thin area having a distinct rim of secondary wall overarching the pit.
**botany.** That branch of biology focusing on plant life.
**bracket fungus.** A basidiomycete that forms shelflike structures (sporophores) as on the trunks of trees.
**Bryophyta.** Nonvascular, land-dwelling green plants, including mosses and liverworts.
**bryophyte.** Nonflowering plants comprising the mosses and liverworts. Possess rhizoids, multicellular archegonia, and antheridia, but lack vascular tissue.
**bud.** In flowering plants, an embryonic shoot. In yeast cells, an outgrowth that produces another cell.
**bud graft.** A form of grafting wherein a bud is united to a stock, bringing about a fusion of the cambiums.
**bulb.** A mass of short, fleshy leaves on a short stem base.
**bundle sheath.** A layer of cells surrounding a vascular bundle. May consist of parenchyma, sclerenchyma, or both.
**button.** An early stage in the growth of a mushroom.

# C

**callose.** A carbohydrate that occurs in cell walls. May obstruct the perforations of sieve plates and commonly develops as a consequence of injury.
**callus.** Undifferentiated tissue that develops over an injured area of stem or root.
**Calvin cycle.** That part of the photosynthetic reaction when carbon dioxide bonds to a 5-carbon sugar and reduces to form other sugars.
**calyptra.** A lid that covers the spore capsule of a moss. Derived from the archegonium after fertilization.
**calyx.** The outermost whorl of flower parts, consisting of the sepals.
**cambium.** A layer of cells that is one cell in thickness, retains meristematic ability, and, by cell division, contributes to the formation of secondary xylem and phloem.
**Cambrian.** The earliest period of the Paleozoic era, when marine algae and cyanobacteria arose.

**capillitium.** A network of noncellular strands formed during the cleavage of spores in the sporangium of slime molds.

**capsule.** In a bryophyte, the sporangium; in an angiosperm, a type of dry, dehiscent fruit.

**Carboniferous.** Approximately three hundred forty-five million years ago during the latter part of the Paleozoic era, when the ferns and gymnosperms arose.

**carboxyl.** A group of atoms (−COOH) that characterizes organic acids.

**carboxylase.** Enzyme involved in the removal of carbon dioxide from a carboxyl group.

**carotene.** A reddish or orange pigment. Occurs in green plants associated with chlorophyll and is the precursor of vitamin A.

**carpel.** The innermost whorl of flower parts. Function as megasporophylls.

**carpogonium.** In red algae, a name for the oogonium, because carpospores arise from it.

**carpospores.** In red algae, spores that arise from the fertilized egg by mitotic divisions.

**Casparian strip.** A waxy band surrounding an endodermal cell along its radial wall.

**catalyst.** An enabling substance that accelerates the rate of chemical reaction without being consumed.

**cell.** A generally microscopically sized unit that composes organisms. Possesses a nucleus and is surrounded by a membrane.

**cell membrane.** Also called the **plasma membrane**, or **plasmalemma**. A selectively permeable, limiting membrane surrounding protoplasm.

**cell plate.** A structure that forms at the equator of the spindle of plant dividing cells. Precedes the formation of the middle lamella.

**cell wall.** The nonliving covering of a cell. Lies outside of the cell membrane and is secreted by the cell. Is usually impregnated with cellulose.

**cellulose.** A complex carbohydrate composed of polymerized glucose units. Chief component of cell walls.

**centriole.** An organelle from which the microtubules of the mitotic spindle appear to radiate. Although present in certain flagellated algal cells and in animal cells, generally absent from most plant cells.

**centromere.** That portion of a chromosome to which a spindle fiber appears to be attached. Also called the **kinetochore**.

**chemoautotrophic.** Capable of oxidizing some inorganic compound as a source of energy. Contrast with photoautotrophic in which plants utilize light energy in synthetic reactions.

**chiasma.** The points where chromatids overlap in homologous chromosomes. Probably the site of crossing over.

**chitin.** A polysaccharide that forms in the cell walls of certain fungi.

**chlamydospore.** From the Greek word *chlamys*, meaning "coat." Here, a heavy-walled covering that envelops a spore.

**chlorenchyma.** Tissue that contains chloroplasts.

**chlorobium chlorophyll.** A type of chlorophyll that is present in green sulfur bacteria and utilizes hydrogen sulfide.

**chlorophyll.** The green pigment that occurs in chloroplasts and catalyzes the light reactions of photosynthesis.

**Chlorophyta.** The green algae.

**chloroplast.** The organelle of plant cells that contains the chlorophylls. The site of photosynthesis.

**chlorosis.** A yellowing condition in chlorophyll-bearing plant cells resulting from the degeneration of chlorophyll, which may be caused by the lack of light or by a deficiency of certain mineral nutrients.

**chromatid.** One of the two daughter strands of a duplicated chromosome. The two strands remain attached at a single point known as the *centromere*.

**chromonema.** A threadlike, DNA-bearing structure that gives rise to chromosomes.

**chromoplast.** A plastid having pigments other than chlorophyll.

**chromosome.** A ribbonlike structure derived from nuclear material, organized from the chromonemata, bearing DNA, and participating in the events of mitosis.

**chrysolaminarin.** A food reserve occurring in organisms of the phylum Chrysophyta.

**Chrysophyta.** An algal group possessing yellowish-green and golden-brown chromatophores and usually having cell walls of overlapping halves.

**chytrid.** One of the chytridiales, aquatic fungi that do not appear to produce mycelium.

**cilia.** Short, hairlike structures that extend from certain special cell types and propel unicellular organisms or gametes through the water by a whiplike motion. Characteristic internal structure is two inner microfibrils surrounded by nine pairs of microfibrils.

**circadian.** Relating to regular rhythms of growth or behavior that occur on an approximately twenty-four-hour basis.

**cisternae.** Flattened sacs characteristic of Golgi bodies.

**cladophyll.** A stem modified in the form of a leaf.

**clamp connection.** In basidiomycetes, a small lateral branch that grows out from a terminal cell and curves back to make contact directly below. Results in maintenance of the proper arrangement of nuclei.

**class.** The rank in the taxonomic hierarchy between *order* and *phylum*; a group of related orders.

**cleistocarp.** An ascus-bearing structure that is completely closed at maturity.

**cleistothecium.** A closed, spherical body containing asci.

**climax.** When plants occupying a given area of land have stabilized. Such plants are said to have reached *climax vegetation*.

**clubroot.** Common name for the infection caused by the slime mold *Plasmodiophora brassicae*. Infects cabbages and related plants. Also called **finger and toe disease**.

**coccus.** A spherical bacterium.

**codon.** A sequence of three adjacent nucleotides that represents a code for a specific amino acid.

**coenobium.** A spherical colony of unicellular organisms surrounded by a common wall.

**coenocytic.** *See* **multinucleated**.

**coleoptile.** The first leaf of a germinating monocot. Sheaths the succeeding leaves.

**collenchyma.** A supporting tissue derived from parenchyma and composed of closely fitting cells having thickened cell walls, particularly at the angles of the cells.

**colloid.** Referring to a state of dispersion wherein the particles are larger than true molecules yet small enough to remain dispersed for a long period of time.

**columella.** In certain fungi, the central, sterile portion of a sporangium.

**companion cell.** In angiosperm phloem, a parenchymatous cell associated with a sieve tube element. Arises from a mother cell, which produces both the companion cell and the sieve tube element.

**complex tissue.** A tissue composed of more than one kind of cell.

**compound leaf.** A leaf composed of a number of leaflets attached to a central rachis or attached at a single point at the end of the petiole.

**conceptacle.** In organisms of the genus *Fucus*, a cavity or chamber in which gametangia are borne.

**conidiospore.** Also known as *conidia*. A fungal spore arising from fragmentation of a hypha or abstriction of the hypha end.

**conifer.** A cone-bearing tree.

**conjugation.** A sexual fusion of isogametes. The transfer and fusion of nuclear material in bacteria, protozoa, and certain algae and fungi.

**continuous phase.** In colloids, the same as the dispersing phase, during which separate particles are dispersed.

**contractile vacuole.** In certain unicellular organisms, a vacuole that enlarges and collapses, expelling its watery contents, often in a regular rhythm.

**cork.** A tissue produced by phellogen or cork cambium. Cells that are nonliving in maturity and have walls impregnated with the waxy material suberin. Considered to be resistant to the passage of gases and water.

**cork cambium.** Also called **phellogen**. A cambial layer outside of the cambium. Produces the cork cells of outer bark, and the phelloderm.

**corm.** A short storage stem surrounded by thin, scaly leaves.

**corolla.** A whorl of floral parts consisting of petals. Counting from the center outward, the second whorl of the series.

**cortex.** A primary tissue of stems and roots. Extends from the primary phloem to the epidermis and is composed chiefly of parenchyma.

**cotyledon.** A seed leaf; one in monocots, two in dicots. Stores or absorbs food.

**Cretaceous.** The most recent period of the Mesozoic era, approximately one hundred thirty million years ago. Angiosperms arose in this period, and gymnosperms declined.

**cristae.** A ridge. Here, used to designate the infoldings of the inner mitochondrial membrane.

**crossing over.** An exchange of corresponding segments between chromatids. Takes place at the site of the chiasma.

**crozier.** A hook formed in the process of ascus development.

**crustose.** Here, used in reference to lichens, which form closely adhering tissue to the substrate of a rock, bark, or soil.

**cryptogam.** Literally, "a hidden marriage"; here, refers to nonflowering plants reproducing by spores, as in ferns and mosses.

**cryptomonad.** A flagellated alga usually included among the Pyrrophyta.

**cuticle.** A waxy layer on the outer wall of an epidermal cell.

**cuttings.** A method of plant propagation whereby stem pieces are placed in the soil and adventitious roots arise from the cut surfaces.

**cyanobacteria.** *See* Cyanophyta.

**Cyanophyta.** Also called **cyanobacteria**. The blue-green algae.

**cycad.** A primitive Cycadopsida plant having unbranched stems and a terminal crown of leathery, compound leaves.

**cystocarp.** In red algae, a fruiting structure. In maturity, has a special protective envelope.

**cytochrome.** An iron-containing, proteinaceous pigment; an enzyme involved in respiration.

**cytokinesis.** The division of a cell.

**cytokinin.** A growth hormone that promotes cell division.

**cytoplasm.** The protoplasm of the cell exclusive of the nucleus.

# D

**day-neutral plant.** Plant in which flowering does not depend on length of day or night.

**deciduous.** Refers to the shedding of leaves, usually during the autumn.

**dehiscent.** Splitting open along predeterminable lines.

**dehydrogenase.** Enzyme involved in the removal of hydrogen atoms from a compound. Plays an important role in biological oxidation-reduction processes.

**dehydrogenation.** The removal of hydrogen atoms from a compound.

**deletion.** A form of chromosomal anomaly wherein part of a chromosome breaks away and is therefore unable to go through the events of mitosis.

**deliquescent.** Becoming dissolved in moisture absorbed from the atmosphere. Also, repeated division of branches.

**deoxyribonucleic acid.** *See* **DNA**.

**deoxyribose.** A 5-carbon sugar having one less oxygen atom than does the sugar ribose.

**dephlogisticated air.** An ancient name for oxygen (*phlogiston* being an ancient name for carbon dioxide).

**deplasmolysis.** Plasmolysis being a process wherein cytoplasm is separated from the cell wall because of a loss of water, this is the reverse process brought about by the intake of water.

**diakinesis.** The final stage of meiotic prophase preceding the formation of the metaphase plate. During this stage, there is a marked contraction of the bivalents.

**dialysis.** The separation of substances in solution by means of a selectively permeable membrane.

**diastase.** An enzyme that converts starch to maltose.

**diatom.** A member of the group of golden-brown algae having siliceous cell walls fitting together much as do the halves of a pill box.

**dichotomous.** Branching into two more or less equal portions.

**dicot.** A plant having two cotyledons in the seed. A shortened form of *dicotyledonous*.

**dicotyledonous.** Having two cotyledons in the seed.

**dictyosomes.** *See* **Golgi bodies**.

**dikaryotic.** Binucleated, the paired nuclei usually deriving from different parents: one male and one female.

**dinoflagellate.** A group of chiefly marine, solitary plantlike flagellates having cells typically enclosed in a cellulose envelope.

**dioecious.** Literally, "in two houses"; here, having separate sexes: male and female. In reference to flowers, staminate and pistillate flowers are on different plants.

**dipeptide.** A union of two amino acids, the linkage made through a nitrogen atom.

**diplococcus.** Bacteria in which the cells occur in pairs or, sometimes, in short chains.

**diploid.** Having two sets of chromosomes, one of maternal origin and the other of paternal origin. Occurs in the sporophyte generation.

**diplotene.** That stage in the meiotic prophase immediately following pachytene and when the homologous chromosomes tend to repel one another.

**disaccharide.** A sugar that yields two monosaccharides upon hydrolysis, most commonly a 12-carbon sugar that yields two molecules of 6-carbon sugar upon hydrolysis.

**dispersed phase.** When the colloidal form (that is, the finely divided particles) is dispersed in a liquid medium.

**distal.** Located away from the site of attachment. The opposite of *proximal*.

**distromatic.** Referring to a thallus two cells in thickness.

**diurnal.** Having a recurrent daily cycle of change.

**DNA.** Deoxyribonucleic acid. That molecule in chromosomes that governs the manufacture of protein and serves as the carrier of genetic information. Composed of phosphate, sugar, and bases, and capable of self-replication.

**dominance.** In genetics, when a gene expresses itself to the entire suppression of its allele.

**double fertilization.** A process occurring in flowering plants and involving two sperm cells: one fertilizing the egg and the other uniting with the polar nuclei to form endosperm.

**drupe.** A one-seeded, fleshy fruit.

**duplication.** A form of chromosomal anomaly wherein a segment of a chromosome is represented twice.

**dwarf male.** Occurs in the Oenogodiaceae. Several cells form an androspore near the oogonium.

# E

**ecology.** The science that focuses on the relationship of organisms to the environment.

**ectotrophic.** Relates to mycorrhiza, growing between but not within root cells.

**egg.** The female gamete.

**elater.** Spindle-shaped cells in the sporangium of a liverwort. Also, the bands attached to the spores of horsetails. Being hygroscopic, they move in varying conditions of humidity. Aid in the dispersal of spores.

**embryo sac.** The female gametophyte of angiosperms. The embryo sporophyte develops within it.

**emulsify.** To obtain a stable dispersion by the use of an emulsifying agent such as soap or detergent.

**encysted.** Enclosed in a protective covering. Results in resistance to desiccation.

**endodermis.** A layer of specialized cells marking the inner margin of the cortex.

**endogenous.** Arising from within, as in the case of a branch root arising from the center portion of the root tissue.

**endoplasmic reticulum.** Also called ER. A network occurring in the cytoplasm, constituted of paired membranes, and often aligned with ribosomes.

**endosperm.** Nutritive tissue within the embryo sac. It often is consumed as the seed matures.

**endospore.** A spore formed within either a sporangium or a cell wall.

**endosymbiosis.** A symbiotic relationship whereby one organism resides within another. Considered by some to be the basis of the origin of eukaryotes.

**endotrophic.** Relating to mycorrhiza, penetrating root cells.

**enucleate.** To remove the nucleus by microdissection methods.

**enzyme.** A protein that, even in low concentrations, is able to accelerate certain chemical reactions without being consumed by those reactions.

**epicotyl.** That portion of a seedling that lies above the cotyledon.
**epidermis.** The outermost layer of cells. A product of primary growth. Present on all parts of the primary plant body.
**epigynous.** In flowers, growing after and appearing to grow from the top of the ovary. Several whorls of flower parts are inserted above the ovary.
**epiphyte.** A plant that grows on another plant.
**epitheca.** The outer and older portion of the half wall of diatoms and dinoflagellates.
**equatorial plate.** In mitosis, the position of the chromosomes at metaphase.
**ER.** *See* **endoplasmic reticulum**.
**era.** A period of time in a geological chart.
**ergot.** A poisonous alkaloid produced in the sclerotia of the fungus *Claviceps purpurea*, which infects rye grasses.
**ergotism.** A disease brought about by eating cereal grains infected with *Claviceps purpurea*. Ingestion causes muscle cramps and, possibly, dry gangrene.
*Escherichia.* The bacterium that dwells in the intestinal tract. Its presence in water can indicate fecal contamination.
**estipulate.** Being without stipules.
**Euglenophyta.** Unicellular, flagellated algae possessing a gullet and either lacking a cell wall or, if a cell wall is present, lacking cellulose. Reproduce asexually by cell division.
**eukaryotic.** Possessing a true nucleus.
**evergreen.** Remaining verdant. Characteristic of pines, holly, and laurel. Contrast with **deciduous**.
**evolution.** A series of changes in organic development resulting in differences in an organism. Changes may be progressive or regressive.
**excurrent.** Having the axis prolonged to form an undivided main stem or trunk, as in spruce and other conifers. Opposite of **deliquescent**.
**exine.** The outer wall of a spore or pollen grain.
**exoenzyme.** An enzyme that is produced by plant cells and that then moves through the cell wall and cell membrane to accomplish its mission outside of the cell.
**exogenous.** Arising from the exterior, as do branches from a stem. Branch roots are endogenous; branch stems are exogenous.
**extraembryonic.** Derived from the fertilized egg but not a part of the embryo.
**eye spot.** Also called a **stigma**. Pigmented, light-sensitive structure in flagellated, unicellular organisms.

# F

**family.** The unit of organism classification that falls between *order* and *genus*.
**far red.** The longest wavelengths of the red portion of the spectrum.

**fermentation.** Anaerobic form of respiration. Produces alcohol or lactic acid as an end product.

**fern.** A vascular plant possessing roots, stems, and leaves. Reproduces by spores, which produce minute gametophytes upon germination.

**fiber.** In vascular plants, an elongated, tapering, thick-walled sclerenchyma cell.

**filament.** A long, slender object. In botany, the anther-bearing stalk of the stamen.

**filamentous.** Long, slender, threadlike.

**filial.** Relating to a son or daughter. In heredity, relating to the offspring of a particular cross.

**filterable virus.** A viral particle consisting of protein and either DNA or RNA and being so minute that it can pass through the pores of a ceramic filter designed to remove bacteria.

**finger and toe disease.** *See* **clubroot.**

**fission.** A division of single-celled forms. No sexual union is involved.

**fixation.** Here, rendering nitrogen into compounds available to plants.

**flagellum** (plural, **flagella**). A slender filament projecting from a cell. Longer but having the same internal structure as cilia. Capable of a different kind of movement than are cilia. Used in locomotion and feeding.

**floridian starch.** A reserve carbohydrate characteristic of the red algae.

**florigen.** The flowering hormone. Has not yet been isolated and identified but is presumed to exist and to stimulate flower production.

**floristics.** The branch of ecology that focuses on distribution and numbers of plants.

**fluorescence.** The emission of light by a substance, as with a chlorophyll extract. Chlorophyll is able to function as a source as well as a receiver of light.

**foliose.** Having similarity to a leaf. In lichens, a flattened, leaflike structure.

**follicle.** A fruit similar to a pod but dehiscing on only one line.

**foot.** In early sporophyte development, the lower portion of the sporophyte that attaches to gametophytic tissue. In bryophytes, the lower part of the embryo sporophyte.

**fret membrane.** The membrane that covers the frets, or lamellae, that interconnect the grana in plastids at irregular intervals.

**fruit.** In angiosperms, the mature ovary containing the seeds and any adjacent parts that may be adhered to the ovary.

**frustule.** The silicious shell of the diatoms. Composed of two valves that overlap.

**fucoxanthin.** A brown pigment occurring particularly in the ova of brown algae.

**fungus** (plural, **fungi**). Saprophytic or parasitic, plantlike structure that lacks chlorophyll and possesses a body composed of mycelia. Included among the fungi are molds, mildews, rusts, smuts, mushrooms, puffballs, and yeasts.

# G

**gametangia.** Organs that bear gametes.

**gamete.** A cell that can fuse with another cell to produce a new individual. Such cells are regularly haploid.

**gametophyte.** A gamete-producing generation that is haploid. The gametes are produced by mitotic divisions.

**gel.** A colloidal substance that tends to be viscous, more or less jelly-like. May result from coagulation.

**gelatinous.** Resembling a jelly in appearance and consistency.

**gemma cup.** In liverworts, a minute cupule in which the gemmae reside.

**gemmae.** Small clusters of vegetative cells. On the thallus of the gametophyte generation of liverworts and in certain fungi, outgrowths capable of developing into new plants. No alternation of generations is involved.

**gene.** The unit of heredity. Constructed of DNA molecules.

**generative nucleus.** In a pollen grain, the one of two nuclei that through its division produces two sperm nuclei.

**genotype.** The genetic constitution, whether latent or expressed. The total of all the genes present in an organism. Contrast with **phenotype**.

**genus** (plural, **genera**). A group of related species. The taxonomic rank between *species* and *family*. The first word of the scientific name.

**germ plasm.** Tissue involved in the production of a new generation. Reproductive cells. Transmitters of hereditary characteristics.

**germinate.** The resumption of growth after a period of dormancy. Applies to a spore or a seed.

**gibberellin.** Growth hormone involved in elongation of stems in a number of higher plants. First isolated in the fungus *Gibberella*.

**girdle.** The overlapping edge of a valve in diatoms. Also, a transverse groove on dinoflagellates.

**glucose.** Also called *grape sugar*. A simple, 6-carbon sugar.

**glycolysis.** An anaerobic step in respiration. A breakdown of sugar to simpler compounds.

**Golgi bodies.** Also called **dictyosomes**. In plant cells, a series of flattened, double lamellae thought to be associated with the production of secretions and cellulose.

**graftage.** The horticultural practice of grafting to unite a scion and a stock. Used to increase the number of clonal plants.

**grana.** Those structures within chloroplasts that under a light microscope look like minute granules and under an electron microscope look like stacked thylakoids. Contain the chlorophylls and carotenoids. Sites of the photosynthesis reactions.

**ground meristem.** Meristematic tissue that gives rise to the fundamental tissue system.

**guard cells.** Specialized, epidermal cells that surround the stomates. Changes in turgor change the shape of the guard cells and serve to open and close the stomates.
**gullet.** A groove present in some dinoflagellates and Euglenids.
**guttation.** The exudation of water and dissolved substances from the leaves of plants.
**gymnosperm.** Seed plant with seeds not enclosed in an ovary. The conifers are the most familiar.
**gynoecium.** The pistil. The female portion of the flower.

# H

**haploid.** Having one chromosome complement per cell. A common characteristic of the gametophyte generation.
**haustorium** (plural, **haustoria**). A cell outgrowth that functions as an adsorbing organ, penetrating a substrate.
**heartwood.** The center of xylem in woody plants. Nonliving, nonfunctional, and commonly dark in color.
**heliotropism.** Positioning in response to sunlight, as in sunflowers turning their heads toward the sun.
**helix.** Spiral in form. Here, describes the form of a DNA or RNA strand.
**hemoglobin.** An iron-containing, protein pigment in red blood cells.
**herbaceous.** Nonwoody and, therefore, persisting only for a single growing season.
**herbarium.** A collection of dried, pressed plants.
**heterocyst.** A transparent, thick-walled cell in the filaments of certain blue-green algae.
**heteroecism.** Relates to parasitic fungi, which require more than one species of host to complete the life cycle.
**heterogamy.** Reproduction requiring two types of gametes as egg and sperm.
**heterothallic.** Refers to organisms that are self-sterile or self-incompatible and, therefore, require two diffcrent compatible strains or individuals to bring about sexual reproduction.
**heterotrophic.** Unable to manufacture organic compounds and, therefore, needing to feed on organic materials. Contrast with **autotrophic** (self-nourishing).
**heterozygous.** Having two different genes at the same locus on a pair of homologous chromosomes.
**hexose.** A 6-carbon sugar.
**holdfast.** Structure at the tips of tendrils that allows the organism to attach to a substrate. Also, the basal part of an algal cell that attaches to a solid object.
**homologous.** Refers to chromosomes that are members of a pair, each member of the pair being derived from different parents.

**homothallic.** Possessing both male and female reproductive structures in the same organism.
**homozygous.** Having a like pair of genes at a specific locus.
**hormogonia.** Short filaments that result from the breaking apart of longer filaments in certain blue-green algae. They fracture at the site of heterocysts.
**hormone.** A chemical substance that is produced in minute amounts and has a marked effect on some other process.
**hornwort.** With the mosses and liverworts, belong to the division Bryophyta. Distinguished from other bryophytes in that each cell has a single chloroplast, mucilage-filled cavities, and an elongated foot.
**host.** A living organism upon which another organism of a different species dwells and from which that other organism derives its nourishment.
**hydathode.** An epidermal structure functioning in the exudation of water.
**hydrogenation.** The union of hydrogen atoms to a compound, a chemical reaction requiring the enzyme hydrogenase.
**hydrolysis.** The chemical breakdown of molecules accomplished by the insertion of the components of water at the point of the breakage.
**hydroxyl.** A radicle consisting of −OH (i.e., one atom of hydrogen and one atom of oxygen). Characteristic of alcohols, glycols, and hydroxides.
**hygroscopic.** Sensitive to variations in moisture. Changes shape in different conditions of humidity.
**hymenium.** In fungi, a mat of mycelium in the floor of the fruiting structure. Bears basidia or asci.
**hypertonic.** Having a concentration of dissolved substance greater than that of a solution serving as a source of comparison.
**hypha** (plural, **hyphae**). In fungi, a single, tubular filament. The hyphae collectively compose the mycelium.
**hypocotyl.** That portion of a germinating seedling that lies below the cotyledons.
**hypodermis.** The tissue immediately beneath the epidermis.
**hypogynous.** Inserted beneath the ovary, as are several whorls of flower parts.
**hypotheca.** The younger of the two valves in the wall of a diatom.
**hypotonic.** Having a concentration of dissolved substance less than that of a solution serving as a source of comparison.

# I

**IAA.** Indole acetic acid. The hormone that is produced at the growing tips of plants and is a cell wall softener.
**icosahedron.** Here, virus having twenty sides.
**idioplasm.** An old term for that part of the protoplasm that functions in the transmission of hereditary properties. It therefore equates with chromatin.
**imperfect.** Asexual reproduction. In reference to flowers, either the pistil or the stamens are lacking, making the flowers unisexual.

**incomplete.** Refers to a flower lacking one or more of the four whorls of flower parts.
**incomplete dominance.** Referring to the phenotype. A blending of traits from each gene of a pair.
**indehiscent.** Not splitting along predeterminable lines.
**independent assortment.** The inheritance of one pair of traits is independent of the distribution of another pair of traits.
**indusium.** On a fern leaf, an epidermal growth that covers the sorus.
**inferior ovary.** An ovary that is attached to the upper part of the calyx. The other whorls of flower parts arise above the ovary.
**inflorescence.** A cluster of flowers, often on a common receptacle.
**infrared.** Lying outside the visible, red end of the spectrum. Wavelengths are longer than those of visible light.
**initial.** A cell of apical meristem that remains meristematic.
**insectivorous.** Said of plants that "capture" insects and "digest" them.
**integument.** A seed coat. An envelope that encloses the nucellus in an ovule.
**intercellular.** Lying between cells.
**internode.** That portion of a stem that lies between two nodes or buds.
**interphase.** In mitosis, the so-called "resting" stage between mitotic divisions; in meiosis, the so-called "resting" stage between the first and second meiotic divisions. The term *resting* is misleading, however, because many changes occur during this stage.
**intine.** An inner layer of the pollen grain wall.
**intracellular.** Within the cell.
**inversion.** Here, refers to when a chromosome is broken, reverses its end-to-end arrangement, and is mended again.
**invertase.** The enzyme that acts on sucrose, converting the sucrose into a mixture of glucose and fructose.
**ion.** That part of a molecule in solution that possesses an electrical charge.
**isodiametric.** Having dimensions that are equal in all directions. Said to be true of parenchyma cells.
**isogamous.** When fused gametes are of the same size and shape.
**isomorphic.** Here, when the sporophyte and gametophyte are of the same shape and size.
**isotonic.** Having the same osmotic concentration as another solution serving as a source of comparison.
**isotopes.** Elements having differing numbers of neutrons in their nuclei, causing them to have differing weights.

# J

**Jarovisation.** Cold stratification. A cold treatment used to render seeds capable of germination.

## K

**karyokinesis.** The division of a nucleus. Contrast with **cytokinesis**.
**kelp.** Large, brown algae, usually consisting of a blade, a stalk, and a holdfast.
**kinetochore.** Same as **centromere**, but this term implies influencing the movement of the chromosome.

## L

**lamella** (plural, **lamellae**). Here, a cell structure resembling a plate. Lamellae appear to lay one upon the other.
**laminarin.** A polysaccharide product occurring in the brown algae.
**late blight.** The disease of potatoes and tomatoes caused by the fungus *Phytophthora infestans*. So-called because it occurs late in the growing season.
**lateral conjugation.** In algae, a sexual union between neighboring cells of the same filament. Contrast with **scalariform conjugation**.
**layering.** A method of inducing adventitious root production in a stem while it is still attached to the parent plant. Soil is mounded over the stem.
**leaf.** An outgrowth from the stem. Involved in photosynthesis. Generally composed of a flattened, green blade attached to the stem by a petiole.
**leaf trace.** Vascular bundle that connects the vascular tissue of the stem with that of the leaf.
**leaflet.** Part of a compound leaf. Distinguished from the leaf by the fact that there is no bud in its axil.
**legume.** A dry fruit that splits along two sutures. The fruit of the bean family.
**lenticel.** A small, lenslike opening in the bark. Said to allow the passage of gases.
**leptotene.** That stage of the meiotic prophase immediately preceding synapsis and during which the chromosomes appear as fine threads.
**leucoplast.** A colorless plastid.
**leucosin.** A white substance produced by numbers of yellow-green algae and thought to be a carbohydrate.
**lichen.** A union between a fungus and an alga living together symbiotically. Grows on rocks and trees.
**lignin.** A constituent of secondary walls, functioning also as an intercellular cement.
**linkage.** Refers to genes being located on the same chromosome and, thus, passing on their characteristics together.
**lip cells.** In the sporangium of a fern, cells that separate in maturity to allow the escape of spores.
**lipids.** Fats or fatty compounds that are insoluble in water but soluble in certain organic solvents.

**lipopolysaccharide.** An organic compound made up of lipid and sugar components.

**litmus.** A compound extracted from several forms of lichens and possessing the quality of being red in acidic solution and blue in alkaline solution, thus used as an indicator.

**liverwort.** A bryophyte. A small, inconspicuous, nonvascular plant. Derives its name from medieval times, when its shape appeared to resemble the lobes of a liver.

**locule.** The cavity of an ovule or anther.

**locus.** That position on a chromosome associated with a particular genetic trait.

**long-day plant.** A plant requiring comparatively prolonged periods of light and relatively brief periods of darkness in order to initiate floral primordia.

**luciferase.** An enzyme that acts upon luciferin to cause the emission of light.

**luciferin.** A pigment that is found in luminescent organisms and furnishes light during oxidation.

**lumen.** A space enclosed by a cell wall. Usually used in reference to dead cells from which protoplast has disappeared.

**luminescent.** Adapted for the production of light.

**lysis.** The disintegration of cells.

**lysosome.** An organelle bounded by a membrane and containing enzymes capable of breaking down proteins and other molecules.

# M

**maize.** Another name for corn.

**marcotting.** Air layering with rooting medium bound to the stem.

**medullary ray.** A parenchymatous connection between the cortex and the pith of the stem.

**megasporangia.** Sporangia bearing megaspores.

**megaspore.** A spore that develops into a female gametophyte.

**megaspore mother cell.** A diploid cell that, through meiotic divisions, gives rise to megaspores.

**megasporophyll.** A modified leaf bearing a megasporangium.

**meiosis.** A type of cell division that results in a reduced number of chromosomes.

**meiospores.** Spores produced by meiotic divisions.

**meristematic.** Being undifferentiated and capable of cell division. Used in reference to tissues.

**mesophyll.** The parenchymatous tissue of the leaf. Located between the layers of epidermis. Possesses chloroplasts.

**Mesozoic.** Geological era that began approximately two hundred twenty-five million years ago and continued until approximately sixty-five million years ago, during which cycads and evergreen trees were dominant. Angiosperms began to rise toward the end of this era.

**metabolism.** The sum of chemical processes occurring in an organism.

**metaphase.** That phase in mitosis when the chromosomes are arranged on an equatorial plate.

**microfibrils.** Minute fibers visible only with the aid of an electron microscope. Can be seen in cross sections of cilia and flagella.

**micron.** $1/1{,}000$ of a millimeter.

**microorganism.** Any organism requiring a microscope or an ultramicroscope to be seen.

**micropyle.** A pore remaining in the seed coat because of the incomplete closure of the integument, the passageway that leads to the nucellus.

**microsporangia.** Sporangia that bear microspores.

**microspore.** A spore that gives rise to a male gametophyte.

**microsporophyll.** A modified leaf to which a microsporangium is attached.

**microtubule.** Structure present in the cytoplasm of many types of eukaryotic cells and in flagella.

**middle lamella.** The thin zone between two adjoining cells that contains a cementing substance that holds the cells together.

**minimal medium.** A nutrient medium containing only sugar and some salts.

**mitochondria.** Organelles possessing double membranes and containing the enzymes associated with respiration.

**mitosis.** The process of duplication of chromosomes which, upon separation, form two genetically identical daughter nuclei. Usually followed by cytokinesis.

**mitospores.** Spores formed by mitotic divisions.

**mole.** Here, designates a gram molecular weight, the molecular weight expressed in grams.

**molecule.** The smallest part of an element or compound that retains the chemical identity of that element or compound. Molecules are often unions of two or more atoms and sometimes contain very large numbers of atoms, although some molecules are monatomic.

**Monera.** A group of prokaryotic, unicellular organisms. In this text, an independent kingdom.

**monocots.** Seed plants whose embryos possess a single cotyledon.

**monocotyledonous.** Having one cotyledon in the seed. Refers to a group of angiosperms.

**monoecious.** Having both male and female reproductive organs in one organism. In reference to flowers, when both staminate and pistillate flowers are present on the same plant.

**monophyletic.** Deriving from a single origin.

**monosaccharide.** Also called a **simple sugar**. Most commonly composed of hydrogen, oxygen, and either five or six carbon atoms. Often classified by number of carbon atoms (e.g., triose, tetrose, pentose).

**monostromatic.** Composed of a single layer of cells. A thallus one cell in thickness.

**morel.** An edible ascomycetous mushroom.

**moss.** A bryophyte in which the gametophyte generation has a leafy appearance. The sporophyte generation grows from the tip of the gametophyte and lacks chlorophyll.

**mound layering.** A form of plant propagation. Adventitious root growth from stems is induced by heaping soil around the base of the plant.

**m-RNA.** The RNA that carries genetic information from the gene in the nucleus to the ribosomes, where it determines the order of amino acids put together in a polypeptide.

**mucoprotein.** Mucopolysaccharides combined with amino acid units or polypeptides. Occurs in body fluids and tissues.

**multicellular.** Consisting of more than one type of somatic cell.

**multinucleated.** Having several nuclei present in a mass of protoplasm without being separated by cell membranes.

**multiseriate.** Arranged in several to many series.

**mushroom.** Fleshy, fruiting bodies of fungi. Mostly basidiomycetes that arise from an underground mycelium.

**mutation.** A change in gene structure that affects the phenotype. Often is naturally occurring and cannot be anticipated.

**mycelium.** A mass of hyphae that composes the body of a fungus.

**mycology.** The study of fungi.

**mycoplasma.** The smallest of the prokaryotic organisms, lying in a sense between the prokaryotes and the viruses.

**mycorrhiza.** A symbiotic association of plant roots and fungi.

# N

**N.** Represents the haploid number of chromosomes.

**naked seeded.** The seed condition in gymnosperms. Rather than being enclosed in an ovary, the seeds lie on the surface of a bract.

**nanometer.** $10^{-9}$ meters; .000000001 meter; one billionth of a meter.

**necrosis.** Death of plant tissue. Caused by low temperature, fungi, etc.

**nematode.** A roundworm.

**net venation.** The arrangement of the veins in the blade of a leaf. Characteristic of the dicotyledonous plants.

**nitrate.** Characterized by the $-NO_3$, the only form of nitrogen available to a plant.

**nitrogen fixation.** The incorporation of atmospheric nitrogen into nitrogen compounds such as nitrate to make them available to green plants. Carried out by certain microorganisms.

**node.** That part of the stem from which leaves arise.

**nondisjunction.** When replicated chromosomes go to the same pole rather than parting and going to opposite poles during the anaphase of mitosis.

**nonseptate.** Not divided by partitions or cell membranes. Allows a multinucleated condition.

**nucellar embryony.** When sporophytic cells, which surround the embryo sac, grow into new sporophytes within the seed.

**nucellus.** The sporophytic tissue immediately surrounding the embryo sac.

**nuclear membrane.** A double membrane surrounding the nucleus and perforated with minute openings.

**nucleic acid.** A combination of a pentose sugar, a phosphate, and a base. The two kinds are DNA and RNA.

**nucleolar organizer.** On certain chromosomes, an area associated with the formation of the nucleolus.

**nucleolus.** A spherical body found within the nucleus of a eukaryotic cell. Made of RNA and protein.

**nucleotide.** A single unit of nucleic acid. Composed of a phosphate, a 5-carbon sugar, and a base.

**nucleus.** That body within a eukaryotic cell that is bounded by a double membrane and contains the genetic material.

**nutation.** Change in position of growing plant parts due to variation in growth rates on different sides of the apex.

# O

**oogamy.** That type of sexual reproduction wherein the gametes that unite are different; one is the egg, and the other is the sperm.

**oogonia.** In certain algae and fungi, unicellular structures each containing an egg cell.

**oospore.** In the oomycetes, a thick-walled zygote.

**operator gene.** An on-off-switch type of gene that acts to stop the production of m-RNA under certain conditions.

**operculum.** In mosses, the lid that covers the sporangium.

**operon.** A group of adjacent genes that are under the control of a single operator gene.

**order.** The taxonomic division below *class* and composed of one or more families.

**organelle.** A body within the cytoplasm of a cell.

**organic.** Formed by living organisms. Used in reference to compounds and the chemistry of carbon compounds.

**organism.** An individual living creature.

**orthogenesis.** The concept that evolutionary changes are predestined, result in progressive trends, and are independent of natural selection.

**osmosis.** The passage of water through a selectively permeable membrane from a solution of lesser concentration of dissolved substance to a solution of greater concentration.

**osmotic shock.** A laboratory procedure wherein cells or bacteriophage are placed in a concentrated salt solution and then suddenly diluted. Causes the particles to lyse, or become separated.

**ovary.** The enlarged base of the pistil. Matures into the fruit.
**ovule.** That structure within an ovary that contains the female gametophyte with its egg cell, and that, upon fertilization, matures into a seed.
**oxidation.** The loss of electrons from an atom or molecule. In biology, an energy-releasing process.

# P

**pachytene.** The stage of the meiotic prophase that follows zygotene and is characterized by the splitting of paired chromosomes into chromatids.
**Paleozoic.** The geologic era that began approximately five hundred seventy million years ago and lasted approximately three hundred million years. Followed the Precambrian era and preceded the Mesozoic era.
**palisade parenchyma.** In a leaf, that layer of mesophyll (photosynthesizing cells) lying immediately under the upper epidermis.
**palmately compound.** A compound leaf having leaflets that radiate out from a central point.
**parallel venation.** When the veins of a leaf lie parallel. Characteristic of the monocots.
**paramylum.** A reserve carbohydrate resembling starch. Produced by various algae and protozoa.
**paraphyses.** Sterile filaments among the reproductive structures. Found in some algae and in ascomycetes and basidiomycetes.
**parasexual.** A form of conjugation between bacteria wherein only a part of the DNA of one cell is moved to the other. Not considered true sexuality because the transfer of genetic material is incomplete.
**parasite.** An organism that lives on or in a host (another organism of a different species) at the expense of the host.
**parenchyma.** An unspecialized plant tissue having thin walls and being loosely put together, thus having intercellular spaces.
**parthenogenesis.** The development of a new individual from an egg cell without the egg cell being fertilized by a sperm.
**passage cell.** Also called a *transfusion cell*. A thin-walled, unsuberized cell found in the endodermis of roots.
**pectin.** A complex, organic compound occurring in primary cell walls and as a cementing substance between cells in the middle lamella.
**pellicle.** The structurally complex outer membrane of organisms such as *Euglena*.
**penicillin.** An antibiotic drug first obtained from the ascomycetous fungus *Penicillium*.
**pentosan.** A polysaccharide that yields 5-carbon sugars upon hydrolysis. Widely distributed in plants.
**pentose.** A 5-carbon sugar.
**pepsin.** A digestive enzyme that acts on proteins.

**peptide.** Two or more amino acids linked by peptide bonds.

**peptide bond.** A bond between two amino acid units. Linkage is between a carboxyl group and an amino group. Accomplished by a dehydrolysis reaction.

**perennial.** A plant that lives longer than two years.

**perfect.** Reproduction wherein the sexual process occurs. In reference to flowers, having both pistil and stamen.

**perianth.** That part of a flower collectively composed of the calyx and the corolla.

**pericycle.** The parenchymatous tissue lying between the endodermis and the vascular cylinder.

**peridium.** In the myxomycetes, the hardened envelope that covers the sporangium.

**perigynous.** A condition wherein the receptacle is concave so that the flower parts seem to be inserted around the ovary.

**peristome.** In mosses, a membrane that covers the mouth of the sporangium.

**peristome teeth.** A fringe of teeth that results when the peristome dries and splits along radial lines.

**perithecium.** In ascomycetes, a spherical or flask-shaped fruiting structure with a small opening.

**permanent.** In relation to cells that have lost the capacity to divide.

**petal.** That part of the corolla that often is conspicuously colored.

**petiole.** The stalk of a leaf.

**PGA.** Phosphoglyceric acid. A 3-carbon compound formed by the interaction of carbon dioxide and the 5-carbon compound ribulose diphosphate. Formed in the first step of the carbon cycle of photosynthesis.

**PGAL.** Phosphoglyceraldehyde.

**phaenerogam.** Literally, "a visible marriage." Applies to the seed plants.

**Phaeophyta.** The brown algae. The chlorophyll is masked by a brown pigment called **phycocyanin**. The group includes the kelps.

**phage.** *See* **bacteriophage**.

**phelloderm.** Cells formed internally by the cork cambium, or phellogen.

**phellogen.** *See* **cork cambium**.

**phenotype.** The visible characteristics directed by heredity.

**phloem.** The vascular tissue that lies peripheral to the vascular cambium and through which the products of photosynthesis are conducted.

**phlogiston.** A residue left over after burning; therefore, an early name for carbon dioxide.

**phosphate.** A salt of phosphoric acid.

**phosphorylation.** Here, an enzymatic reaction that attaches a phosphate group to a carbohydrate.

**photon.** A unit of light. A quantum of light.

**photoperiodism.** The effect of alternating light and dark periods on the growth of plants and the formation of floral primordia.

**photophosphorylation.** Here, adding a phosphate group to ADP (adenosine diphosphate), thus producing ATP (adenosine triphosphate), the energy coming from light.
**photosynthesis.** The process whereby light is used to bring about the reaction between carbon dioxide and water, resulting in the formation of carbohydrate and the release of oxygen.
**phycocyanin.** A bluish-green pigment occurring in blue-green algae.
**phycoerythrin.** A reddish pigment occurring in red algae.
**phyllotaxy.** The arrangement of leaves on a stem.
**phylum.** The primary taxonomic division.
**phytochrome.** A pale-blue, proteinaceous pigment sensitive to red and far-red light.
**pileus.** The cap of a fleshy fungus, such as the cap of a mushroom.
**pinnate.** The arrangement of either veins in a leaf or leaflets on the rachis wherein there is a single midrib from which smaller veins or leaflets arise.
**pinnately compound.** A compound leaf having the pinnate arrangement of leaflets.
**pistil.** The central organ of flowers. Consists of the ovary, style, and stigma.
**pistillate.** Having a pistil but lacking stamens.
**pit.** A recess or cavity in the cell wall where the secondary wall is interrupted.
**pith.** The parenchymatous tissue that occupies the central portion of the stem.
**placenta.** The portion of the interior of an ovary to which the ovules are attached.
**plant succession.** Plants occupying a certain terrain giving way to other plants over a range of time.
**plaque.** A cleared area in a lawn of bacterial cells. Results from the lysis of the bacterial cells.
**plasma membrane.** *See* **cell membrane**.
**plasmalemma.** *See* **cell membrane**.
**plasmodesmata** (singular, **plasmodesma**). Minute, cytoplasmic threads that extend through openings in cell walls and connect the protoplasts of adjacent living cells.
**plasmodium.** A multinucleated mass of protoplasm surrounded by a membrane. A stage in the life cycle of a myxomycete.
**plasmolysis.** The pulling away of the cell from the cell wall. A consequence of water loss.
**plastid.** An organelle in eukaryotic cells. Serves various functions including food storage and, for chloroplasts, photosynthesis. Bounded by a double membrane.
**pleiotropy.** The capacity of a gene to affect more than one characteristic.
**plumule.** That portion of the young seedling that lies above the cotyledons.
**plurinucleated.** *See* **multinucleated**.
**pneumococcus.** A round bacterium that causes pneumonia.
**polar nuclei.** The two centrally located nuclei in an eight-nucleated embryo sac.

**polarized light.** Light in which the waves occur in a single plane. Such light is obtained by allowing ordinary light to pass through Nicol prisms, which are made of clear calcite.

**pole.** Here, the positioning of chromosomes at the conclusion of mitosis.

**pollen tube.** Tube formed following the germination of a pollen grain, when the grain resides on the stigma of a flower. The tube carries the male gametes to the ovule.

**pollination.** The transfer of pollen from the anther to the stigma of a flower.

**pollination drop.** An exudate of a gymnosperm ovule to which pollen grains adhere and, by way of the shrinkage of the drop, are drawn into the micropyle.

**pollinia.** A mass of waxy material containing pollen grains.

**polycotyledonous.** Having more than two cotyledons in the seed. Characteristic of most gymnosperms.

**polygenic inheritance.** Inheritance wherein several to many genes influence the development of a single trait.

**polymer.** A complex, organic compound composed essentially of repeating structural units (e.g., a starch made of repeating glucose molecules).

**polypeptide.** A polymer of a number of amino acids united through the amino group.

**polyphyletic.** Here, refers to the theory that different kinds of organisms arose independently from separate origins.

**polyploidy.** A cell having more than two sets of chromosomes.

**polysaccharide.** A polymer of repeating sugar units.

**Precambrian.** The geological era that began approximately six hundred million years ago and during which primitive marine life is believed to have originated.

**primary growth.** The growth of a stem or root that arises from the cells at the apex.

**primary pit connection.** In red algae (among the Floridiophyceae), strands of cytoplasm that connect adjacent cells.

**primary wall.** The first cell wall that is formed during the time of cell growth. Because it is formed first, it lies outside of the secondary wall.

**primordial.** Earliest formed.

**procambium.** The primary meristematic tissue that gives rise to primary vascular tissues.

**prokaryotic.** Being without a true nucleus and other organelles such as plastids, Golgi bodies, and mitochondria.

**prop organ.** In bacteriophage, a slender stalk.

**prophase.** An early stage in mitosis during which the chromonemata give rise to chromosomes.

**proplastid.** A cytoplasmic body that develops into a plastid. The precursor of a plastid.

**protein.** A polymer of amino acids.

**prothallium.** A small, flat, green thallus attached to the soil by rhizoids. The gametophyte generation of a fern.

**protist.** Unicellular, nucleated organism.

**protoderm.** The outer cell layer of primary meristem. Gives rise to the epidermis.

**proton.** The nucleus of a hydrogen atom. Also, a particle within the nuclei of other types of atoms.

**protonema.** A first-formed thread. Produced by the germination of a moss spore.

**protoplasm.** A general name for living substance.

**protoplast.** The organized living unit of a single cell.

**provascular tissue.** Tissue located just below the cells of the apical meristem and destined to develop into vascular tissue.

**proximal.** At the site of attachment. Contrast with **distal**.

**pteridophytes.** In an old system of classification, the vascular plants excluding the seed plants.

**purple bacteria.** Free-living bacteria that contain bacteriochlorophyll. Tend to be purplish or, sometimes, reddish in color.

**purple sulphur bacteria.** Bacteria that require hydrogen sulfide as a raw material and give off sulphur. Use light and manufacture carbohydrate.

**pycnium** (plural, **pycnia**). *See* **spermogonium**.

**pycnospore.** Spore produced by a pycnium. Of two kinds: plus and minus.

**pyrenoid.** A body inside a chloroplast and associated with the deposition of starch.

**pyriform.** Pear shaped.

**Pyrrophyta.** The yellow-green algae. Includes the dinoflagellates and cryptomonads.

# R

**rachis.** The central axis of a compound leaf.

**radicle.** The portion of an embryo plant that lies below the cotyledons and forms the embryonic root.

**radioactive.** Exhibiting radioactivity spontaneously, emitting alpha or beta rays and sometimes gamma rays by disintegration of atomic nuclei.

**raphe.** A longitudinal line on the valve of a diatom.

**ray.** Parenchyma cells arranged in a radial formation and running out from the center of the stem. Living ray cells conduct laterally.

**receptacle.** The structure to which the parts of a flower are attached. In organisms of the genus *Fucus*, the bulbous enlargement that is at the end of the stalk and contains conceptacles.

**recessive.** A gene that does not express itself in the phenotype unless homozygous.

**reciprocal cross.** When a cross of A as a pollen source and B as an egg source is reversed, with A becoming the egg source, and B becoming the pollen source.

**reciprocal translocation.** When two chromosomes are broken, and the parts are traded in the mending process.

**recognition site.** The part of a t-RNA molecule that determines which particular amino acid will attach to the molecule.

**reduction division.** The division in the process of meiosis wherein the chromosome number is reduced to the haploid.

**reniform.** Kidney shaped or bean shaped.

**replicate.** Here, the production of a second DNA molecule of the same constitution.

**repressor gene.** A gene that may obstruct the function of an operon, a genetic unit composed of a structural gene and an operator gene.

**reservoir.** A vesicle in cytoplasm that may store resin, oil, or a product of metabolism.

**resin duct.** A channel lined with cells that secrete resin into the channel. Present in pines.

**respiration.** In cells, the process whereby food is broken down and energy is thus liberated. Both anaerobic and aerobic steps are involved.

**resting stage.** A name once given to the interphase of mitosis and meiosis. In a sense a misnomer because many changes occur during this stage.

**rhizoid.** In lower organisms, rootlike extensions leading into the soil or water and capable of functioning in the sense of roots.

**rhizome.** A horizontal stem that grows underground.

**Rhodophyta.** The red algae.

**ribonucleic acid.** *See* **RNA.**

**ribose.** A 5-carbon sugar.

**ribosome.** A particle that lies along the endoplasmic reticulum and is composed of a protein portion and RNA. The site of protein synthesis.

**RNA. Ribonucleic acid.** A nucleic acid that yields the 5-carbon sugar ribose upon hydrolysis. Occurs in cytoplasmic structures and some nuclei.

**root.** Descending axis of a plant. Normally but not always below ground. Serves as an anchor and absorbs water and mineral nutrients.

**root cap.** A mass of dead cells covering and protecting the growing meristematic cells of the root tip.

**root hair.** In a root, a part of an epidermal cell in the zone of maturing cells. The only part of the root capable of taking up water.

**rough endoplasmic reticulum.** That part of the reticulum along which ribosomes are arrayed.

**RuDP.** Ribulose diphosphate. A phosphorylated 5-carbon sugar that has a role in photosynthesis.

**rust.** A basidiomycete fungus disease that must infect two different species of hosts in order to complete its life cycle. Many are economically destructive. Frequently produce a reddish or brownish discoloration on the infected host.

## S

**saprophyte.** An organism that lives on dead organic matter.

**sapwood.** The part of the secondary xylem that is functional in conducting water.

**scalariform conjugation.** Form of conjugation wherein the cells of one filament unite with the cells of another filament that lies parallel to the first. Ladderlike conjugation.

**Schizomycetes.** Class that includes the bacteria, mycoplasma, and rickettsiae.

**Schizophyta.** Phylum of unicellular organisms, each having a relatively simple nuclear structure, lacking a nuclear membrane and nucleolus, and dividing primarily by asexual means.

**scion.** That portion of the shoot that is used in grafting.

**sclereid.** Also called a **stone cell**. A short sclerenchyma cell having heavy lignified cell walls and tubular pits.

**sclerenchyma.** Elongated cells having thick secondary walls. A supporting tissue that is not living at maturity.

**secondary growth.** Growth primarily in diameter and achieved by the cambium.

**secondary pit connection.** In the red algae (among the Floridiophyceae), a second channel, in addition to the primary pit connection, created to a neighboring cell.

**secondary wall.** Cell wall formed after the maturity of the cell and, thus, deposited on the inside of the primary wall. Most often impregnated with cellulose.

**segregation.** The separation of chromosomes during the anaphase of meiosis. The separation of genes.

**selective permeability.** Having the capacity to allow certain substances to pass through and to disallow the passage of others. Characteristic of an osmotic, living membrane.

**self-incompatible.** The pollen of a flower is incapable of fertilizing the ovules of the same flower.

**senescence.** A decline of vigor related to aging.

**sepal.** Part of the outermost floral envelope. Collectively compose the calyx.

**separation layer.** A layer of cells in the abscission zone, the breaking of which causes a leaf to fall.

**septate.** Divided by partitions.

**septum.** A wall or membrane that separates two plant cells or two cavities.

**serpentine layering.** A form of plant propagation whereby a stem is laid partially in a trench, some segments of the stem being under the soil, and other adjacent segments extending into the air.

**seta.** In liverworts and mosses, the stalk that bears the capsule.

**sexual.** The type of reproduction involving a union of cells. The gametes that fuse may be the same in size and shape (isogamy) or different, as with egg and sperm (heterogamous).

**short-day plant.** A plant requiring comparatively prolonged periods of darkness and relatively brief periods of light in order to initiate floral primordia.
**sieve cell.** A long, slender cell having perforations in the wall. One of the elements of phloem.
**sieve plate.** Perforations in the end walls of two connecting sieve cells.
**sieve tube.** Sieve-cell elements arranged end to end to make a conducting tube.
**sieve-tube element.** In flowering plants, one of the components of the sieve tube.
**simple layering.** The method of plant propagation whereby the stem is bent over and the upper portion is placed in contact with the soil. The portion of the stem that touches the soil then takes root.
**simple leaf.** A lateral outgrowth from a stem consisting typically of a single, flattened, green blade joined to the stem by a petiole and often having stipules at the base of the petiole.
**simple sugar.** See **monosaccharide**.
**simple tissue.** A plant tissue composed of only one type of cell.
**Siphonous.** In the green algae, that one of a series of presumed lines of evolution characterized by a multinucleated condition.
**sirenin.** A chemical attractant secreted by female gametes of certain fungi.
**slime mold.** Another name for *myxomycetes*.
**smooth endoplasmic reticulum.** Endoplasmic reticulum that does not have ribosomes lying along its borders.
**smut.** A basidiomycetous fungus. A disease of cereal grasses. Changes plant organs into masses of black spores.
**sol.** A nonviscous colloidal dispersion.
**somatoplasm.** All the cells of a body excluding those in the reproductive tissues.
**soredia.** In lichens, a type of reproductive body. Composed of a number of algal cells surrounded by fungal hyphae.
**sori.** Clusters of sporangia.
**species.** All organisms of one kind. A division of *genus*. The second word of the scientific name.
**sperm.** The male reproductive cell.
**spermatia.** Nonmotile male reproductive gametes.
**Spermatophyta.** The seed-bearing plants.
**spermogonium.** In rust fungi, the structure that produces spermatia.
**spindle.** The cluster of filamentous threads lying within a cell and apparent at the time of mitosis.
**spirillum.** An elongated form of bacterium having tufts of flagella at one or both ends.
**spongy mesophyll.** Also called **spongy parenchyma**. A chloroplast-bearing parenchyma lying in the lower portion of a leaf and having conspicuous intercellular spaces.
**spongy parenchyma.** See **spongy mesophyll**.

**spontaneous generation.** An ancient theory purporting that life arose spontaneously from nonliving precursors.
**sporangia.** Spore-bearing organs.
**sporangiophore.** The stalk on which a sporangium is borne.
**spore.** An asexual reproductive body.
**spore mother cell.** Diploid cell that produces meiospores by meiotic divisions.
**sporophyte.** A plant that produces spores.
**sporophytic budding.** When the sporophytic cells that surround the embryo sac grow to new sporophytes within the seed.
**stalk cell.** In the developing pollen grains of gymnosperms, cells produced by the division of the generative cell.
**stamen.** That part of a flower that bears the anthers and pollen sacs.
**staminate.** Having stamens but lacking a pistil.
**staphylococcus.** A form of coccus bacterium in which the cells occur in small clusters.
**stem.** The main body of the aboveground portion of a plant. Also, the ascending axis, whether above or below the ground.
**sterigma.** On a basidium, the pointed projection that bears the basidiospores.
**stigma.** That portion of a flower that is receptive to pollen. Also used to indicate a light-sensitive spot in an algal cell.
**stipe.** A supporting stalk such as the stalk of a gill fungus or the leaf stalk of a fern.
**stipule.** An appendage at the base of the petiole of a leaf.
**stolon.** A horizontal stem growing above the ground.
**stomate.** An opening in the surface of a leaf and bounded by guard cells.
**stone cell.** *See* **sclereid.**
**streptococcus.** A coccus form of bacteria in which the cells occur in short chains.
**strobilus.** A cone. Composed of a number of modified leaves that become ovule-bearing scales.
**stroma.** The ground substance of plastids.
**structural gene.** One of the two genes that contribute to the formation of an operon.
**style.** An elongated portion of the pistil. Bears the stigma at its upper end.
**suberin.** A waxy substance that occurs in the walls of cork cells.
**sulcus.** A groove or furrow that occurs in the cell walls of organisms of the phylum Pyrrophyta.
**superior ovary.** A floral ovary that is positioned above the insertion point of other floral parts.
**suspensor cells.** In the development of the gymnosperm sporophyte within the seed, extraembryonic cells derived from the divisions of the fertilized egg.
**swarm spore.** A stage in the life cycle of a myxomycete. When the sporangium breaks open, the spores develop into swarm spores.

**symbiont.** An organism that lives in a symbiotic relationship.
**symbiosis.** A close association between two different species that is mutually advantageous.
**synapsis.** A pairing of homologous chromosomes in the early stages of meiosis.
**synecology.** That branch of ecology that focuses on plant communities.
**synergid.** In an angiosperm gametophyte, the cell that lies at the sides of the egg cell.
**synnema.** A cluster of conidiophores having short chains of conidiospores at their tips. Found among the Fungi Imperfecti.

# T

**taproot.** A root having a prominent central portion growing vertically downward.
**taxonomy.** The study of classification.
**teliospore.** In rust fungi, a thick-walled spore in which meiosis follows the union of gametic nuclei.
**telophase.** That phase in mitosis following anaphase, when chromosomes arrive at the poles.
**template.** A pattern upon which another thing is formed. Used here in reference to a molecule.
**terrestrial.** Dwelling on land.
**test cross.** *See* **back cross**.
**tetrad.** A group of four spores formed from a spore mother cell by meiotic divisions.
**Tetrasporine.** A theoretical line of evolution among the green algae comprising a uniseriate series of uninucleated cells. Presumed to be the ancestral form of the vascular plants.
**tetrasporophytes.** In the life cycle of red algae, spores that are produced by diploid carpospores and reproduce by the production of tetraspores.
**thallophyte.** Member of the phylum Thallophyta. Characterized by a thallus, a plant body lacking specialized conducting tissues and displaying simplicity of form.
**thallus.** A plant body that lacks specialized conducting tissues, roots, stems, and leaves.
**thylakoid.** The lamellae of the grana that occur within the chloroplast.
**torus.** A thickened portion of primary wall lying at the center of a pit.
**totipotency.** When all the cells of a body (the somatic cells) have a complete set of genes. If any one of them recovers the capacity to divide, it can result in the formation of a complete individual.
**tracheid.** An elongated, thick-walled, conducting and supporting cell of xylem.
**Tracheophyta.** The vascular plants.

**translocation.** A segment of a chromosome breaking away from its parent member and attaching to a different chromosome.
**transpiration.** The loss of water on the part of a plant.
**trench layering.** A vegetative method of plant reproduction wherein a cane or stem is placed in a trench and covered with soil.
**trichogyne.** A receptive emergence from an oogonium and through which spermatia may travel to the egg nucleus. Occurs in both ascomycetes and red algae.
**tripeptide.** Three amino acids linked by peptide bonds.
**triplet.** A codon. The three bases that represent instruction to use a particular amino acid in the construction of a protein.
**t-RNA.** Transfer RNA. A type of RNA that functions as a carrier molecule to bring amino acids to the ribosomes, where they may then be assembled into polypeptides.
**truffle.** A subterranean mushroom belonging to the Ascomycetes class.
**tube nucleus.** The one of two nuclei that result when a pollen grain germinates on the stigma of a flower and that then leads down the pollen tube.
**tuber.** A thick, fleshy terminal part of a stem, usually formed underground.
**turgor.** The state of turgidity, or tension, in living cells. The distension of the protoplasm against the cell wall, caused by the cell's fluid content.
**turgor pressure.** The pressure that develops because of the fluid in a turgid plant cell. Results from water flowing into the cell.
**tylose.** The intrusion of the cytoplasm of one cell into a neighboring cell, resulting in obstruction of the cavity of the second cell.

# U

**ultracentrifugation.** Very high-speed centrifugation that makes possible the sedimentation of colloidal and other small particles. Used in the stratification of cell components.
**ultraviolet.** Beyond the visible spectrum at the violet end. Having a wavelength shorter than that of visible light.
**unicellular.** Composed of a single cell.
**uninucleated.** Having a single nucleus.
**uniseriate.** In a single row of cells.
**unit characters.** One of Gregor Mendel's laws of heredity, which declares that each characteristic is transmitted as a single and unchanging unit.
**universal veil.** Veil that surrounds a mushroom when the mushroom is still in a button stage.
**urediniospore.** Also called a **uredospore**. In the red-cell stage of the life cycle of a rust fungus, a one-celled, summer spore.
**uredospore.** See **urediniospore**.

## V

**vacuole.** A cavity that is within the cytoplasm, filled with a watery fluid, bound by a membrane, and considered to be nonliving.

**vascular bundle.** A strand of tissue containing both primary xylem and primary phloem and frequently enclosed in a bundle sheath of parenchyma or fibers.

**vascular cambium.** A one-cell-thick zone that is capable of mitotic divisions and gives rise to secondary xylem and secondary phloem.

**vegetative propagation.** The means of reproduction that uses body tissues. Reproduction without sex.

**venation.** The arrangement of the veins in a leaf.

**venter.** The basal end of an archegonium containing an egg.

**vernalization.** Exposing seeds to an interval of cold with the goal of inducing growth and flowering.

**vesicle.** Small, membrane-bound sacs that make up the Golgi complex. The vesicle functions as a packing station.

**vessel.** A tubelike structure in the xylem. Composed of cells positioned end to end and having perforated end walls.

**virulent.** Capable of causing disease.

**virus.** An ultramicroscopic particle consisting of protein and either DNA or RNA.

**Volvacine.** One of three theoretical lines of evolution of the green algae. Characterized by *Chlamydomonas*-type cells, which may cluster into colonies.

## W

**wheat rust.** A basidiomycete disease that infects wheat and barberry plants. Requires different hosts to complete its life cycle.

**wood.** The secondary xylem produced by the vascular cambium.

**wound hormone.** Theoretical hormone thought to help enable cells that have become permanent to regain the capacity to divide following wounding.

## X

**xanthophyll.** A carotene-like, yellow pigment present in chloroplasts associated with chlorophyll.

**xylem.** A conducting tissue of plants through which most of the water and minerals move.

## Y

**yeast.** An ascomycete fungus. Generally reproduces by budding and often occurs as single cells.

## Z

**zeatin.** A cytokinin isolated from corn.

**zoospore.** A motile spore.

**zygomorphic.** Capable of being divided into similar halves by a single plane. Bilaterally symmetrical. Said with regard to flowers.

**zygospore.** A zygote that results from the fusion of isogametes, possesses a thick wall, and goes into a resting stage. Also, a zygote that becomes encysted. Found in organisms of the genus *Spirogyra*.

**zygote.** A fertilized egg resulting from the fusion of two gametes.

**zygotene.** The synaptic stage of meiosis during which homologous chromosomes are paired.

**zymase.** A complex of enzymes present in yeasts, bacteria, and higher plants. Facilitates glycolysis, the breakdown of sugar.

# Index

**Note:** Page numbers in **bold type** reference non-text material

Abscisic acid (ABA), 272
Abscission layer, leaf, 312
*Acetabularia*, 166, **167**
*Acetabularia crenulata*, 167
*Acetabularia mediterranea*, 167
Acetic acid, chemical formula for, **40**
Achene, described, 327
Acids
  amino, chemical formula for, 39, **40**
  organic, chemical formula for, 39, **40**
Actinomorphic flowers, 321, **322**
Active transport, defined, 87
Adaptation, ecology and, 353–54
Adder's tongue, 229
Adenosine triphosphate. *See* ATP
*Adiantum*, 229
Aerobic
  bacteria, 110–11
  pathways, respiration and, 103–5
*Albugo candida*, 180, **181**
Alcohol, chemical formula for, 39
Algae. *See* Specific type of algae
*Allomyces arbuscula*, 173, **175**
Alpha–glucose, chemical formula for, **44**
Amide dinucleotide phosphate (NADPH), photosynthesis and, 96
Amino acids
  chemical formula for, 40, **41**
  DNA and, 67–68
Ammonia, chemical formula for, 37
*Amorphophallus titanum*, 320, 321
Amphimixis, 333
*Anabaena*, 122
Anaerobic pathways, respiration and, 103–5
Anaphase, mitosis and, 23

*Anchusa officinalis*, doctrine of signatures and, 213
Aneuploidy, cellular aberrations and, 31
Angiosperms, 255–62
  gymnosperms compared to, 261
  life cycle of, 255–59
  lilies and, 260–61
  xylem/phloem in, 238–42
Animal cell, illustrated, **6**
Anisogamous sexual reproduction, 129
Antheridiophores, *Marchantia* and, 222
Antheridium, 133
*Anthoceros*, 225
Anthocyanins, 18
Antipodal cells, angiosperms and, 256
Apetalous flowers, 321
Apical meristem tissue, 235
Apogamy, 334
Archegoniophores, *Marchantia* and, 222
Aristotle, on creation of life, 343
Ascomycetes, 169, 183–93
  ergot and, 192–93
  fruiting bodies and, 186–87
  morels and, 191–92
  pathogenic, 189–91
  reproduction in, 184–85
  truffles and, 191–92
  yeasts and, 187–89
Asparagus, *chadophylls* and, 297
ATP (Adenosine triphosphate)
  manufacture of, 8
  molecule, 102
  photosynthesis and, 96
Attachment site, t-RNA and, 70
Autumn crocus, 57
Auxin, 263–69

  obtaining, 269
  uses of, 268
Avery, O.T., search for DNA and, 63

Bacillariophyceae, 160–63
*Bacilli*, 111
Back cross, described, 55
Bacteria, 361–62
  aerobic, 110–11
  benefits of, 112
  characteristics of, 111
  growth of, 113–15
  hazards of, 112–13
  identifying, 113
  original, 110
Bacterial transformation, Fred Griffith and, 61–62
Bangiophycidae, 149–51
Basal body, described, 17
Basidiomycetes, 169, 195–205
  mycorrhiza, 204–5
  puffballs, 203
  rusts, 195–200
  smuts, 200–203
Bateson, William, on segregation and gene interaction, 346
Beal, W.J., seed longevity research by, 331
Beans seed, longevity of, 331
*Berberis vulgaris*, 197
Berry, described, 327
Beta-glucose, chemical formula for, **44**
Biological clocks
  circadian rhythms, 275–77
  *Gonyaulax polyedra* and, 277
Bivalents, meiosis and, 27
Black, Joseph, photosynthesis research and, 90
Black bread mold, 175–76

Blackmann, F.F., photosynthesis research and, 93–95, 98
Bladderwort, 356
Blast fibers, 238
Blue-green algae, 119–22, 346
  characteristics of, 119–20
  types of, 120–22
Bordered pits, gymnosperms and, 242
Boysen-Jensen, Peter, oat seeding experiments by, 263–64
Bread mold, black, 175–76
Brown algae, 119, 143–48
  parthenogenesis and, 334
  products of, 144
  reproduction in, 144–48
Brown rot, 112
  plum, 204
Bryophytes
  green algae or, 140
  hornworts, 225
  liverworts, 221–25
  mosses, 225–27
Bud graft, 340
Buffon, Comte de Georges Louis LeClerc, on evolution, 343
Bugloss, doctrine of signatures and, 213
Bulbs, described, 295
Butane, chemical formula for, **38**
Butcher's broom, cladophylls and, 297

C4 plants, photosynthesis and, 99–100
Calvin, Melvin, photosynthesis research and, 98–99
Calvin cycle, photosynthesis and, 98–99
Calyptra, mosses and, 227
CAM carbon pathway, photosynthesis and, 100
Cambium tissue, 235
*Camptosorus*, 229
Carbohydrates, chemical formula for, 42–45
Carbon cycle, 106
Carbon dioxide, chemical formula for, **37**
Carboniferous period, 245
*Carina*, defined, 233

Carolus Linnaeus
  circadian rhythms and, 275
  on creation of life, 343
  plant classification and, 215–17
Carotenes, 18
Casparian strip, 303
Catkins, 321, **322**
*Caulerpa floridana*, described, 20
Cavendish, Henry, photosynthesis research and, 90
*Celandine*, doctrine of signatures and, 213
Cells
  aberrations in composition of, 31–35
  chloroplasts and, 15–17
  cilia and, 17
  doctrine of, 6–8
  endoplasmic reticulum and, 11–12
  Golgi bodies and, 10–11
  illustrated, **6, 7**
  membrane, 13–14
  nuclear, 12–13
  mitochondria and, 8–10
  plastids and, 18
  vacuoles and, 18–20
  walls of, 14–15
Cellulose, plant cells and, 10
Centrales, 161, 162
Centromere, mitosis and, 23
*Cercospora*, 208
*Cetraria islandica*, 213
*Chamaesiphon*, 122
Chemistry
  formulas,
    alcohol, 39
    amino acids, 40, **41**
    carbohydrates, 42–45
    empirical/structural, 37–38
    lipids, 45–47
    organic acids, 39, **40**
    polymers, 40, **41**
    proteins, 41, **42**
Chiasmata, meiosis and, 28
*Chlamydomonas*, 125–28
*Chlorella*, photosynthesis research and, 94–95
Chlorenchyma, 236
*Chlorococcum*, 212
Chlorophyll

chloroplasts and, 15, 17
  laboratory experiments and, 3
  photosynthesis and, 93
Chlorophyta, 119
Chloroplasts, cells and, 15–17
Chromatids, meiosis and, 27
Chromoplasts, 18
Chrysolaminarin, defined, 161
*Chrysophyceae*, 160
Chrysophyta, 119, 160–63
  algae, 19
Chytridiomycetes, 169, 173–75
*Chytridium sphaerocarpum*, 173, **174**
Cilia, cells and, 17
Circadian rhythms, 275–77
*Cladonia mitis*, 213
Cladophylls, butcher's broom/asparagus and, 297
Clamp connection, puffballs and, 203
*Claviceps purpurea*, 362
Climate, ecology and, 354–55
*Clostridium*, food poisoning and, 362
Club
  fungi, 195–200
  mosses, 230–32
  root, 171
*Cocci*, 111
*Codium*, 139
*Colchicum autumnale*, 57
Cold stratification, germination and, 333
Collenchyma, 235, 236
Colloids, protoplasm and, 82–83
Columella, black bread mold and, 176
Complex tissues, 238–43
Coniferales, 246–51
Conifers, 245
Continuity of germ plasm, 50
Cork, 235
Cork cambium tissue, 235
Corms, described, 295
Corn smut, 200–201
Corolla, 255
Correns, Karl Erich, Mendel's experiments and, 51
*Corynebacterium sepedonicum*, 112
Crick, F.H.D., DNA research and, 63

Cristae, described, 9
*Croococcus*, 212
Crossing over, meiosis and, 28–29
*Cryptomonads*, 166
*Cryptophyceae*, 166
Cultivated plants, 360–61
*Cuscuta gronovii*, 356
Cuttings propagation, 338–39
Cuvier, Georges, on evolution, 343–44
*Cyanobacteria*, 120
Cyanophyta, 119
Cyanophytes, 119–22, 346
Cyclosporine, 362
Cytokinins, 271–72

Dark reactions, photosynthesis and, 95
Darwin, Charles
  on evolution, 344
  heliotropism and, 263
  *On the Origin of Species*, 217
de Candolle, Alphonse, origins of cultivated plants and, 360
de Mairan, Jean Jacques, circadian rhythms and, 275
de Saussure, Nicholas Theodore, photosynthesis research and, 90–91
DeBary, Henrich, *Puccinia graminis fungus* and, 195–96
Dehiscent, mosses and, 227
Dehydrogenase, dehydrogenation and, 105
Deletion, cellular aberrations and, 31
Deoxyribonucleic acid (DNA). See DNA (Deoxyribonucleic acid)
Deoxyribose, described, 60
DeVries, Hugo
  Mendel's experiments and, 51
  theory of mutations and, 345
Diakinesis, meiosis and, 28
Dialysis, defined, 87
Diastase enzymes, 19
Diatomaceous earth, 163
*Diatoms*, 160–63
Dicot stems
  herbaceous, 292
  woody, 283–91
Dictyosomes, plant cells and, 11

Diffusion, protoplasm and, 84–87
*Dinoflagellates*, 164–65
*Dinophyceae*, 164–65
Dioecious, 131
*Dionaea muscipula*, 356
Dioscorides, plant classification and, 215
*Diplococci*, 111
Diploid number, 21
Diplotene, meiosis and, 28
DNA (Deoxyribonucleic acid)
  amino acids and, 67–68
  enzymes and, 71–72
  functions of, 65–67
  gene repression and, 73–74
  laboratory experiments and, 3
  mutations in, 72–73
  search for, 59–63
  structure of, 63–65
  transfer RNA and, 69–71
Doctrine of signatures, lichens and, 213
Dodder, 356
Double fertilization, angiosperms and, 259
*Drosera rotundifolia*, 356
*Drosophila melanogaster*, genetic mapping and, 30
Drupe, described, 327
*Dryopteris*, 229
Duckweed, 319
Duplication, cellular aberrations and, 31, 32
Dyes, lichens and, 213

*E. coli*, 114–15
  T-2 bacteriophage and, 76
Ecology
  adaptation and, 353–54
  climate and, 354–55
  environment and, 354
  global-warming and, 355–56
  interrelationships, 356–57
  natural recycling in, 357
  plant, 353
  plant succession and, 357–58
*Ectocarpus*, 144, **145**
Ectotrophic association, Mycorrhiza and, 204
Eimer, Theodor, straight line evolution and, 345

Electron transfer, photosynthesis and, 96–98
Elm, seed longevity of, 331
Endodermis, 301
Endoplasmic reticulum, 10–11
Endosperm, Coniferales and, 248
Endospores, described, 113–14
Endosymbiosis, described, 9–10
Endotrophic association, Mycorrhiza and, 204
Englemann, T.W., photosynthesis research and, 93
*Enquiry into Plants*, 215
Environment, ecology and, 354
Enzymes, DNA and, 71–72
Epidermis, 235
Epigynous flowers, 322, **323**
Epiphytes, 356
Epitheca, *Bacillariophyceae* and, 160
Equatorial plate, mitosis and, 23
*Equisetum arvense*, 233
*Equisetum gigantium*, 232
Ergot, 192–93
Ergotism, 362
*Escherichai coli*. See E. coli
Ethane, chemical formula for, **38**
Ethyl alcohol, chemical formula for, **39**
Ethylene, 269–70
*Eudorina*, 128
*Euglena*, 158–60
Euglenophyta, 119, 157–60
Eukaryotic
  cells,
    described, 8–9
    Golgi bodies and, 10
  life, 347
*Euonymus europaeus*, plant classification and, 215
*Euphrasia officinalis*, doctrine of signatures and, 213
Evergreen, described, 249
*Evernia prunastri*, 213
Evolution, 346–50
  changes in early thought about, 343–44
  Charles Darwin and, 344
  theories of, 344–46
  theory of, 217

Exoenzymes, described, 113
*Experiments on Air*, 90
*Experiments on Vegetables, Discovering Their Great Power in Purifying the Common Air in Sunshine, but Injuring it in the Shade or at Night*, 90
Extraembryonic cells, seeds and, 328
Eyebright, doctrine of signatures and, 213

Fat molecule, chemical formula for, **45**
Fatty acid, chemical formula for, **45**
Fermentation, 101–6
Ferns, 229–30
Finger and toe disease, 171
Fission, bacteria and, 113
Flagella, mitochondria and, 9
Floridiophyceae, 151–53
Florigen, 272–73
Flowers
  evolutionary modifications in, 324
  formation of, 317–19
  variations in, 319–24
Follicle, legumes and, 327
Formic acid, chemical formula for, **39**, **40**
Formulas, chemistry. *See* Chemistry, formulas
Four O'clock's, incomplete dominance and, 56
Fret membranes, grana and, 15, 17
Fructose, chemical formula for, 43
Fruit
  fly, gene mapping and, 30
  forms of, 327
Frustules, *Bacillariophyceae* and, 160
*Fucus*, 147–48
  parthenogenesis and, 334
Fungi, 169–70, 361–62
  Ascomycetes, 183–93
  Basidiomycetes, 195–205
  Imperfecti, 207–10
  Phycomycetes, 173–82
Fungi Imperfecti, 169

Galactose, chemical formula for, 43
Gel, defined, 83

Gene, repression, 73–74
Generative nucleus, angiosperms and, 257
Genetic mapping, A.H. Sturtevant and, 30
Genetics
  applying, 55–57
  Gregor Mendel and, 50–51
    experiments by, 52–55
    genetic laws of, 51
  pre-Mendelian, 49–50
Geologic time chart, 348
  plant forms and, 350
Germ plasm, 50
Germination, seed, 332–33
Gibberellins, 270–71
*Ginkgo biloba*, 252–53
Global-warming, ecology and, 355–56
*Gloeocapsa*, 121, 212
Glucose, chemical formula for, **43–45**
Glycerol, chemical formula for, **45**
Glycine, chemical formula for, **41**
Glycolysis, respiration and, 103
Golgi, Camillo, Golgi bodies discovered by, 10
Golgi bodies, 6, 10–11
*Gonium*, 128–29
*Gonyaulax*, 165
*Gonyaulax polyedra*, 277
Grafting propagation, 339–40
Gram, Hans Christian, bacteria identification and, 113
Grana
  chlorophyll and, 15
  fret membranes and, 15, 17
Grasses, emergence of, 349
Great Potato Famine, 181, 362
Green algae
  Acetabularia, 166, **167**
  Bryophyta or, 140
  Siphonous line, 139
  Tetrasporine line, 131–38
  Volvacine line, 125–31
Griffith, Fred, bacterial transformation and, 61–62
Ground meristem, 283
Guard cells, 308–10
Guttation, leaves and, 310
Gymnosperms, 245–53

angiosperms compared to, 261
classifying, 246–53
*Coniferales*, 246–51
*Ginkgo biloba*, 252–53
xylem/phloem in, 242–43

Hammerling, Joachim, *Acetabularia* and, 167
Haploid number, 21
Haustoria, black bread mold and, 176
Heliotropism, 263
  defined, 263
Hemlock, germination of, 333
Herbaceous dicot stems, 292
Heteroecism, 199
Heterozygous, defined, 53
Homozygous, defined, 53
Honeysuckle, germination of, 333
Hooke, Robert, cells and, 5
Hormones, 263–74
  Abscisic acid (ABA), 272
  auxin, 263–69
  Cytokinins, 271–72
  Ethylene, 269–70
  Florigen, 272–73
  Gibberellins, 270–71
  Photoperiodism, 272–73
  Phytochrome, 273
  Wound hormone, 273
Hornworts, 225
Horsetails, 232–34
Human life, emergence of, 349
Hybrids, described, 52
Hydathodes, 310
Hydrogenation, respiration and, 105, **106**
Hypertonic solution, 85
Hyphae, black bread mold and, 176
Hypogynous flowers, 322, **323**
Hypotheca, *Bacillariophyceae* and, 160
Hypotonic solution, 85

Idioplasm, 345
Indian Pipe, 356
Indusium, ferns and, 229
Inflorescent flowers, 321
Ingenhousz, Jan, photosynthesis research and, 90

Inheritance of acquired characteristics, 49
Insectivorous plants, 356–57
Interphase of mitosis, 22
Interrupted fern, 229
Inversion, cellular aberrations and, 31
Invertase enzymes, 19
Irish potato, 295
Irish Potato Famine, 181, 362
iso-butane, chemical formula for, **38**
Isodiametric, 236
Ivanoski, D., virus research and, 75

Janssen, Zacharias, 5
  microscope invented by, 1
Janzen, Daniel, on epiphytes, 356
Jarovisation, described, 333

Kelp, 143
Kepler, Johannes, microscope improved by, 1
Kinetochore, mitosis and, 23
Krebs, Hans
  Krebs cycle, 105

Lamarck, Jean
  on evolution, 343
  genetics and, 49
Lamellae, chlorophyll and, 15
*Laminaria*, 145–47
Laminarin, 143
Lavoisier, Antoine Laurent, photosynthesis research and, 90
Layering propagation, 338, **339**
Leaf trace, 289
Leaves
  classification/identification of, 311–14
  guttation and, 310
  shot hole disease of, 112–13
  simple versus compound, 307
  structure of, 311
  transpiration and, 308–10
  transplanting, 310
*Lecanora tartarea*, 213
Legume, described, 327
*Lemma minor*, 319

Leptotene, meiosis and, 28
Leucoplasts, 18
Lichens, 169, 211–14
Life
  modern-day theory of, 2–3
  over time, 349
  spontaneous generation of, 1
  viewed through a microscope, 1–2
Light
  photosynthesis and, 93–95
  reactions, photosynthesis and, 95
Lignin, cells and, 14
Lilies, 260–61
Lipids, chemical formula for, 45–47
*Liriodendron*, 324
Liverworts, 221–25
*Lobaria pulmonaria*, doctrine of signatures and, 213
Luciferin, dinoflagellates and, 165
*Lycopodium*, 230–32
Lysosomes, described, 11

Magnolia, seed longevity of, 331
Maidenhair
  fern, 229
  tree, 252
Maize, *Ustilago zeae* and, 200–201
Mannose, chemical formula for, 43
*Marchantia*, 221–25
Margulis, Lynn
  blue-green algae research and, 119
  on mitochondria, 9–10
*Marsilia*, 229
*Materia Medica*, 215
Matthei, Heinrich, m-RNA and, 67–68
Medullary rays, 292
Megasporangia, Coniferales and, 247
Megaspore mother cell, Coniferales and, 247
Megasporophylls, Coniferales and, 247
Meiosis
  basics of, 26–30
  daughter nuclei and, 21
Membranes, nuclear, 12–13
Mendel, Gregor, 50–51, 346

experiments of, 52–55
genetic laws of, 51
Mendelian genetics, 49–58
Meristematic tissue, 235
Metaphase, mitosis and, 23
Methane, chemical formula for, **38**
Methyl alcohol, chemical formula for, **39**
Micropyle, angiosperms and, 257
Microsporangia, Coniferales and, 246
Microsporophylls, club mosses and, 230
Middle lamella, cells and, 14
Miescher, Friedrich, nucleic acid and, 59
Milkweed flowers, pollinia and, 324
Miller, Stanley, life experiment by, 3
Minerals, plant nutrition and, 279–80
*Mirabilis jalapa*, incomplete dominance and, 56
Mistletoe, 356
Mitochondria, 6
  described, 8–10
Mitosis
  aberrations in, 31–35
  basics of, 21–23
  plant, 24–26
Modified stems, 295–97
Monera, 119–22
Moniliales, 208
Monocot stems, 293–94
*Monotropa uniflora*, 356
Morgan, Thomas Hunt, on genes and chromosomes, 346
Mosses, 225–27
m-RNA, 67–68
Mutations
  in DNA, 72–73
  term coined by Hugo DeVries, 51
Mycoplasm, described, 7–8
Mycorrhiza, 204–5
Myxomycetes, 170

NADPH (Nicotine amide dinucleotide phosphate), photosynthesis and, 96
Nageli, Karl Wilhelm, theory of evolution by, 344–45
n-butane, chemical formula for, **38**

Needham, John, on proving spontaneous generation, 2
*Nelumbo nucifera*, seed longevity of, 331
Nematodes, 361
Net venation, leaf, 312
Nirenberg, Marshall, m-RNA and, 67–68
*Nitrobacter*, 112
*Nitrosomonas*, 112
*Noctiluca*, 165
Nondisjunction, cellular aberrations and, 31
*Nostoc*, 122, 212
Nucellar embryony, 333
Nuclear membranes, 12–13
Nucleic acid
 Friedrich Miescher and, 59
 nucleotides and, 60
Nucleolar organizer, mitosis and, 23, **24**
Nucleotides, nucleic acid and, 60
Nucleus, defined, 6
Nutrition, plant, 279–80

*Ochrolechia tartarea*, 213
*Oedogonium*, 133–35
*Oenothera biennis*, 33
*Oenothera lamarckiana*, Hugo DeVries experiments with, 51
*On the Causes of Plants*, 215
*On the Origin of Species*, 217
*On the Purification of Air by Plants*, 90
Oogamy, 131
Oomycetes, 169, 178–81
Operculum, mosses and, 227
*Ophioglossum*, 229
*Ophioglossum vulgatum*, 21
Organelles, cells and, 6
Organic acids, chemical formula for, 39, **40**
Orthogenesis, 345
*Oscillatoria*, 121
Osmosis, 85–87
*Osmunda*, 229
Ovules, described, 255

Paal, Arpad, plant curvature and, 265

Pachytene, meiosis and, 28
*Pandorina*, 128–29
Parallel venation, monocot stems and, 293
Parasexual process, described, 115
Parenchyma, 235
Parenchyma cells, 236
Parthenogenesis, 334
Pasteur, Louis, on proving spontaneous generation, 2
Pathenogenetic egg cell development, 334
Pathogenic
 ascomycetes, 189–91
 fungi, 362
Pears, germination of, 333
Pectin, cell walls and, 14
*Pectobacterium carotovorum*, 112
*Peltigera canina*, doctrine of signatures and, 213
Pennales, 161
Pentane, chemical formula for, **40**
Perianth, flowers and, 320
Pericycle, 301
Perigynous flowers, 322, **323**
Petals, described, 255
Phaeophyta, 119, 143–48
 products of, 144
 reproduction in, 144–48
Phelloderm, cork cambium and, 235
Phenotype, defined, 53
Phloem
 angiosperms and, 238–42
 in gymnosperms, 242–43
*Phoradendron sp.*, 356
Phosphorylation, 103
Photoperiodism, 272–73
Photosynthesis
 C4 plants and, 99–100
 Calvin cycle and, 98–99
 CAM carbon pathway and, 100
 chlorophyll and, 93
 electron transfer and, 96–98
 light and, 93–95
 research on,
  early, 89–92
  modern-day, 82
 respiration and, 102, **103**
Phycomycetes, 169, 173–82
 Chytridiomycetes, 173–75

Oomycetes, 178–81
 Zygomycetes, 175–77
Phyllotaxy, leaf, 314
Phytochrome, 273
*Phytophthora infestans*, 180–81, 362
*Pinus aristata*, leaves of, 249
*Pinus strobus*, leaves of, 249
Pistal, described, 255
Pistillate, flowers and, 320
Pitcher plant, 356
Pits, tubular, 237
Plant cells
 cellulose and, 10
 dictyosomes and, 11
 illustrated, **6**, **7**
Plantae, 119
Plants
 classification of, 215–19
  Carolus Linnaeus and, 215–17
  early efforts of, 215
  problems in, 217–18
  system of, 218–19
  theory of evolution and, 217
 cultivated, 360–61
 ecology of, 353
 human uses of, 360
 mitosis in, 24–26
 nutrition for, 279–80
 succession of, 357–58
 viruses, 79–80
Plasmalemma, described, 13
Plasmodesmata, described, 14
*Plasmodiophora brassicae*, 171
Plasmodium, myxomycetes and, 170
Plastids, cells and, 18
Pleiotropy, described, 57
*Pleodorina*, 128
*Pleurococcus*, 131
Plum brown rot, 204
Pneumonia, *Lobaria pulmonaria* and, 213
Polar nuclei, angiosperms and, 256
Pollen tube, Coniferales and, 248
Pollination drop, Coniferales and, 248
Pollinia, milkweed flowers and, 324
Polygenic inheritance, described, 56
Polymers, chemical formula for, 40, **41**
Polyploidy, plant mitosis and, 26

*Polysiphonia*, 152, **153**
*Polytrichum*, 226
Population, feeding an increasing, 359–60
*Porphyra*, 149–51
Potatoes, 295
 Mycorrhiza and, 204
 Phytophthora infestans and, 180–81
Priestley, Joseph, *On the Purification of Air by Plants*, 90
Primary
 growth, described, 249
 pit connections, Floridiophyceae and, 151
Procambium, 238–39
Prokaryotic, 347
 described, 8
Propagation
 cuttings, 338–39
 division and, 337
 grafting, 339–40
 layering, 338, **339**
Propane, chemical formula for, **38**
Propionic acid, chemical formula for, **40**
Propyl alcohol, chemical formula for, **39**
Proteins, chemical formula for, 40, **41**
Protoderm, 283
Protonema, mosses and, 226
Protoplasm
 colloids and, 82–83
 composition of, 81
 defined, 6
 diffusion and, 84–87
Provascular tissue, 283
*Pseudomonas solanacearum*, 112
Pteridophytes
 club mosses, 230–32
 ferns, 229–30
 horsetails, 232–34
*Puccinia graminis* fungus, 195–97
Puffballs, 203
Purple sulphur bacteria, C.B. van Neil and, 91–92
Pycnia, *Berberis vulgaris* and, 198
Pyrrophyta, 119, 164–67

Rabies, *Peltigera canina* and, 213
*Rafflesia*, 320
Recognition site, t-RNA and, 70
Red algae, 149–53
Redi, Francesco, spontaneous generation and, 1
Respiration, 101–6
 aerobic pathways and, 103–5
 anaerobic pathways and, 103–5
 hydrogenation and, 105, **106**
 photosynthesis and, 102, **103**
Resting stage of mitosis, 22
Rhizomes, described, 295
*Rhizopus nigricans*, 175–76
Rhodophyceae, 149–53
Rhodophyta, 119
*Rickettsiae*, food poisoning and, 362
Ring rot, 112
RNA, DNA patterning of, 67
*Roccella*
 *fuciformis*, 213
 *tinctoria*, 213
Rocky Mountain Spotted Fever, 362
Root
 Casparian strip, 303
 growth, 304
 contributors to, 299
 hairs, 300–301
 structure of, 301–2
Rough endoplasmic reticulum, 11
Rusts, 195–200

*Salmonella*, food poisoning and, 362
*Salvinia*, 229
*Saprolegnia*, 178–80
*Sarracenia flava*, 356
Scheiner, Christoph, microscope improved by, 1
Schleiden, Matthias, 7
 cell doctrine and, 6
Schwann, Theodor, cell doctrine and, 6
Sclereids, 237
Sclerenchyma, 235
 kinds of, 237
Secondary growth, described, 249–50
Secondary pit connections, Floridiophyceae and, 151

Seed
 functions of, 330–31
 germination of, 332–33
 longevity of, 331, **332**
 plants, emergence of, 347, 349
 reproduction of, 333–34
 structure/characteristics of, 327–30
 variations in, 331
Sepals, 255–56
Separation layer, leaf, 312
Sexual reproduction, anisogamous, 129
Shield fern, 229
Shoestring fern, 229
Shot hole, described, 112–13
Siberia, winter wheat in, 333
Sieve plates, 289
Sieve tubes
 angiosperms and, 241
 elements of, 241
Silkworm disease, 204
Simple tissue, 235–38
Singer, Fred, on global-warming, 355
Siphonous line of green algae, 139
Slime mold, *Plasmodiophora brassicae*, 171
Smooth endoplasmic reticulum, 11
Smuts, 200–203
Snake bites, *Anchusa officinalis* and, 213
Soft maple, seed longevity of, 331
Sol, defined, 83
*Solanum tuberosum*, 180–81, 204, 295
Spallanzani, Lazzaro, on proving spontaneous generation, 2
Species, defined, 219
Spermatia, *Porphyra* and, 151
Spermogonia, *Berberis vulgaris* and, 198
Spindle fibers, mitosis and, 23
*Spirilli*, 111
*Spirogyra*, 135–37
Spontaneous generation, 1
Sporanglophores, black bread mold and, 176
Sporophytic budding, 333
Stamens, described, 255
Staminate, flowers and, 320

Stanley, Wendell, virus research and, 75
*Staphylococci*, 111
*Staphylococcus*, food poisoning and, 362
Starch, chemical formula for, **44**
Stemonitis, 170
Stems
  dicot,
    herbaceous, 292
    woody, 283–91
  modified, 295–97
  monocot, 293–94
Stinking corpse lily, 320
Stomates, 308, **309**
Stone cells, 237
*Streptococci*, 111
Stroma, chlorophyll and, 15
Sturtevant, A.H., genetic mapping and, 30
Suberin, cells and, 14
Sugars, chemical formula for, 42–45
Sundew, 356
Suspensor cells, Coniferales and, 248
Swarm spores, Myxomycetes and, 170
Synergid, angiosperms and, 256
Synnema, Moniliales and, 208

T-2 bacteriophage, 76–78
Telophase, mitosis and, 23
Test cross, described, 55
Tetrads, meiosis and, 27
Tetrasporine line of green algae, 131–38
Tetrasporophytes, *Polysiphonia* and, 152
Theophrastus, plant classification and, 215
Thylakoids, 15
Time chart
  geologic, 348
    plant forms and, 350
Tissues
  complex, 238–43
  meristematic, 235
  simple, 235–38
Tobacco mosaic virus (TMV), virus research and, 75

Tomatoes, Phytophthora infestans and, 180–81
Torus, bordered pits and, 242
Tracheids, xylem and, 239
Translocation, cellular aberrations and, 31, 32–33, **34–35**
Transpiration, leaves and, 308–10
Transplanting, leaves and, 310
*Trebouxia*, 212
Trichogyne, *Porphyra* and, 151
t-RNA, 69–71
Tsutsugamuchi Fever, 362
Tube nucleus, angiosperms and, 257
Tuberculosis, *Lobaria pulmonaria* and, 213
Tubers, described, 295
Tubular pits, 237
Tugor, 84
  pressure, 84
Tulip tree, 324
Turtle grass, described, 20

*Ulothrix*, 132–33
*Ulva*, 137–38
Unsaturated fat, chemical formula for, **47**
*Ustilago maydis*, 200–201
*Ustilago zeae*, 200–201, 362
*Utricularia vulgaris*, 356

Vacuoles, cells and, 18–20
*Vallecula*, defined, 233
van Helmont, Jan B.
  photosynthesis research and, 89
  spontaneous generation and, 1
van Leeuwenhoek, Anton
  bacteria and, 109
  microscope "wee beesties" and, 2
van Niel, C.B., photosynthesis research and, 91–92
Vascular bundle, pine leaf and, 249
Vascular cambium tissue, 235
*Vaucheria*, 155–57
Venalization, germination and, 333
Venation, leaf, 312
Venter, *Marchantia* and, 222

Venus flytrap, 356
*Verrucaria maura*, 211
Vessels, xylem and, 239
Virchow, Rudolf, on life, 1
Viruses, 75–80, 361–62
  plant, 79–80
  T-2 bacteriophage, 76–78
*Vittoria*, 229
Volvacine line of green algae, 125–31
*Volvox*, 130–31
von Seysenegg, Tschermak, Mendel's experiments and, 51

Walking fern, 229
Watson, James D., DNA research and, 63
Watson-Crick DNA model, 63
Waxy cuticle, epidermal cells and, 235
Weismann, August, genetics and, 49–50
Wheat
  *Puccinia graminis* fungus and, 195–97
  rust, 195
White pine, leaves of, 249
Willstatter, Richard, photosynthesis research and, 93
Winter wheat, germination of, 333
Wohlor, Friedrich, urea synthesized by, 3
Wood, described, 286, 288–89
Woody dicot stems, 283–91
Wound hormone, 273

*Xanthomonas*, 362
Xanthophylls, 18
Xanthophyta, 155–57
Xylem
  in angiosperms, 238–42
  in gymnosperms, 242–43

Yellow-green algae, 155–57

Zygomorphic flowers, 321, **322**
Zygomycetes, 169, 175–77
Zygotene, meiosis and, 28